Reclaiming Space

Reclaiming Space

Progressive and Multicultural Visions of Space Exploration

Edited by

James S. J. Schwartz, Linda Billings, and Erika Nesvold

OXFORD
UNIVERSITY PRESS

Oxford University Press is a department of the University of Oxford. It furthers
the University's objective of excellence in research, scholarship, and education
by publishing worldwide. Oxford is a registered trade mark of Oxford University
Press in the UK and certain other countries.

Published in the United States of America by Oxford University Press
198 Madison Avenue, New York, NY 10016, United States of America.

© Oxford University Press 2023

CIP data is on file at the Library of Congress

ISBN 978–0–19–760479–3

DOI: 10.1093/oso/9780197604793.001.0001

1 3 5 7 9 8 6 4 2

Printed by Sheridan Books, Inc., United States of America

For the dispossessed—of this planet, and of the stars

Contents

PART 3   CULTURAL NARRATIVES AND SPACEFLIGHT

PART 4   BEING ACCOUNTABLE IN THE PRESENT

PART 5 VISIONS OF THE FURTHER FUTURE

Foreword

Lori Garver

Humanity's first steps beyond Earth's atmosphere were to essentially advance military objectives. The National Aeronautics and Space Administration (NASA) was established in 1958 for expressed peaceful purposes but was in reality an instrument of the Cold War. The premise put forward in my memoir, *Escaping Gravity*, is that while this linkage increased NASA's budget immensely, it also drove the fledgling space agency's culture toward episodic large engineering projects and away from investments in scientific research and technical innovation that could have returned more broad societal benefits. In turn, this focus attracted a workforce most interested in conducting daring exploits for a few White men, and the massive scale of such endeavors could only be conducted by the world's two superpowers of the time, the United States and the Soviet Union.

Remembrances of Apollo focus on the bold, daring dreams of a young President and romanticize the era and purpose of the mission. "We choose to go to the Moon in this decade and do the other things, not because they are easy, but because they are hard" is a justification still repeated in an attempt to increase NASA's funding today. Rarely acknowledged are the recordings of Kennedy telling NASA Administrator James Webb in November 1962 that if we can't beat the Russians, "we ought to be clear, otherwise we shouldn't be spending this amount of money, because I'm not that interested in space." The historic record is clear, but still this narrative prevails, reinforced by historians and institutions committed to perpetuating the legend. Mythology sells.

Nostalgia for the early "manned" space program depicts it as a period that was thrilling and wonderful for everyone. In reality, that was primarily the case for Anglo-Saxon Caucasian men in the United States. NASA's overwhelming early success narrowed its field of vision and prevented the US space program from being oriented towards more fundamental and inclusive goals. By putting these past achievements on a pedestal, we limited our future. The massive institutional bureaucracy and industry interests developed for Apollo required exorbitant fixed costs just to be maintained, and once in place, legacy interests were naturally conditioned to seek similar missions and goals that could use the same infrastructure and workforce.

The space community has been frustrated by a lack of US government investment in space exploration but hasn't wanted to accept the fundamental source of the problem. Too many stakeholders are invested and incentivized to protect the current way of doing business. Apollo's mandate was unique, but because of its success, it set a culture ill-suited to carry the hopes and dreams of the global citizenry.

Space endeavors can contribute to the greater good by connecting the world and reaching beyond, but fundamentally, we still have a system that creates programs to suit its own needs—doing the things the people who already work there want to do—instead of creating programs that address the public's needs. Public polls on space issues consistently rank studying Earth Science and detecting asteroids as the top two priorities for space exploration, while sending a few humans to walk on the Moon and Mars rank at the bottom of the list. Aligning space exploration programs to address today's challenges would expand interest and talent beyond its existing narrow constituency and encourage a more diverse workforce. An important justification for human spaceflight is its ability to inspire. Instead of trying to refuel a Cold War hysteria, we should be designing missions to inspire those we have yet to reach.

In reality, there are as many motivations to advance space development as there are people on the planet. That is the point. Instead of benefiting a few, space should be fully utilized to benefit humanity and society. Future space activities can help us not only to thrive, but also to survive.

Space is a transformative location with unique characteristics—like Earth's own atmosphere and oceans. Earlier societal expansions that exploited the oceans and the atmosphere grew to support vastly different uses, including transportation, communications, scientific research, recreation, and warfare. Early explorers who learned to operate in these otherwise hostile, alien environments, carried their own cultural bias, and foreign—sometimes deadly—biology.

Society's expansion and exploitation of space are evolving to support similarly diverse uses. Early successes in human spaceflight enhanced prestige and reinforced the intent for global space dominance by competing spacefaring nations. Advancing technologies and reducing barriers to entry for space exploration now require us to evaluate and adjust our intentions. As capabilities and industries mature, government regulations related to public safety, environmental stewardship, and shared resources must also evolve.

Exploring space has allowed us to begin to understand the mysteries of the universe, including how life began and whether it exists elsewhere in any form. Space missions have opened our eyes to what lies beyond our atmosphere, as well as on the Earth below. Operating in space provides us instantaneous global delivery of voice, data, and video information; precision measurement of time and location; and Earth observations that measure the interactions between the atmosphere, land, ice, and oceans that affect us all.

We've learned that recent human activities are changing our home planet in ways that aren't easy to see from our own backyards. To understand what is happening here, we need to see ourselves from a new perspective, one that shows how we are connected—7.7 billion people and 8.7 million species—to the only living planet in our own galaxy or in the known universe. It is only from this perspective that we can fully understand what can be done to allow planet Earth to remain a vital home to future generations.

The Industrial Age fueled population expansion across the globe and our first steps beyond. In the Digital Age, we now gather and instantaneously access massive amounts of data from space that inform Earth systems models, revealing how unprecedented amounts of greenhouse gases being released into our atmosphere are causing a climate crisis that threatens our existence.

Temperatures in the atmosphere, land, and oceans are increasing; glaciers are melting, and seas are rising. These changes are fueling extreme weather events, catastrophic storms, flooding, and drought that affect every aspect of our environment—air quality, water availability, food supply, biodiversity, and disease. All of life as we know it is under stress. Data shows that over the next few decades—a blink of an eye in Earth's story—the damage we have caused will accelerate uncontrollably, making it even more difficult, if not impossible, for us to reverse. We are facing the tipping point for human life on our home planet. What we do now will determine the rest of our story.

Thanks to advancements in space development, we live in a moment in history when scientific and technological progress allows us to understand the negative impacts of previous inventions, offering us a rare and fleeting opportunity to recover. Armed with the knowledge of what is happening and why, our new perspective offers solutions. Improved sensor technologies, data accessibility and distribution can provide critical, timely information to more precisely measure, model, predict, and adapt to the climate crisis, limiting human suffering.

Sixty-three years ago, NASA was challenged to reach farther and, in successfully achieving this goal, provided a new perspective of ourselves and our beautiful, fragile home planet. Apollo 8 astronaut Bill Anders, who took the Earthrise photo capturing the Earth as it rose behind the Moon, said, "We came all this way to explore the Moon, and the most important thing is that we discovered the Earth."

President Kennedy's speech that set the stage for Apollo explained the challenge poetically: "We set sail on this new sea because there is new knowledge to be gained, and new rights to be won, and they must be won and used for the progress of all people."

Space activities can make meaningful contributions to sustaining humanity on the Earth and eventually beyond. Our combined investments have led to advancements that now allow us to tap the collective genius of humanity to find solutions to the previously insurmountable problems we face today. Future space voyages can lift all people by more fully utilizing atmospheric and space-based science and technology to address society's challenges.

We now have the knowledge, understanding, and capability to chart a course that utilizes the realm of space sustainably for the benefit of us all. Expanding our presence beyond our own planet shouldn't be designed to escape or give up. It should be part of a larger strategy to assure all our survival.

Reclaiming Space: Progressive and Multicultural Perspectives on Space Exploration offers new and valuable insights for more equitable and sustainable space exploration initiatives. Linda, Jim, and Erika have curated an impressive range of diverse topics and perspectives for this volume that have rarely been considered. It is my hope that *Reclaiming Space* will begin a global conversation about the future of space for humanity.

The most important lesson we learned from our first forays beyond our thin atmosphere was that we are in this together. We overcame gravity by working collectively toward an aligned goal. That same force must now allow us to overcome political and policy differences that are driven by circumstances such as what we look like, where we live, and who we love. Keeping our shared goals in mind can give us the ability to explore space for the benefit of all.

Preface

The idea for this volume was first conceived in the autumn of 2016, at the inaugural meeting of the Society for Social and Conceptual Issues in Astrobiology (SSoCIA). It was there that two of us (Linda Billings and James Schwartz) wondered aloud to one another about putting together an edited volume dedicated to progressive, radical, and "counter-culture" perspectives on space exploration and the culture of spaceflight. Perhaps, we thought, we could assemble a spiritual successor to Stewart Brand's 1977 edited collection *Space Colonies (A Coevolution Book)*, which platformed a variety of environmentalist critiques (and defenses) of space exploration during the public commotion surrounding Princeton physicist Gerard O'Neill's advocacy for orbital habitats in the 1970s.

Beginning in the 2010s, however, it was no longer O'Neill's "high frontier" that was captivating spaceflight culture. Instead, it was Richard Branson's and Virgin Galactic's efforts to grow the space tourism industry. It was Jeff Bezos's and Blue Origin's plans to create space-based industrial workforces. It was Elon Musk's and SpaceX's mission to make humanity a "multiplanetary" species. In other words, we are heirs to a spaceflight culture, ensconced as the "NewSpace" movement, that by all appearances is dedicated to promoting the interests of the most wealthy and powerful Earthlings.

This volume is a testament to the critical mass that has been reached in response to the NewSpace movement. In 2020, when Schwartz at last followed up with Billings about the idea for *Reclaiming Space*, the volume came together remarkably—breathtakingly—quickly. Erika Nesvold joined as coeditor within a few short weeks. After a month, we had a dozen confirmed contributors as well as proposals under consideration at several publishers. After only another two or three months our contributor list had grown to nearly thirty authors, and we had secured a publication contract for the volume. Throughout this process, as editors, we have been overwhelmed by the enthusiasm of our contributors. We are honored to have constructed a platform from which so many individuals from so many different backgrounds have chosen to speak, and we are deeply grateful for their diligent and artful contributions, each written under the many stressors inherent to life during a deadly global pandemic. While our contributors are not individuals who are opposed to space exploration, they all have concerns about how it is proceeding, and they have many good ideas about how we can aim for a better future, in space and on Earth.

We recognize at the same time that this book is long overdue, in the sense that too few attempts have been made to invite culturally nondominant perspectives into the space exploration conversation (an error to which we editors, as privileged members of the dominant culture, are not immune). We also recognize that this book is inadequate, in the sense that no book, nor any proliferation of books, could ever be sufficient for effecting positive changes on this world and in the space and satellites that surround it, where changes are badly needed.

Reclaiming Space: Progressive and Multicultural Visions of Space Exploration is for a wide audience. Our aim has been to curate a volume that is accessible to general space-interested readers, but that also remains novel, meaningful, and informative to space exploration researchers, employees of national spaceflight programs, and members of the space industry. Its essays should be no more difficult to read than something you might be asked to read in a first-year humanities or social science course at university. We hope it inspires younger minds just as much as it challenges the assumptions of experienced experts.

A review of the biographies of our contributors will show a great variety of home countries, backgrounds, expertise, and lived experiences. While many of our contributors are academics who work and teach at universities, some teach at other kinds of institutions, some are artists, some are dancers, some are actors, and some are award-winning authors of science and speculative fiction. At the same time, all of our contributors are curious and passionate about space. And all are eager to help us figure out how to explore space in ways that substantially benefit all of humanity, as opposed to benefiting only a privileged few.

Our academic contributors hail from disciplines as varied as anthropology, astrobiology, astronomy, astrophysics, communication science, law, linguistics, philosophy, and physics. This multidisciplinarity appears even among the editors, as one (Nesvold) earned her doctorate in astrophysics, another (Billings) earned her doctorate in mass communication, and yet another (Schwartz) earned their doctorate in philosophy. Let this highlight that there are *many* ways to contribute to conversations about space exploration and space policy—these are not domains that are reserved exclusively for the STEM (Science, Technology, Engineering, Mathematics) disciplines. If indeed, as is commonly assumed in spaceflight culture, we are obliged to extend human society into space, it is not clear how we will do that without the help of *all* of the disciplines, vocations, and artistries (as well as hobbies and pastimes) that we require to maintain societies here on Earth. It might be difficult to justify ruling anyone *out* of the conversation.

Certainly, there are many more voices and perspectives that deserve a hearing than can fit in an edited volume of modest length. Although we did not seek to exclude any families of perspectives from this volume, the result is neither fully nor perfectly inclusive. There are many other perspectives that we ought to appreciate and understand. Tentatively, we hope to assemble sequels to this volume in order to platform a greater variety of voices and perspectives. We also would like to encourage others to pursue related ideas for anthologies and special issues, meetings and informal gatherings, and conferences and symposia, for there is so much important work to be done! That said, we would be remiss if we did not point the reader to other anthologies which have helped to inspire and pave the way for the book you are reading right now, including (but not limited to): Isaiah Lavender III's *Black and Brown Planets: The Politics of Race in Science Fiction* (University Press of Mississippi, 2014); Grace Dillon's *Walking the Clouds: An Anthology of Indigenous Science Fiction* (University of Arizona Press, 2012); Ed Finn and Joey Eschrich's *Visions, Ventures, Escape Velocities: A Collection of Space Futures* (Center for Science and the Imagination, Arizona State University, 2017); and Eugene Hargrove's *Beyond Spaceship Earth: Environmental Ethics and the Solar System* (Sierra Club Books, 1986).

We have many individuals to thank for helping bring this volume into existence. First and foremost, we are beyond grateful to each of our contributors, who have filled this volume with compelling and insightful ideas about space exploration. We encourage readers to learn more about their unique and fascinating novels, art projects, research, and other work. Should you peruse the contributor biography section of this book, no doubt you will discover more than one topic that intrigues you! We also thank Jeremy Lewis, Bronwyn Geyer, Michelle Kelley, and Oxford University Press for their warm support and encouragement throughout the process of editing this book. Lastly, we thank Tony Milligan for insightful conversations very early in the editing process.

Jim thanks everyone who has helped them to shape their values for the better, as well as those who played important behind-the-scenes roles in bringing this volume to life. In no particular order, I thank: Cindy Johnson, Zach Pirtle, Susan Sterrett, Susan Castro, Jeff Smith, Mukesh Chiman Bhatt, Sheri Wells-Jensen, Jonathan Trerise, Daniel Capper, Daniela de Paulis, Margaret Race, Lucas Mix, Ted Peters, Keith Abney, Wendy Whitman Cobb, Chelsea Haramia, William R. Kramer, Deana Weibel, John Traphagan, Kelly Smith, Lucianne Walkowicz, Chyla Pugh, Ashley Bergman, Katie Klein, Rebecca Nicole,

Jessica Garner, Deb Michling, Rebecca Teich-McGoldrick, Kate Nance, Glenn Baughman, Jay Price, Kathryn Denning, Steve Dick, Roger Launius, Chinyere Okafor, Neal Allen, A. J. Link, Brian Green, Tony Milligan, Charles Cockell, Gonzalo Munévar, David Hewitt, Les Johnson, Marcus Cooper, Susan Vineberg, Matt McKeon, Debra Nails, Jamie Lindemann Nelson, John Corvino, Ben Ragan, Zach Weinersmith, Steve Baxter, Ben Kitley-Hassenger, Corey Kitley-Hassenger, Maüz, Jonathan Natale, Alexandria Mason, David Cheely, Jennifer Blount Goosen, Travis Figg, Tom, Josh, and the fabulous community at J's Lounge, Sarah Lundin, Crystal Chitwood, Greg Grillot, Ginnie Johnson, Kristin Platt, Conrad Johnson, Marna Cole, Sean Johnson, Addie Carter, Indigo Challender, Ila Phelps, Em White, Hannah Erickson, Lee Jones, Sam Anderson, Samara Bergman Murphy, and my students. Also, my deepest gratitude to Erika and Linda, for being phenomenal coeditors!

Linda thanks: My NASA funders—Mary Voytek, head of the NASA Astrobiology Program; Lindley Johnson and Kelly Fast with NASA's Planetary Defense Program. Roger Launius and Steve Dick, former NASA chief historians, who gave me opportunities to research, write about, and publish critiques of the dominant ideology of space exploration.

Erika thanks: Linda and Jim, for inviting me to this project and making it a delightful experience; my invaluable JustSpace cofounder, Lucianne Walkowicz; all of the scholars and activists who have shared their time and expertise to improve me understanding of these issues, particularly Margaret Newell of The Ohio State University and Sarah Newell of the International Labor Rights Forum; and all of the tireless (and tired) activists working to show us visions of a better future on Earth, without which we cannot hope to imagine a better future in space.

With apologies to those we have forgotten to acknowledge,
Jim, Linda, and Erika

Contributors

Mukesh Chiman Bhatt is a Kenyan polymath physicist, linguist, and space lawyer of Indian descent living in the United Kingdom. He has previously worked and published on defense and aerospace materials and currently researches and writes on space law (and law in space) with respect to evolution and human agency in the settlement of outer space. This also intersects with his interests in migration, cultures, and philosophies from around the world. Currently (2021) he is affiliated to Birkbeck, University of London in the United Kingdom, and his work may be found on LinkedIn, Academia.edu, and Researchgate along with the usual books and journals.

Linda Billings is a consultant to NASA's astrobiology and planetary defense programs in the Planetary Science Division of the Science Mission Directorate at NASA Headquarters in Washington, DC. She also is director of communications with the Center for Integrative STEM Education at the National Institute of Aerospace in Hampton, Virginia. She earned her PhD in mass communication from Indiana University. She worked for more than thirty years in Washington, DC, as a researcher; communication planner, manager, and analyst; policy analyst; and journalist. She served as a member of the staff for the National Commission on Space (1985–1986), and an officer of Women in Aerospace (WIA) for fifteen years. She was elected a Fellow of the American Association for the Advancement of Science in 2009. She blogs about space matters at doctorlinda.wordpress.com.

Daniel Capper is a Professor in the School of Humanities at the University of Southern Mississippi, where he teaches Asian religions, comparative religions, and research methods. Trained in the field of religion and science dialogue at the University of Virginia and the University of Chicago, his interdisciplinary studies explore environmental ethical interactions with the nonhuman natural world, especially among American Buddhists. Capper's many publications include the books *Guru Devotion and the American Buddhist Experience, Learning Love from a Tiger: Religious Experiences with Nature,* and *Roaming Free like a Deer: Buddhism and the Natural World.* Currently his research probes the interfaces between Buddhism, environmental ethics, and extraterrestrial ecological issues such as the disposition of space junk or the imminent mining of our moon. Please visit his research website at www.space-environments.org.

Octavio Chon-Torres holds a PhD in Philosophy and is a doctoral candidate in Education at the Universidad Nacional Mayor de San Marcos, Peru. He also holds a master's degree in Epistemology from the same university. He is currently a Professor at the Universidad de Lima, Peru, associate member of the International Astronomical Union in the commission F3 for Astrobiology, and member of the Working Group on Astrobioethics. He is the President and founder of the Peruvian Association of Astrobiology. His publications focus on the philosophy of astrobiology, astrobioethics, and transdisciplinarity. In addition, he is the Director of the Stratosphere Project, a transdisciplinary research initiative in astrobiology. He is also an amateur radio operator (OA4-DWK).

Edward C. Davis IV has conducted anthropological research in Africa, Latin America, Europe, Asia, and North America. As an historical linguistic anthropologist, Davis earned his BA at the Gallatin School at New York University; both MA and PhD degrees in African American studies at the University of California, Berkeley; and an MPhil in Anthropology at University of Cambridge as a Gates Scholar and member of St. John's College. Dr. Davis honors his Underground Railroad roots along Illinois' Trail of Tears Highway as a volunteer with the Angola Partnership Team. As a mission within the Illinois conference of the United Church of Christ, APT rebuilds postwar medical, educational, and agricultural infrastructure at the request of the *Igreja Evangelica Congregacional em Angola*. Davis remains the only descendant of 1619 Virginia working in his African ancestral homeland, which inspired the name for his company *Uloño GPS, Inc.* from the Ovimbundu word for intergenerational knowledge. For eight years, Davis directed Africana studies and Anthropology as a tenured professor at a college named for Malcolm X. Today, Davis teaches at Southland College Prep High School in Richton Park, Illinois.

Kathryn Denning is a Canadian anthropologist and archaeologist at York University, Toronto. She specializes in the long view of humanity's history, and our ethical engagements with life that is extinct, extant, and as-yet-unknown. She researches humanity's cultural expansion into space, how we imagine our future beyond Earth, and how we anticipate other life not yet discovered. Kathryn was on Crew #78 at the Mars Desert Research Station, has collaborated with scientists of the NASA Astrobiology Institute, guest lectures at International Space University, and is on the Science Advisory Board of the SETI Institute and the Board of the Just Space Alliance. She is passionate about fair and accurate space education for all, and regularly speaks with writers, documentary-makers, and at public events.

Francesca Ferrando (*pronouns*: they/them) teaches Philosophy at NYU-Liberal Studies, New York University. A leading voice in the field of Posthuman Studies and founder of the Global Posthuman Network, they have been the recipient of numerous honors and recognitions, including the Sainati prize with the Acknowledgement of the President of Italy. They have published extensively on these topics; their latest book is *Philosophical Posthumanism* (Bloomsbury, 2019). In the history of TED talks, they were the first speaker to give a talk on the topic of the posthuman. US magazine *Origins* named them among the 100 people making change in the world (www.theposthuman.org).

Lori Garver is the Founder and CEO of Earthrise Alliance, a philanthropy that creates meaningful earth science content from satellite data to address climate change. She serves as an Executive in Residence at Bessemer Venture Partners, and is a nonresident Senior Fellow at the Belfer Center for Science and International Affairs at Harvard's Kennedy School. Garver served as Deputy Administrator of the National Aeronautics and Space Administration (NASA) from 2009–2013 and led the agency's transition team for President-Elect Obama. Previous senior positions have included General Manager of the Air Line Pilots Association, Vice President at the Avascent Group, and Associate Administrator for Policy and Plans at NASA. Lori is a cofounder of the Brooke Owens Fellowship, an internship and mentorship program for collegiate women and gender minorities pursuing degrees in aerospace fields. Her memoir, *Escaping Gravity*, was published in the summer of 2022. Garver is the recipient of numerous honors and awards including the 2020 Lifetime Achievement Award from Women in Aerospace, three NASA Distinguished Service Medals and an Honorary Doctorate from Colorado College. She holds a BA in Political Economy from Colorado College and an MA in Science, Technology and Public Policy from the George Washington University.

Alice Gorman is an internationally recognized leader in the field of space archaeology and author of the award-winning book *Dr Space Junk vs the Universe: Archaeology and the Future* (MIT Press, 2019). Her research focuses on the archaeology and heritage of space exploration, including space junk, planetary landing sites, off-earth mining, and space habitats. She is an Associate Professor at Flinders University in Adelaide and a heritage consultant with over twenty-five years' experience working with Indigenous communities in Australia. Gorman is also a Vice-Chair of the Global Expert Group on Sustainable Lunar Activities and a member of the Advisory Council of the Space Industry Association of Australia. In 2021, asteroid 551014 Gorman was named after her in recognition of her work in establishing space archaeology as a field.

Chelsea Haramia is an Associate Professor of Philosophy at Spring Hill College. Her current research focuses on astrobiology ethics, the ethics of science and technology, the ethics of space exploration, and feminist philosophy. She is especially active in debates concerning the Search for ExtraTerrestrial Intelligence (SETI) and the activity of Messaging ExtraTerrestrial Intelligence (METI). She has recently published multiple journal articles on the morality of messaging extraterrestrial intelligence, and she has a chapter forthcoming in the *Routledge Handbook of Social Studies of Outer Space* outlining the philosophical and ethical dimensions of current METI and SETI projects. She is also coeditor of the open-access journal *1000-Word Philosophy*, which houses a growing set of original, philosophical essays aimed at experts and nonexperts alike.

Evie Kendal is a bioethicist and public health researcher at the Department of Health Sciences and Biostatistics, Swinburne University of Technology. Evie completed her Bachelor of Biomedical Science in 2008, specializing in human pathology and behavioral neuroscience, before completing a Bachelor of Arts in Comparative Literature and Cultural Studies, Classical Studies (Latin) and Philosophy, with First Class Honors in English Literature. She completed her Master of Bioethics in 2012 and her PhD in Bioethics from Monash University in 2018. Evie also completed a Master of Public Health and Tropical Medicine at James Cook University in 2017, specializing in reproductive and environmental health. Evie's research interests include ethical dilemmas in emerging biotechnologies, space ethics, and public health ethics.

Mary Robinette Kowal is the author of the *Lady Astronaut Universe* and historical fantasy novels: *The Glamourist Histories* series and *Ghost Talkers*. She's a member of the award-winning podcast *Writing Excuses* and has received the Astounding Award for Best New Writer, four Hugo awards, the RT Reviews award for Best Fantasy Novel, the Nebula, and Locus awards. Stories have appeared in Strange Horizons, Asimov's, several Year's Best anthologies and her collections *Word Puppets* and *Scenting the Dark and Other Stories*. Her novel *Calculating Stars* is one of only eighteen novels to win the Hugo, Nebula and Locus awards in a single year. As a professional puppeteer and voice actor (SAG/AFTRA),

Mary Robinette has performed for LazyTown (CBS), the Center for Puppetry Arts, Jim Henson Pictures, and founded Other Hand Productions. Her designs have garnered two UNIMA-USA Citations of Excellence, the highest award an American puppeteer can achieve. She records fiction for authors such as Seanan McGuire, Cory Doctorow, and John Scalzi. Mary Robinette lives in Nashville with her husband Rob and over a dozen manual typewriters.

William R. Kramer has degrees in biology, environmental policy, public administration, and political science (University of Hawaii). His doctoral dissertation, "Bioethical considerations and property rights issues associated with the discovery of extraterrestrial biological entities—Implications for political policy in the context of futures studies," grew from his experiences as a wildlife biologist with the US Fish and Wildlife Service, where he was head of the Endangered Species Recovery Program. He was a consultant to the US Navy for over a decade on environmental issues, taught bioethics and biology at Hawaii Pacific University, worked at HDR, Inc. as their extraterrestrial environmental analyst, and now lectures on futures studies, bioethics, and environmental ethics at the International Space University. William lives in Frederick, Maryland, and can be reached through his website: OuterSpaceConsulting.com.

Ingrid LaFleur is a curator, artist, Afrofuture theorist, pleasure activist, and founder of The Afrofuture Strategies Institute (TASI). As a former candidate for the mayor of Detroit, LaFleur has made it her mission to ensure equal distribution of the future. She explores the frontiers of social justice through emerging technologies and science, and new economies and modes of government. Through TASI, LaFleur implements Afrofuture foresight and approaches to empower Black bodies and oppressed communities. As a thought leader, social justice technologist, public speaker, teacher, and cultural advisor she has led conversations and workshops at Centre Pompidou (Paris), TEDxBrooklyn, TEDxDetroit, Ideas City, New Museum (New York), Harvard University, Oxford University, and Museum of Modern Art (New York), among others. Her work has been featured in the *New York Times*, NPR's *This American Life*, and *Hyperallergic* to name a few.

William Lempert is an Assistant Professor of Anthropology at Bowdoin College in Brunswick, Maine. Drawing on years of ethnographic fieldwork with Indigenous media organizations in the Kimberley region of Northwestern Australia, his research engages the dynamic process of filmmaking as a critical mode of political transformation. This work informs his current writing on ways in which Indigenous futurisms reimagine the proliferation of virtual reality and outer space colonization.

Alan Marshall is a specialist in both the global history of technology and also the long-term future global impact of technology, receiving his doctorate in these topics from the University of Wollongong (Wollongong, Australia) in 1999. Since 2011, Alan has been a full-time Visiting Professor in Thailand's largest tertiary institution, Mahidol University. Before that, Alan held numerous research fellowships across the world, from the Institute of Advanced Studies in Science, Technology and Society in Graz, Austria to the Curtin University of Technology in Perth, Australia. His published books include *Ecotopia 2121: A Vision of Our Future Green Utopia*, *Dangerous Dawn: Our New Nuclear Age*, and *Wild Design: The Ecomimicry Project*.

Tanja Masson-Zwaan is an Assistant Professor and Deputy Director of the International Institute of Air and Space Law at Leiden University, and President Emerita of the International Institute of Space Law. She advises the Dutch Government on space law issues and was cofounder of the Hague International Space Resources Governance Working Group. She publishes on diverse space law topics and coauthored *Introduction to Space Law* (2019). She is full Faculty at International Space University. Tanja is closely associated with associations and organizations worldwide, including the International Academy of Astronautics, International Law Association, Académie de l'Air et de l'Espace, Open Lunar Foundation, ASU Interplanetary Initiative, Netherlands Space Society, and Secure World Foundation. She received several awards including a Royal decoration for her work in space law.

Tony Milligan is a Senior Researcher in the Philosophy of Ethics with the Cosmological Visionaries project, based out of the Department of Theology and Religious Studies at King's College London. Publications include *The Ethics of Political Dissent* (2022); *Pravda v době populismu/Truth in a Time of Populism* (2019); the coedited white paper on *Astrobiology and Society in Europe* (2018); the coedited volume *The Ethics of Space Exploration* (2016); *The Next Democracy?* (London: Roman and Littlefield, 2016); *Animal Ethics: the Basics*, (2015); *Nobody Owns the Moon: The Ethics of Space Exploitation* (2015); the coedited volume *Love and its Objects* (2014); *Civil Disobedience: Protest, Justification and the Law* (2013); *Love* (2011); and *Beyond Animal Rights: Food, Pets and Ethics* (2010).

Prathima Muniyappa is a designer, conservator, and a research assistant for the Space Enabled research group. She is a master's student in the Media Arts and Sciences at the Media Lab. She is interested in addressing issues of social justice, democratic access for historically marginalized communities, and enabling Indigenous agency. Her research investigates alternative cosmologies and cultural ontologies for their potential to contribute to emerging discourse on technoimaginaries in the realm of space exploration, synthetic biology, and extended intelligence. Prior to coming to MIT, she completed a master's in Design Studies in Critical Conservation at the Graduate School of Design, Harvard, under a Fulbright Scholarship. In investigating traditional Indigenous knowledge, practices, and folkways for conservation, her design research on sacred groves and alternative forest practices' assumes complete synergy between the constructed categories of "nature" and "culture," and leverages the philosophy of nonduality as an operational paradigm. It achieves this synergy by investigating cultural identity as an ecological manifestation. She holds a BDes in Spatial Design from the National Institute of Design, India, and is a Young India Fellow 2013–2014.

Erika Nesvold has a PhD in Physics from the University of Maryland, Baltimore County, and has performed computational astrophysics research at NASA Goddard Space Flight Center, the Carnegie Institution for Science, NASA Ames Research Center, and the SETI Institute. She is currently an Astrophysics Engineer for the Universe Sandbox astronomy simulation software. In 2017–2018, she produced and hosted a limited podcast series, *Making New Worlds*, about the ethics of space settlement. In 2018, she cofounded a nonprofit, the JustSpace Alliance, to advocate for a more inclusive and ethical future in space and to harness visions of tomorrow for a more just and equitable world today.

Jacque Njeri is a multidisciplinary design creative cum visual artist whose work addresses culture, feminism, and female empowerment through projected extraterrestrial realities. She has a bachelor's degree in Art in Design from the University of Nairobi. She has released four art projects so far, namely, the Stamp Series, the Mau-mau Dream, and the Maasci Series, which saw her receive global acclaim from featuring in CNN and BBC to participating in multiple exhibitions in her home country, Kenya, and notably the 2018 Other Futures Festival in Amsterdam and as keynote speaker in the 2019 Kikk Festival in Belgium. Her most recent project titled is Genesis, a modern futuristic interpretation of the Matriarch of the Kikuyu people of Kenya. Her website is located at www.jacquenj eri.com.

Andrea Owe is a Norwegian environmental moral philosopher working in the domain of catastrophic risk and future studies. She is a Research Associate at the Global Catastrophic Risk Institute—a think tank working on the risk of events that could significantly harm or even destroy human civilization at the global scale. Owe is deeply committed to the cause of nature and its nonhuman inhabitants, working particularly on risks associated with climate change, environmental destruction, and artificial intelligence, as well as long-term future trajectories that include expansion of life and civilization into outer space. She wrote her thesis about environmental ethics in space. Owe is also an artist, musician, and avid hiker, living in an old farmhouse by the forest and mountains north of the capital of Norway, Oslo.

Daniela de Paulis is a former contemporary dancer, a licensed radio operator, and a media artist exhibiting internationally. From 2009 to 2019 she was artist in residence at the Dwingeloo radio telescope, where she developed the Visual Moonbounce technology and a series of innovative projects combining radio technologies with live performance art and neuroscience. She is member of the International Academy of Astronautics Search for Extraterrestrial Intelligence (SETI) Permanent Committee. She is the recipient of the Baruch Bloomberg Fellowship in Astrobiology at the Green Bank Observatory in West Virginia. She is a regular host for the *Wow! Signal* Podcast. She has published her work with the *Leonardo MIT Journal*, Inderscience, Springer, Cambridge University Press, and RIXC, among others.

David Colby Reed is a systems designer, educator, and technologist. He is currently a PhD student at the MIT Media Lab, where he's a member of the Space Enabled research group and a cofounder of the Space Governance Collaborative. His research focuses on designing participation, voice, and equity into the architectures of complex systems like law, economies, and technologies. The research vision is to design systems that create patterns of social relations that are compatible with democracy. Prior to coming to MIT, he founded, with Lee-Sean Huang, Foossa, a service design, strategy, and futuring practice based in NYC. His Foossa projects have ranged from one-day design thinking and problem-framing workshops, to years-long service innovation projects in partnership with clients and other stakeholders. He's worked to design public services for the City of New York, storytelling experiences for the Kigali Genocide Memorial, technology tools for multilateral organizations, and financial instruments to advance economic security, among other projects. He's also worked on projects to advance the ethical governance of artificially intelligent systems, and he's a past Fellow of the Assembly program at the Berkman-Klein Center at Harvard and the MIT Media Lab. He's a Fellow of the Royal Society of Arts (RSA), and a former trustee of the NYC chapter of the Awesome Foundation. He studied cognitive science at Harvard (BA), public policy and economics at NYU (MPA), and at the Media Lab (MSc '20, PhD in progress).

Ruvimbo Samanga is a lawyer specializing in commercial law and exploring creative synergies between her expertise and chosen interests in human rights, space law, and international law. She has supported policy development and awareness with organizations such as the United Nations Economic Commission for Africa, the Open Lunar Foundation, Space in Africa, and UNOOSA Space4Water Initiative. An avid space advocate, she manages her time between her research, consultancy work, space education projects, and entrepreneurial pursuits. Her current research focus includes GIS policy, Water Management Policy, Remote-Sensing and Data Policy, Lunar Resource Development, Property Rights in Outer Space, and Capacity-Building in Emerging Space Economies.

James S. J. Schwartz (*pronouns*: they/them) is an Assistant Professor of Philosophy at Wichita State University, where they research and teach courses on a variety of issues related to the ethics of space exploration. They are author of *The Value of Science in Space Exploration* (Oxford University Press, 2020), and editor (with Tony Milligan) of *The Ethics of Space Exploration* (Springer, 2016). Their publications have appeared in numerous academic journals, including *Space Policy*, *Acta Astronautica*, *Advances in Space Research*, *Astropolitics*, and *Environmental Ethics*, among others. They have also contributed chapters to over a dozen edited collections, most recently *The Human Factor in Settlement of the Moon* (Springer, 2021), *Astrobiology: Science, Ethics, and Public Policy* (Wiley, 2021), and *The Expanse and Philosophy: So Far Out Into the Darkness* (Wiley, 2021).

Vandana Singh is an Indian science fiction writer inhabiting the Boston area, where she is a Professor of Physics and Environment at Framingham State University. A former particle physicist, her academic work for over a decade is on a transdisciplinary, justice-centered understanding of climate change. Her science fiction short stories have been gathered into two collections, most recently the Philip K. Dick award finalist, *Ambiguity Machines and Other stories*, and several of her works have been reprinted in Year's Best anthologies. In 2021 she was a Climate Imagination Fellow at Arizona State University's Center for Science and the Imagination. More about her can be found on her website, vandana-writes.com.

Saskia Vermeylen is a reader in law at the University of Strathclyde in Glasgow, Scotland. She works as a legal theorist in the area of property frontiers and has a specific interest in outer space law. She studies space law as an aesthetic and in her Leverhulme-funded research she started to question space law from an Afrofuturist and African-centric perspective. Her collaboration in the coauthored chapter with the Kenyan artist, Jacque Njere, is part of a wider dialogue about the role of African art in developing a hopeful space law manifesto that not only criticizes the colonial legacy but also voices the dreams of an inclusive and diverse space program for the benefit of humanity and beyond. Saskia has published widely on this topic and is also experimenting with the performativity of the law through a decolonial curatorial practice.

Janet de Vigne is an academic, actor, and singer working at the University of Edinburgh in the field of language education. She's interested in process-oriented ideas about existence and is a fan of Otto Scharmer, Donna Haraway, and Gilles Deleuze. Her current research focuses on organizational development, embodiment and human-terra, space place, and memory. A lifelong interest in science fiction and the way humans imagine their and Terra's future has led her to this point.

Deana Weibel is a professor of anthropology at Grand Valley State University in Allendale, Michigan, and also teaches in GVSU's Religious Studies program. Her exploration of sacred sites, sacred objects, and people's experiences of them has taken her to a variety of pilgrimage centers from Chimayó in New Mexico to Lourdes in France and Boudanath Stupa in Kathmandu, Nepal. Her most recent work has focused on the idea of outer space as sacred space and the religious lives and motivations of astronomers, flight surgeons, engineers, technicians, and astronauts. Weibel has conducted ethnographic research on religion and space at various NASA flight centers around the United States, the Mojave Air and Spaceport, and at the Vatican Observatory in Castel Gandolfo, Italy.

Sheri Wells-Jensen is an associate professor in the Department of English at Bowling Green State University in Bowling Green, Ohio. Her research interests include xenolinguistics, astrobiology, phonetics, braille, teaching English to speakers of other languages (TESOL), language creation, and disability studies. She is on the boards of Messaging Extraterrestrial Intelligence International, SciAccess, and the Society for Social and Conceptual Issues in Astrobiology. She was beyond thrilled to have been selected to fly aboard the first parabolic flight of Mission: AstroAccess, which took twelve disabled people into a zero gravity environment, and she is currently at work on a book documenting that experience.

Danielle Wood serves as an Assistant Professor in the Program in Media Arts & Sciences and holds a joint appointment in the Department of Aeronautics & Astronautics at the Massachusetts Institute of Technology. Within the Media Lab, Professor Wood leads the Space Enabled Research Group which seeks to advance justice in Earth's complex systems using designs enabled by space. Professor Wood is a scholar of societal development with a background that includes satellite design, earth science applications, systems engineering, and technology policy. In her research, Professor Wood applies these skills to design innovative systems that harness space technology to address development challenges around the world. Professor Wood is also the Faculty Advisor at MIT for African and African Diaspora Studies. Prior to serving as faculty at MIT, Professor Wood held positions at NASA Headquarters, NASA Goddard Space Flight Center, Aerospace Corporation, Johns Hopkins University, and the United Nations Office of Outer Space Affairs. Professor Wood studied at the Massachusetts Institute of Technology, where she earned a PhD in engineering systems, SM in aeronautics and astronautics, SM in technology policy, and SB in aerospace engineering.

1

An Introduction to *Reclaiming Space*

James S. J. Schwartz, Linda Billings, and Erika Nesvold

Does Space Need to be Reclaimed?

Space, to use a worn metaphor, is in the mind of the beholder. When we contemplate the seemingly limitless universe, we tend to project onto space our own hopes and dreams (as well as our fears and anxieties). But like responses to Rorschach inkblots, there are many different hopes, dreams, fears, and anxieties that one can project onto the night's sky. To those who approach it with a thirst for profits, space appears as a resource-rich goldmine, beckoning to anyone with enough wealth and privilege to take advantage of untapped markets. To those who approach it with a yearning for human expansion, space appears as a frontier that is humanity's birthright to conquer, its new manifest destiny. To those who approach it with a passion for knowledge and understanding, space appears as a tantalizing and pristine laboratory for scientific exploration. In these ways, our *visions* for humanity's future in space—what planets and moons we hope to visit, what we hope to accomplish when we get there—are more products of our *perspectives* about space (and our underlying worldviews and value systems) than anything else.

But the approaches listed above, which tend to dominate modern conversations about human plans for space, only represent a small portion of the human perspectives that exist. To those who are mindful of humanity's regrettable treatment of terrestrial ecosystems, for instance, space appears as a true wilderness, intricately structured, and worthy of protection. Meanwhile, to those aware of humanity's long and ongoing history of exploitation, injustice, and greed, space appears alarmingly as a harsh setting for history's next colonialist chapter. And to those who appreciate its unique and fragile beauty, space appears as a grand, sublime gallery of awesome creation. And there are of course individuals who hold several of these perspectives at the same time—they are not in every case mutually exclusive.

This book provides a platform for expressing perspectives on space that have not yet received a fair or adequate hearing. That is, this book exists to

James S. J. Schwartz, Linda Billings, and Erika Nesvold, *An Introduction to* Reclaiming Space In: *Reclaiming Space.*
Edited by: James S. J. Schwartz, Linda Billings, and Erika Nesvold, Oxford University Press. © Oxford University Press 2023.
DOI: 10.1093/oso/9780197604793.003.0001

incubate, illuminate, and illustrate a more diverse and inclusive conversation that highlights nondominant and counter-mainstream perspectives on space exploration. In the process of soliciting contributions, we (the editors) deliberately sought writers and thinkers known for questioning the space exploration *status quo*. Rather than soliciting the perspectives of astronauts or space entrepreneurs, we instead prioritized searching for contributors with more tangential, unconventional, and/or "outsider" relationships with spaceflight culture.

This is not to denigrate astronauts or space entrepreneurs, but to acknowledge that representatives of these groups already have strong voices in spaceflight culture, in politics, and in popular culture more generally, and that it has taken too long for the dominant voices in spaceflight culture to invite and heed any other perspectives on space. In this respect, we editors are obliged to acknowledge that this volume, which is being edited by members of the dominant cultural group, for an academic publisher highly respected in Western intellectual culture, risks perpetuating our own biases as editors, and risks reinforcing power structures that effectively serve to exclude rather than include marginalized voices. As editors of this volume, then, we must acknowledge we have privileges, including those which led to the three of us to being in a position to create this platform in the first place. So, beyond the need to contextualize the contents of this volume, we have also written this introduction in a way that outlines the methods and values we used in bringing this volume to life. We acknowledge the validity of perspectives and lived experiences according to which books such as these are distractions from the necessary and difficult work of challenging unjust, discriminatory societal institutions.

The space exploration *status quo* is decidedly W.E.I.R.D. (Western, Educated, Industrialized, Rich, and Democratic), male, white, politically libertarian, and nondisabled. So, our goal as editors was to find contributors who represented various departures from this *status quo*. What would space exploration be like if we prioritized—or even simply acknowledged—the perspectives or value systems of individuals who are disabled, or aren't white, or aren't male, or aren't libertarian, or aren't Western? What can these perspectives teach us all about space exploration and its value (or potential harm) that cannot be easily recognized or appreciated under the *status quo*? What kinds of changes should we advocate for?

As we reached out to prospective contributors, we asked them: What do *you* think is most important about space exploration? What perspectives on space exploration and spaceflight culture do *you* think are the most important for everyone to consider? How has *your* path through life highlighted what you take to be especially valuable (or troubling) about the way humans are currently pursuing the exploration of space?

By characterizing this volume as a platform for progressive as well as multicultural visions of space exploration, we mean to be inclusive of many kinds of departures from the space exploration *status quo*. But it is worth clarifying specifically what we mean by "progressive," since traditionally, the idea of "progress" is ideological, referring to the widely discredited belief that human societies develop in predictable, linear ways, with identifiable, objectively desirable end-states. As professor of world arts and Indigenous studies researcher David Delgado Shorter observes in the context of the Western search for extraterrestrial intelligence (SETI), this understanding of "progress" is also associated with settler colonialism, for "[p]revious well-known colonial encounters form much of the intellectual foundation for current ways of thinking about 'civilization,' often leading to uncritical assumptions about progressive linearity in the development of life and what counts as intelligence" (2021, 27). Afrofuturist writer and artist Ingrid LaFleur raises a similar criticism of the idea of "progress" in her contribution to this volume (Chapter 13).

So, the sense of "progressive" that we have in mind is not its role as an adjective for a set of beliefs. We mean something else by the term. In particular, we suggest using "progressive" as an adjective describing a family of *methodologies* or *practices* for identifying, framing, and resolving societal problems, for seeking the full participation at all levels of society of marginalized groups, and for repairing the injustices visited upon marginalized peoples by dominant cultural groups. In this sense, to be "progressive" starts from recognizing *all* of what needs to be taken into consideration when making decisions that affect others, in space or on Earth.

As we hope the chapters of this volume demonstrate, being interested in space exploration—including being deeply and passionately interested in space—is compatible with being severely critical of the way that it is currently being conducted by space-faring nations and corporations. At the same time, we would like to acknowledge that, in a volume of less than thirty chapters, it is not possible to provide a complete or representative sample of nonmainstream perspectives on space. So, the reader should not expect this book to provide a comprehensive, maximally inclusive collection of progressive and multicultural approaches to space exploration. It will not teach you everything you need to know about the diversity of thought about space. But more than likely it will teach you about more than a few perspectives you hadn't previously considered. And, we hope, the experience will helpfully guide your future engagements with the perspectives of others. We more than welcome further efforts to communicate a greater share of the variety of perspectives on space, through whatever channels are the most effective for, and responsive to, all those who are still considered as "others" or "outsiders" within spaceflight culture.

However, we aren't suggesting, through this book, that by surveying various perspectives, we will eventually discover who has the "correct" vantage point on space, if such a thing even exists. As we described moments ago, this book is more about *methodology* than it is about *ideology*. That is, this is less a book about *what you should believe* about space exploration and more a book about *how you can learn to think more critically and responsibly* about space exploration. What we, along with our contributors, are calling for, is wider appreciation of the *diversity* of perspectives on space, and for wider inclusion of these perspectives in spaceflight culture, popular culture, and in space policy research and decision-making. It is only through earnestly pursuing such diverse engagements that we can begin to identify and prioritize spaceflight goals in a way that respects the dignities and interests of all peoples and cultures.

Think of this volume, then, as one small part of a necessarily long and very overdue *inclusive* conversation about space exploration. Think of it as an early attempt to communicate some of the great diversity of thought about space that exists around the world. Think of it as a means for acquiring a broader cultural, ethical, and philosophical understanding of space exploration. And, perhaps most importantly, think of this volume as a source of examples and insights into how you can use *your own values and life experiences* to create *your own visions and dreams* for the future of space.

Recurring Themes

Despite the diversity of the professional and personal backgrounds of our contributors, there are nevertheless several patterns that emerge in the chapters that follow. In particular, four themes stand out as common reference points for significant and oftentimes problematic aspects of the space exploration status quo. Moreover, it should not be terribly surprising to learn that there are common themes across a volume such as this. The perspectives of the contributors to this volume, while interestingly varied, are all perspectives affected by the same underlying geopolitical realities.

One common theme comes from the cultural narratives and metaphors of spaceflight propounded by global space powers, such as the United States' embrace of Manifest Destiny in space. A second theme is the observed reality that contemporary spaceflight activities take place in a political sphere dominated by libertarian sensibilities and in an economic environment that aspires to free-market capitalism. A third theme is space exploration's rich and eventful legal history that includes not only national space legislation but also legally binding international treaties such as the 1967 United Nations Outer Space Treaty. The fourth theme is the need to consider ethics when balancing the interests of the many different stakeholders in space exploration.

Here we discuss the significance of each theme. However, we advise the reader that these discussions are not intended to provide anyone with a comprehensive introduction to the history and culture of spaceflight,[1] but only to contextualize this volume and its contents.

Western Grand Cultural Narratives of Space Exploration

"History doesn't repeat itself, but it often rhymes" is an apt observation often attributed (without evidence) to Mark Twain. Space is not a blank page where we can effortlessly start fresh, because every human working toward our future in space, whether on or off the Earth, carries with them the weight of history and the norms and values of the societies in which they were raised. If we do not deliberately identify and interrogate the cultural narratives that influence our motivations, we risk recreating the atrocities of our past. This process of identification and interrogation is akin to moral philosopher Alasdair MacIntyre's (2000) plea that we focus more on accumulating "self knowledge," that is, knowledge of the origins of our beliefs, and of the broader societal interests and agendas that those beliefs serve, so that we do not inadvertently help to advance causes that work against our sincerely held values.

Today's space advocates often deliberately invoke the history of European colonization on Earth when describing their plans for our future in space. This framing of space as the next "New World" to be conquered ignores the widespread and ongoing negative consequences of colonization here on Earth. Terrestrial empire-building was often motivated by the pursuit of new resources for profit, with devastating effects for Indigenous peoples and their environments. Even if there is no indigenous life to be harmed in space—an assumption not yet proven—many of our contributors acknowledge that there is danger in embracing a *colonialist attitude* toward space, a policy of never-ending expansion on a ceaseless quest for more territory and resources.

Space advocacy aimed at the public, rather than investors, tends to appeal to cultural motivations for colonization rather than economic ones, but even these attempts echo the propaganda of European colonization, particularly in North America. Many space advocates claim that space is humanity's "destiny," with some (including former US President Donald Trump) explicitly naming "America's manifest destiny in the stars" (Armus 2020). Manifest Destiny, the mid-19th-century idea that Americans had a God-given right (and responsibility) to expand across North America, was used to justify the displacement and genocide of the continent's Indigenous people.[2]

For space advocates who believe that space is our destiny as a species, failing to meet this destiny carries the threat of societal and technological stagnation. If we don't test ourselves against the final frontier of space, the argument goes, we will miss out on our only remaining opportunities for challenge, innovation, and adventure. These ideas parallel the "frontier thesis" of American historian Frederick Jackson Turner, who argued in the late nineteenth century that the recently closed western frontier in the United States had been responsible for shaping the country's character, including its love of democracy, independence, and egalitarianism (Turner 1920).[3]

It's not surprising that the cultural narratives that dominate much of today's public conversations about space rhyme, not with history as a whole, but with a specific era and perspective. Many of these conversations have been taking place in the "Western world": Europe and the regions of the world, once colonized by Europeans, where ethnic and cultural makeups are still significantly influenced by the colonizers. If we only consider Western philosophies, values, and perspectives into our plans for space, we will continue to perpetuate the horrific legacy of colonialism. Of course, there is no single anticolonialist approach to space: even within this volume, for which colonialism is a common theme, we find a diverse range of responses to its presence in our history and its influence on our future.

Economic Frameworks and Political Ideologies

Space exploration's cultural context cannot be understood independently from its political and economic context. *Capitalism* is the dominant form of political economy in much of the world, especially in the West. It embraces private ownership and profit-making of the means of production. Many advocates of the human exploration, exploitation, and colonization of space embrace free-market capitalism as the way to advance their aims. This way of thinking is especially prevalent in the United States, which still claims, arguably, to be *the* world leader in space. Capitalism does not necessarily imply "free," as the governments of capitalist economies not only regulate commercial activity but also subsidize it.

The eighteenth-century economist and moral philosopher Adam Smith's book, *An Inquiry into the Nature and Causes of the Wealth of Nations*, is considered to be a foundational text of capitalist political economy. However, many if not most practicing capitalists have not fully embraced Smith's philosophy. For example, in his discussion of taxes, Smith wrote,

> The necessaries of life occasion the great expense of the poor. They find it difficult to get food, and the greater part of their little revenue is spent in getting it. The luxuries and vanities of life occasion the principal expense of the rich, and a magnificent house embellishes and sets off to the best advantage all the other luxuries and vanities which they possess. A tax upon house-rents, therefore, would in general fall heaviest upon the rich; and in this sort of inequality there would not, perhaps, be anything very unreasonable. It is not very unreasonable that the rich should contribute to the public expense, not only in proportion to their revenue, but something more than in that proportion. (Smith 1776)

Too many practicing capitalists today—especially the ultra-rich, such as billionaire space exploration and exploitation advocates Jeff Bezos, Richard Branson, and Elon Musk—show little inclination to "contribute to the public expense." In fact, they have accepted public subsidies to build their various businesses, and thus their own wealth.

Neoliberalism is a twentieth-century reconceptualization of capitalism, calling for free-market capitalism that reduces or eliminates regulation of prices, trade, and banking. Neoliberalism has no concern for the poor. Nonetheless, neoliberalism in the United States accommodates corporate subsidies and tax breaks, which are anathema to free-market capitalism.

Libertarianism, an ideology embraced by many space exploration and exploitation advocates, is an even more extreme flavor of neoliberalism, promoting private property rights, individual freedom, and unlimited growth as key priorities.

A number of contributors to this volume express concerns about extending these ideologies and exploitative economic frameworks into outer space, as humans expand their presence beyond Earth. By doing so, they argue, space explorers, exploiters, and colonizers would continue practices that, on Earth, have done great harm to people and their environments. However, you should not expect to find a unified, contrasting political or economic ideology, to which each of our contributors subscribes. Each of our contributors has a distinct understanding of what is problematic about the spaceflight status quo, and each has their own suggestions for what to do about it.

Space as an Evolving Legal Domain

Space law, meanwhile, is an outgrowth and reflection of spaceflight culture, especially the cultures of the major space powers. The 1967 United Nations Treaty on Principles Governing the Activities of States in the Exploration and Use of Outer Space, Including the Moon and Other Celestial Bodies, known as the Outer Space Treaty (OST), is the backbone of international space law. This treaty establishes that

> The exploration and use of outer space shall be carried out for the benefit and in the interests of all countries and shall be the province of all mankind; outer space shall be free for exploration and use by all States; [and] outer space is not subject to national appropriation by claim of sovereignty, by means of use or occupation, or by any other means. (UN General Assembly 1967)

This treaty was ratified at a time when only the United States and the Soviet Union had access to space. Now China, Europe, India, and Japan have access to space, as do several commercial launch companies such as the New Zealand-based Rocket Lab. Consequently, the customers of all of these launch service providers also have access to space. While an increasing number of nations are now involved in space exploration and development—one of the latest to enter being the United Arab Emirates—the effort is by no means inclusive.

The 1979 United Nations Moon Agreement was an attempt by non-spacefaring nations to establish their rights in space via international treaty, and it

> reaffirms and elaborates on many of the provisions of the Outer Space Treaty as applied to the Moon and other celestial bodies, providing that those bodies should be used exclusively for peaceful purposes, that their environments should not be disrupted, that the United Nations should be informed of the location and purpose of any station established on those bodies. In addition, the Agreement provides that the Moon and its natural resources are the common heritage of mankind and that an international regime should be established to govern the exploitation of such resources when such exploitation is about to become feasible. (UN General Assembly 1979)

The Moon Agreement, however, has been ratified by only eighteen U.N. member states, including none the major space powers, leading some to consider it a "failed treaty." The reality today is that rights in space—to orbital slots and other resources—are in the hands of those who can get to space.

In 2015, U.S. President Barack Obama signed the Commercial Space Launch Competitiveness Act (CSLCA) into law, directing the federal government "to facilitate a pro-growth environment for the developing commercial space industry" (2015). More controversially, the CSLCA also calls for facilitating "commercial exploration for and commercial recovery of space resources by United States citizens." Extraterrestrial resource exploitation—while further into the future than some advocates may claim—is controversial. On Earth, mining is conducted for profit. There is no reason to believe that the enterprise of mining in space will be any different.

In 2020, under the Trump Administration, NASA produced the so-called Artemis Accords—"principles for cooperation in the civil exploration and use of the Moon, Mars, comets and asteroids for peaceful purposes" (NASA, n.d.). Artemis is NASA's plan to establish a permanent human presence on the Moon. According to NASA, the accords aim to "establish a common vision via a practical set of principles, guidelines, and best practices to enhance the governance of the civil exploration and use of outer space with the intention of advancing the Artemis Program." NASA composed these accords unilaterally, without input from other space agencies, and then invited others to sign on. While several nations have signed the accords, China and Russia—among the top three space-faring nations in the world—notably declined the invitation. They have expressed an intent to collaborate on exploring and developing the Moon and cislunar space (the space between Earth and the Moon). While the United States continues collaboration with Russia on the International Space Station, US federal law prohibits most forms of collaboration between NASA and China in space.

The European Union's Treaty on the Functioning of the EU establishes that the EU "shall draw up a European space policy [and] coordinate the efforts needed for the exploration and exploitation of space . . . excluding any harmonisation of the laws and regulations of the Member States" (European Union 2012). Interestingly, the Grand Duchy of Luxembourg has taken an interest in space mining, and in 2017, enacted a law governing the exploration and use of space resources (2017). Again, space mining is controversial and is unlikely to occur for years into the future.

This summary of current space law is by no means a comprehensive listing. What it does demonstrate is that space-faring nations—especially the United States—are dominating the legal and regulatory discourse about space development and exploitation. Nonspace-faring nations are largely left out. Several contributors to this volume offer ways of making space exploration and development more equitable and inclusive through the consideration of legal principles and regulatory frameworks that are inspired by nondominant cultures and cultural values.

Ethics and The Many Faces of "Space Exploration"

"What should I do?" is among the oldest questions of ethics, the branch of philosophy devoted to analyzing right and wrong action. As this book is, broadly speaking, about what is or should be the right things for us to do in (or in relation to) space exploration, it is also therefore a book that touches on the subject of *space ethics*. While readers do not need to have taken a course in ethics in order to understand any of the contributions to this volume, their understanding of certain chapters may be enhanced by some familiarity with some of the major concepts of ethics, for instance:

- Consequentialist views of ethics, according to which an action is right or wrong based on whether it produces good or bad *consequences*.
- Duty-based views of ethics, according to which an action is right or wrong based on whether it conforms to the moral obligations (the moral duties) we have as individuals.
- Virtue-based views of ethics, which hold that the key to living a good life is to cultivate a good character through moral and practical education.
- Environmental ethical views, which expand the moral sphere from just humans to include animals, species, ecosystems, and even non-living things like rocks and canyons.[4]

Different perspectives in ethics are like different frameworks, or different lenses, for asking and answering questions about what we should do in space. However, space exploration is not a monolith—it is not one individual thing, but instead includes a great variety of different kinds of activities: satellite services (including Earth observation, telecommunications, weather and climate monitoring, resource prospecting, and national security); space tourism; scientific exploration (including crewed and robotic missions to Earth orbit, the Moon, and beyond); space resource exploitation (including solar power generation, asteroid mining, and lunar mining); as well as human expansion into and settlement of space.

This means that when we ask the question "What should we do in space?" there are many possible answers to select from, and some of the answers might be incompatible. For instance, if we build hotels on the Moon, we might not be able to build as many science laboratories there. If we settle Mars before we have satisfied ourselves that Mars lacks endemic life, human contamination could forever interfere with our ability to resolve whether Mars is or was previously home to a second genesis of life.[5]

The upshot is that there are at least as many ways to be *in favor* of space exploration as there are types of spaceflight activities. One does not have to support *the entirety* of existing and proposed spaceflight activities in order to be someone who values space exploration. Indeed, many of the chapters of this volume demonstrate that sometimes our passionate interests in space demand that we *avoid pursuing* types of spaceflight activities that would effectively destroy or damage the space environment, such as the strip-mining of the lunar surface for helium-3 or mining the rings of Saturn.

Ultimately, we cannot really have a productive discussion about the value or importance of "space exploration," because the term "space exploration" is ambiguous. But we *can* engage in productive discussions about which specific spaceflight activities we should pursue, and how and when we should pursue them. Space ethics is an important tool for organizing these discussions.[6]

To deny, for instance, that we are presently ready to attempt the settlement of Mars, is not to deny the importance or wisdom of pursuing sending science missions to learn more about the red planet. To question, for instance, whether we should turn the Moon into a mining district is not to question the significance of traveling to the Moon for artistic reasons. To doubt that space resource exploitation will benefit all of humanity is not to doubt that other spaceflight activities are worth supporting, all things considered.[7]

Volume Overview

This volume is divided into five sections of chapters, grouped thematically. Part 1 includes chapters that provide the reader with overlooked and underappreciated aspects of spaceflight's historical context. Part 2 includes chapters that discuss the impact of creative projects, such as works of science fiction and space art, on our thinking about space exploration. Part 3 contains chapters addressing and demonstrating by example why it is vital to engage with diverse cultural perspectives on space. Part 4 contains chapters with a distinctively pragmatic orientation, which explore pathways for effecting change in the space sector. Lastly, Part 5 contains chapters seeking to advise our exploration of space over the long-term, including discussions of reproductive rights and educational rights for future members of space societies. However, you should not read too much into how we have partitioned this volume, as the partitioning was done in response to, not in anticipation of, the chapters contained in this book. Moreover, the distinguishing features of each part are fuzzy; many chapters would not seem out of place appearing in other places in the volume.

Part 1: The Evolution and History of Spaceflight

In Part 1, a diverse collection of scholars examines belief systems that, one way or another, have been driving space exploration throughout its history. They address ideas, beliefs, and values that have been largely excluded from the mainstream discourse about the human future in space, but that nevertheless provide important prompts for critical reflection.

In Chapter 2, social scientist and volume coeditor Linda Billings critiques the theory of political economy known as neoliberalism. This ideology, or belief system, demanding free-market capitalism, private property rights, and individual liberty, has been driving US space exploration efforts from their beginning. Billings argues that neoliberalism promises to export Earthly practices of colonization and exploitation into space, to the exclusion, and even to the detriment of much of humankind. Billings is not alone in this critique—several other contributors offer their own reasons for being wary of the conquest-and-exploitation mode of space exploration and development.

In Chapter 3, Mukesh Chiman Bhatt, a physicist, linguist, and space lawyer, joins several other contributors to the volume in calling for the decolonization of the way that space-faring nations are planning for and going about space exploration. Bhatt explains that the Western scientific method, and the Western scientific world view, are not the only valid ways of learning about the world and the universe. For millennia before humans gained access to space, he notes, non-Western cultures around the world—African, Arab, Chinese, Indian, Indigenous, and Japanese—have been exploring and using the vastness of space and envisioning human space travel and habitation. This non-Western knowledge plays little to no role in present planning for the human future in space. And he argues that it should. He reports on contributions of nations in the Global South to space exploration and says they should be included in planning for the extension of human presence into space.

Next, in Chapter 4, from a "contrarian" perspective, Edward C. Davis IV, an historical linguistic anthropologist, explores links in US history between slavery, racism, McCarthyism, Cold-War "us versus them" thinking, generational divides, and the television drama series *Lovecraft Country*. In his chapter, he "seeks liberation from systems of knowledge rooted in binary oppositions, such as East-West, North-South, or Black-white." Davis writes of what he calls a religion of space racism, explaining that without contrarian revisions, upholding the myth of US exceptionalism will extend white supremacy beyond Earth into space.

In Chapter 5, Alan Marshall, a specialist in the global history of technology and the long-term future global impact of technology, explains how space exploration, development, and settlement as currently envisioned are not inclusive. Advocates claim that colonizing outer space and exploiting extraterrestrial resources will benefit humanity. Marshall explains that no mechanisms are in place to allow global participation in planning for space development. Wealthy nations and billionaire "space barons" are leading the effort, assuming that free-market economics will govern space development and exploitation. What this means, as Marshall explains, is that parties to the 1967 United Nations Outer Space Treaty established that space is the province of "all mankind." But it does not provide for equal access to space.

Part 1 closes in Chapter 6, in which anthropologist William Lempert demonstrates how remnants of scientific racism, represented here in the form of the now-discredited science of phrenology, have shaped the discourse of scientists engaged in SETI. Their thinking assumes that intelligent extraterrestrial life will be technological and exploratory. But this way of thinking ignores knowledge about Indigenous peoples and cultures here on Earth— cultures that are not necessarily technological or exploratory—but that are nonetheless not inferior to Western expansionist cultures, as they often have been deemed by the West. Lempert explores how, why, and by whom "intelligence" is defined, and he shows how misconceptions about intelligence have been used, and could again be used, to exploit certain groups of people. He makes it clear that while he is not deeming SETI scientists racist, he is concerned about the possible misuse of erroneous conceptions of intelligence.

Part 2: The Art of Envisioning Space

Part 2 includes chapters that focus substantially on the impact of human creativity on our exploration and conceptualization of space. The section opens with the perspectives of two authors of science and speculative fiction, Mary Robinette Kowal and Vandana Singh. Authors of these genres of fiction, as experts in the art of crafting stories, have a kind of advantage over scholars when it comes to contemplating the social, political, ethical, and other quandaries associated with spaceflight: they have been exploring these issues for as long as science and speculative fiction have existed! While much of Western science fiction literature is either unreflectively (or unabashedly) disposed toward Western, colonialist, libertarian, and masculine values, there is an appreciable body of alternative, subversive, and counter-mainstream science fiction, written by authors who use stories to challenge, rather than reinforce, dominant cultural forces, values, and assumptions.

In Chapter 7, Mary Robinette Kowal calls attention to the fact that, because this volume was written and published in the English language and will be accessible only to English language speakers, it thereby serves to reinforce Western cultural hegemony in space. For Kowal, languages do not exist merely to communicate information about the facts, they also help us preserve cultural identities and traditions. She poses a challenging dilemma: The success of space missions depends on crewmembers sharing a common language, but cultural representation cannot be produced in space without active attempts to foster diversity of spoken, living languages among space travelers.

In Chapter 8, Vandana Singh reconciles the Western, colonial history of Earth and space exploration with the author's own fascination with science and exploration, formed during her childhood in India. Singh, who is also a professor of physics, holds that it is critical for us to understand and interrogate the *stories* we tell about space. "By immersing us in entirely different ways of being, thinking, and exploring" through diligently crafted speculative fiction, Singh claims, "we can stretch the boundaries of the imagination past the mechanistic, Eurocentric, colonialist ways of thinking about Earth and Space."

Next, in Chapter 9, Danielle Wood, Prathima Muniyappa, and David Colby Reed of the Space Enabled Research Group at the Massachusetts Institute of Technology offer anticolonial and antiracist designs for human societies in space, as alternatives to extending our historical legacy of colonization, exploitation, racism, and environmental degradation beyond Earth. The authors look to Black feminism, Indigenous cultures, poetry, and science fiction for ideas about how to design human settlements in space that are inclusive and environmentally sustainable. "There is no guarantee that the inspiring novelty of space will push humans to design liberatory pathways, but there is an opportunity to start," they conclude.

Of course, our artistic, creative, and inspirational interactions with space are not limited to the stories we read in novels and anthologies—much of our appreciation of space is influenced by imagery, including artistic imagery and visual media, as well as many other forms of artistic creativity. But conventional or stereotypical space art and imagery, which emphasizes and depicts spacecraft, space habitats, and suited spacefarers (usually of Western origin),[8] belies the great variety of artistic engagement with space, which often envisions radical and progressive alterations to the cultures and purposes of spaceflight.

In Chapter 10, multidisciplinary artist Jacque Njeri and sociolegal scholar Saskia Vermeylan discuss the cultural and legal lessons to be learned from the works of African space artists, including Njeri's own *MaaSci* project, which envisions the Maasai people of Kenya and Tanzania as space explorers. As part of their analysis, they highlight the need for an *Africanfuturist* perspective (i.e., one centering African peoples and cultures) in contrast with an *Afrofuturist* perspective (the focus of Chapter 13 by Ingrid LaFleur). Their chapter concludes by encouraging the use of Africanfuturist values and cultural narratives in the interpretation of legal principles such as the "Common Heritage of Mankind" principle that appears often in international space law.

Meanwhile, in Chapter 11, media artist Daniela de Paulis and philosopher Chelsea Haramia reflect on photographic representations of space. As they argue, images of planetary surfaces taken from space rovers and landers can provide "examples of a new form of knowledge produced through remote embodiment," for instance, knowledge of what it is like to stand on and view the landscape of another planet. Importantly, de Paulis and Haramia stress the need to make room for "deviant" or "outlaw" responses to space imagery— that is, responses which depart from the senses of awe and feelings of accomplishment that normally attend space imagery. By engaging more with deviant responses, we can more readily and more clearly notice injustices associated with space exploration, especially injustices inflicted on non-dominant groups.

Part 2 concludes with Chapter 12, in which philosopher and volume editor James Schwartz reflects on the aesthetic appreciation of the solar system, and on how our appreciation of the beauty of space influences how we craft visions of space futures. They argue that our engagement with space will be more rewarding, and our future in space more secure, if we prioritize sooner rather than later a respect for the *existing* and pristine beauty of space, as opposed to our desiring aesthetic engagement with altered, perturbed, conquered, or terraformed places in space.

Part 3: Cultural Narratives and Spaceflight

Part 3 contains perspectives on space exploration from an array of scholars who are concerned about the ways in which space-faring nations are going about planning for extending human presence into space. As we discussed earlier in this introductory chapter, the dominant way of thinking about and planning for this activity is Western. It is a product of a white male–dominated community. It is capitalistic, pro-growth, and pro-expansion. Women, people of color, and Indigenous peoples may succeed in this community by embracing this way of thinking—but do they have to? Authors in this section of the book offer different ways of thinking about the human future in space, presenting visions of human societies in space that are not discriminatory or exploitative, that do not perpetuate the harmful practices of Earthly exploration and colonization, that accommodate feminist, Indigenous, and Afrofuturist visions of how we might live in space.

Like many contributors to this volume, Ingrid LaFleur—a curator, artist, Afrofuture theorist, and pleasure activist—is concerned about the current direction of plans for extending human presence into space. In Chapter 13, she calls for decolonizing the way we think about doing it. Billionaire space industrialists are promoting a vision of space conquest, exploitation, and colonization. Afrofuturism, she claims, revisits precolonial and ancient African relationships to space in order to envision a human future in space that does not perpetuate any of the harmful "-isms" that have plagued human societies on Earth—racism, sexism, and colonialism, among others. The Western obsession with "progress"—which has come to mean growth and expansion—is an element of current thinking about extending human presence into space. Though we seem to be heading in the wrong direction, LaFleur concludes on a hopeful note. "We can still change course," but in order to do so, Afrofuturists need to be included in planning for a human future in space.

In Chapter 14, cultural anthropologist Deana Weibel writes of the historical practice of predominantly Western white male explorers using so-called native guides to help them navigate unfamiliar terrain. If human explorers land on other planetary bodies, there will be no natives to aid them. Thus far, the human exploration of space has been an exclusive enterprise, limited to a small number of people with a particular set of skills and expertise that suits them for space flight. As the human exploration of space proceeds, Weibel says, it would be useful to engage with people of Indigenous cultures, which largely have been excluded from space exploration. Indigenous people have skills, expertise, and knowledge that could be useful in exploring harsh and rugged extraterrestrial terrains. And importantly, as she observes in her conclusion, "humans from more than just a few of Earth's populations should have a say as to whether, how, and where we should go."

In Chapter 15, philosopher Tony Milligan explains that space ethics does not consist in, and should not strive to produce, a set of ethical principles that stays fixed for all time. He describes three pathways to drawing on Indigenous knowledge in planning for the human future in space, and for developing a more expansive space ethics. We should anticipate that this process will be disruptive, and Milligan argues that it will be. Human expansion into space will be a multigenerational project. Drawing on Indigenous knowledge will not lead to the establishment of a stable center of space ethics, and it will not lead to the controversy-free resolution of problems in space ethics. But this process could deepen understanding of the difficult nature of ethics in general and space ethics in particular, he concludes.

In Chapter 16, Daniel Capper, a scholar of philosophy and religion, offers a Tibetan perspective on space exploration, urging readers to consider loving what he calls "nonliving stuff" in space—extraterrestrial craters and mountains, planetary rings, and other features of the extraordinarily beautiful extraterrestrial environments in our solar system. While it is not peculiar to Western cultures, many of us have been acculturated to value life over nonlife, he says, and he notes that this traditional lack of valuation of nonlife is "glaringly obsolete" as humans move into the solar system. Tibetans, whose religiosity, he explains, is a blend of Buddhism and traditional Tibetan spirituality, have close relationships with their mountains. The mountains are respected members of Tibetan communities. If we are to be responsible citizens of space, he argues, we will need to develop a sense of kinship with what's in our solar system, to extend the idea of stewardship of Earth into outer space.

Chapter 17 closes Part 3 and sets salient context for Part 4. In this chapter, anthropologist Kathryn Denning strikes at the heart of the emotional core of the volume by discussing *hope* in the context of the increasingly devolving disconnections between the spaceflight priorities of ordinary people and marginalized groups (who have never really been consulted) and the space ambitions of the super wealthy. Denning draws parallels between the grief expressed in the various chapters of this book and the ecological grief felt by many in light of the ongoing ecological crises. She notes that while present-day youth may be more motivated than older generations to solve global (and solar systemic) problems, that does not excuse inaction on behalf of those currently in power. Denning's chapter provides a powerful statement of what we can *realistically* hope for in the first place, and what we *should* hope for from among the genuinely available possibilities.

Part 4: Being Accountable in the Present

In Part 4, contributors explore the risks of perpetuating social inequalities and injustice in space and ask how we can apply lessons from our past and present on Earth to improve the lives of humans living and working in space, as well as those of us who remain here on Earth. They also ask how we can share the use of the space environment and its resources equitably, while also protecting space from irreparable harm due to human activities.

In Chapter 18, space archaeologist Alice Gorman summarizes the evolution of space habitats from early fictional concepts through to today's reality on the International Space Station, noting in particular the strong resistance to designing space environments and technology for women's bodies. Reframing today's space habitats as a "contact zone" between male and female cultures, Gorman asks "How can space habitat design support rather than undermine equality?"

Meanwhile, in Chapter 19, xenolinguist Sheri Wells-Jensen asks, "Will disabled people fly?" She examines the justifications that have traditionally been given for excluding disabled people from space. She argues that the real barrier to including disabled astronauts is the lack of accessible spaceflight systems. Wells-Jensen describes her first-hand experience with a new project aimed at collecting data from disabled participants in parabolic flights. These participants can identify the challenges of inhabiting a zero-gravity environment specific to their disabilities and propose modifications to space technologies to increase accessibility.

In Chapter 20, astrophysicist, space ethics advocate, and volume coeditor Erika Nesvold explores the prospects of labor rights for space workers, interrogating three commonly referenced historical analogs for the growing space industry: the North American transcontinental railroad, the California Gold Rush, and today's international fishing industry. She argues that if proponents of the space economy ignore the labor rights abuses in these industries, they risk reproducing similar forms of labor exploitation in space.

The final three chapters of Part 4 turn to the use (and abuse) of the space environment. In Chapter 21, Ruvimbo Samanga, a lawyer specializing in commercial law and a self-described "avid space advocate" addresses the topic of rights to extraterrestrial resources. Thus far, China, Russia, and the United States have collected lunar materials and returned them to Earth for study. Now companies are proposing to mine lunar and other extraterrestrial resources for profit. Samanga details the intricacies and vagaries of the existing legal regime that is governing activities in space, and cites a need to strengthen this regime. As Samanga argues, this should be informed by African nations' historical and continuing experiences as sites of Western mineral exploitation, and the legal principle of subsidiarity, to ensure fair and equal access to the benefits of space resource extraction for all nations, not just those wealthy ones with space-faring capabilities.

In Chapter 22, space law professor Tanja Masson-Zwaan moves us back toward low Earth orbit, considering the potential conflict growing there between companies launching large satellite constellations and astronomers trying to preserve a dark night sky for their scientific observations. She argues that the astronomical community's initial suggestions for resolving this conflict were not legally tenable and proposes more realistic action to balance the desires of the commercial space industry with the needs of publicly funded science.

Part 4 concludes in Chapter 23, in which bioethicist and environmental scientist William R. Kramer continues this exploration of the environmental impacts of our actions in space. He warns that ignoring these impacts "will establish a precedent of disregard that will continue as we expand beyond Mars." Instead, he proposes applying lessons learned from our environmental history on Earth, including the use of impact assessments and the inclusion of non-Western and Indigenous perspectives to develop a cooperative relationship with the space environment.

Part 5: Visions of the Further Future

Our final section looks beyond our near-future plans for space, in some instances *far beyond* the near future, imagining the birth and childhood of future generations, the discovery of extraterrestrial life, the expansion of terrestrial life out into the universe, and the world of our distant future descendants, possibly so far evolved from us that they can no longer be defined as humans. How can we ensure a better world for these future generations we will never meet? What will they think of us as they look back into history toward our time, when their ancestors were just starting to move beyond their home planet?

Bioethicist and public health researcher Evie Kendal, recognizing that a sustainable human presence in space would require healthy reproduction practices in space, asks "Is there an ethical way to select humans for Noah's Ark?" In Chapter 24, she responds with an analysis of different proposals for ethically preserving diversity in a population of humans headed to space to live there permanently (rather than to visit there momentarily). Along the way, she ponders scenarios in which humanity's survival depends on the willingness and ability of individuals to procreate.

Reproduction in space will mean raising and educating children in space, so in Chapter 25, education researcher Janet de Vigne explores various forms of future education represented in science fiction, including the educational system of Orson Scott Card's 1985 novel *Ender's Game*. She argues for the de-colonization and reinvention of the curriculum as we prepare to move beyond the Earth. Our descendants in space, she argues, will benefit from a child-centered educational system that positions Indigenous and Western ways of knowing as equal.

Meanwhile, in Chapter 26, philosopher and astrobioethicist Octavio A. Chon-Torres considers our responsibilities toward life, both terrestrial and extraterrestrial. He asks whether we, as humans, have an obligation to protect and maintain life, and whether this obligation would extend to any extrater-restrial life we might find. He also considers whether the discovery of extra-terrestrial life, or even simply the expansion of humanity to multiple planets, would improve the human condition.

Along this theme, in Chapter 27 environmental philosopher Andrea Owe argues for space expansion on the grounds that it will help to ensure the con-tinuation of terrestrial life. She sees the protection of the Earth's ecosphere and its continued existence as our primary obligation, and suggests that taking Earth's ecosphere as our starting point for space expansion "will enable future worlds that can sustain and optimize human flourishing." Owe's chapter is evidence that space expansion, if combined with ecological conscience, may serve admirable as opposed to either ambiguous or morally unsavory goals.

The volume's final chapter provides a fitting conclusion to *Reclaiming Space*, in the form of a message to the far future. In Chapter 28, philosopher and posthuman researcher Francesca Ferrando reaches far into the future to our posthuman descendants in space, who over enough time may have evolved beyond today's conception of humanity. Ferrando's message to these future space inhabitants, which includes a letter as well as a poem, is also a message they wish to send to today's terrestrial readers. Their message concerns the enduring connections we have with other humans, both with humans alive today and with the many generations yet to come.

Notes

1. But see (Dick and Launius 2007) for a more comprehensive overview.
2. For reflections on the impact of these events on Indigenous cultures, see Shorter and TallBear (2021).
3. For a professional historian's perspective on the impact of Turner's frontier thesis, see Limerick (1992).
4. For an accessible introduction to ethics that discusses consequentialism, duty-based views, and virtue ethics, see (Fieser, n.d.); for an equally accessible introduction to environmental ethics, see (Cochrane, n.d.).

5. See (Haramia 2021) for further discussion about space ethics in the context of Mars exploration.
6. For accessible introductions to space ethics, see (Green 2021) and (Schwartz and Milligan 2021); see also the academic volume *The Ethics of Space Exploration* (Schwartz and Milligan 2016).
7. We thank Anthony Chan and Chelsea Haramia for helpful suggestions on previous drafts of this introductory chapter.
8. See (Newell 2019) for an engaging discussion of the history of American space art.

References

Armus, Teo. "Trump's 'Manifest Destiny' in Space Revives Old Phrase to Provocative Effect." *The Washington Post*, February 5, 2020. https://www.washingtonpost.com/nation/2020/02/05/trumps-manifest-destiny-space-revives-old-phrase-provocative-effect/.

Cochrane, Alasdair. n.d. "Environmental Ethics." *Internet Encyclopedia of Philosophy*. https://iep.utm.edu/envi-eth/.

Dick, Steven and Roger Launius, eds. 2007. *The Societal Impact of Spaceflight*. NASA SP-2007-4801.

European Union. 2012. "Consolidated Version of the Treaty on the Functioning of the European Union." *Official Journal of the European Union* C 326, no. 47, 131–132.

Fieser, James. n.d. "Ethics." *Internet Encyclopedia of Philosophy*. https://iep.utm.edu/ethics/.

Green, Brian. 2021. *Space Ethics*. Lanham, MD: Rowman and Littlefield.

Haramia, Chelsea. 2021. "Why We Need Space Ethics." *Noema*, November 23. https://www.noemamag.com/why-we-need-space-ethicists/.

Limerick, Patricia. 1992. "Imagined Frontiers: Westward Expansion and the Future of the Space Program." In *Space Policy Alternatives*, edited by Radford Byerly, 249–261. Boulder, CO: Westview Press.

MacIntyre, Alasdair. 2000. "The Recovery of Moral Agency." In *The Best Christian Writing 2000*, edited by Joseph Wilson, 111–136. San Francisco, CA: Harper Collins.

NASA. n.d. "The Artemis Accords: Principles for a Safe, Peaceful, and Prosperous Future." Accessed January 25, 2022. https://www.nasa.gov/specials/artemis-accords/index.html.

Newell, Catherine. 2019. *Destined for the Stars: Faith, the Future, and America's Final Frontier*. Pittsburgh, PA: University of Pittsburgh Press.

Official Newspaper of the Grand Duchy from Luxembourg. "Law of July 20, 2017, on the Exploration and Use of Space Resources." July 2017. https://legilux.public.lu/eli/etat/leg/loi/2017/07/20/a674/jo.

Schwartz, James and Tony Milligan, eds. 2016. *The Ethics of Space Exploration*. New York: Springer.

Schwartz, James and Tony Milligan. 2021. "'Space Ethics' According to Space Ethicists." *The Space Review*, February 1, 2021. https://www.thespacereview.com/article/4117/1.

Shorter, David. 2021. "On the Frontier of Redefining 'Intelligent Life' in Settler Science." *American Indian Culture and Research Journal* 45: 19–44.

Shorter, David and Kim TallBear, eds. 2021. "Settler Science, Alien Contact, and Searchers for Intelligence." *American Indian Culture and Research Journal* 45, no. 1: 1–167.

Smith, Adam. 1776. "Taxes upon the Rent of Houses." *The Wealth of Nations*, Book V, Chapter II, Part II, Article I. https://www.adamsmithworks.org/documents/chapter-ii-of-the-sources-of-the-general-or-public-revenue-of-the-society.

Turner, Frederick. 1920. *The Frontier in American History*. New York: Henry Holt and Company.

UN General Assembly. 1967. "Treaty on Principles Governing the Activities of States in the Exploration and Use of Outer Space, including the Moon and Other Celestial Bodies." 2222 (XXI). http://www.unoosa.org/oosa/en/ourwork/spacelaw/treaties/outerspacetreaty.html.

UN General Assembly. 1979. "Agreement Governing the Activities of States on the Moon and Other Celestial Bodies." RES 34/68. https://www.unoosa.org/oosa/en/ourwork/spacelaw/treaties/moon-agreement.html.

U.S. Commercial Space Launch Competitiveness Act. 2015. H.R. 2262. 114th Congress. https://www.congress.gov/bill/114th-congress/house-bill/2262.

PART 1
THE EVOLUTION AND HISTORY
OF SPACEFLIGHT

2
Neoliberalism

Problematic. Neoliberal Space Policy? Extremely Problematic

Linda Billings

Introduction

Neoliberalism and its kissing cousins, American exceptionalism and libertarianism (a more extreme form of neoliberalism), have been at the heart of the American ideology of human space flight since the United States created a space program in 1958. Until the end of the Cold War, keeping ahead of the Soviets seemed to be sufficient public rationale for the National Aeronautics and Space Administration's (NASA) costly human space flight program, though neoliberal ideology was always present. Once the Cold War ended, neoliberalism came to the fore in public advocacy, and policy, for human space flight. I will expound upon the ideologies of neoliberalism, libertarianism, and American exceptionalism, as well as the ideology of manifest destiny, later in this chapter, but first I will offer some observations of the current state of US space policy and advocacy.

While official US national space policy documents may be light on overtly neoliberal rhetoric, other official statements—speeches by presidents and their appointees, advisory reports, and especially pronouncements by various human space flight advocacy groups—are more explicit in advocating for characteristically neoliberal ideas such as free enterprise, private property rights, and unlimited growth in space. In a September 2021 interview, NASA administrator Bill Nelson reiterated the idea of American exceptionalism and also the idea of manifest destiny, a deeply Christian belief (more on that later). He told political scientist Larry Sabato that space exploration "is fulfilling the destiny of the American character . . . We've always been explorers and adventurers; we've always had a frontier . . . It's part of who we are, as explorers . . . We don't ever want to give it up, that's part of our character as a people" (Nelson 2021).

Linda Billings, *Neoliberalism* In: *Reclaiming Space*. Edited by: James S. J. Schwartz, Linda Billings, and Erika Nesvold, Oxford University Press. © Oxford University Press 2023. DOI: 10.1093/oso/9780197604793.003.0002

Why are these ideologies, these beliefs, dangerous? It's because they represent the interests of a largely white, male, Western contingent of human exploration advocates who are bent on the conquest and exploitation of space, for political and economic reasons, and apparently are indifferent to the voices, views, and concerns of other groups. This volume was conceived to present different voices and views on the human future in space.

Many of the advocacy groups courted by NASA—such as the National Space Society (NSS)—are most explicitly neoliberal, claiming in their mission statements that US space policy should enable private property rights in space, unfettered private-sector exploitation of solar system resources, and colonization of other planetary bodies.[1] That this belief system has been driving US policy for human space exploration and development for so long, and continues to do so today, is disturbing. What the NSS mission statement advocates—and what NASA tacitly if not explicitly endorses—does not and will not serve the welfare and interests of all humankind, and not even the welfare and interests of all Americans. What would be served are the interests of a small, elite, and influential group of wealthy people and the interests of aerospace corporations—which undoubtedly will continue to seek government support, financially and otherwise, for their proposed endeavors.

In my view, "commercialization" of space is a misleading term. All companies/corporations are "commercial"—that is, they are profit-driven, whether they make their money in government contracting or sales in the private sector, or both. Even during the Apollo years, when NASA and its contractors appeared to be working closely together, the contractors were still earning profits.

Nonetheless, pro-space neoliberals and libertarians have been promoting the "commercialization" of space, achieved by a space exploitation and colonization agenda, on Capitol Hill and at the White House for decades. In 2015, an Alliance for Space Development formed to advocate for expanding permanent human presence beyond low-Earth orbit (Alliance for Space Development, n.d.). Members of the alliance include the NSS, the Space Frontier Foundation, the Mars Society, the Tea Party in Space, the Space Development Foundation, and the Space Tourism Society. The founders and leaders of Alliance members are overwhelmingly white American men.

Key Concepts in the Ideology of the Human Exploration of Space

The American ideology of human space exploration and exploitation encompasses some key beliefs: American exceptionalism, neoliberalism, libertarianism, and manifest destiny. To lay a foundation for this critique, I will establish some definitions for these key terms. (To my mind, libertarianism is a more extreme form of neoliberalism, but I will discuss both here.)

American Exceptionalism

In his 1996 book, *American Exceptionalism: A Double-Edged Sword*, the influential political scientist Seymour Martin Lipset described what he called the ideology of "Americanism," which he said could be described in five words: liberty, egalitarianism, individualism, populism, and laissez-faire (Lipset 1996). Lipset explained how this belief system shaped, and continues to shape, US organizing principles and political institutions.

The bright-and-shiny edge of Lipset's double-edged sword of exceptionalism represents freedom and opportunity, leadership, and setting a good example to the world. NASA has promoted these values as key aims of the agency's activities. The dark-and-jagged edge represents the drive for US global dominance—to be the world's only superpower in the military, economic, technological, and aerospace arenas, to be *the* leader. NASA and other US government organizations continue to advocate for US dominance in space. This jagged edge of exceptionalism pushes forward capitalism and development into the space domain, whenever and wherever possible, according to the principle that those who get there first get the most.

American exceptionalism, as it appears in space exploration rhetoric and policy, looks bright and shiny—it's about the U.S. leading in space exploration for the benefit of humankind. Beneath that shiny surface, though, lies neoliberal/libertarian ideology. This ideology declares outer space a wide-open frontier, open to exploitation and colonization, ripe for "commercialization," unfettered by government oversight.

Neoliberalism

In his book *A Brief History of Neoliberalism*, anthropologist David Harvey observes that neoliberalism has become a hegemonic—that is, culturally dominant—mode of discourse in the global political economy. Indeed, it has. It is a Western mode of discourse and thinking. This theory of political economy, according to Harvey:

> Proposes that human well-being can best be advanced by liberating individual en-
> trepreneurial freedoms and skills within an institutional framework characterized
> by strong private property rights, free markets, and free trade . . . It has pervasive
> effects on ways of thought to the point where it has become incorporated into the
> common-sense way many of us interpret, live in, and understand the world . . . The
> process of neoliberalization [deregulation, privatization, and withdrawal of the
> state from many areas of social provision] has, however, entailed much 'creative
> destruction' . . . of prior institutional frameworks and powers. (Harvey 2005, 2–3)

Deregulation, privatization, and withdrawal of the state from oversight of the human exploration and exploitation of space is exactly what has been taking place in recent decades. Though exploration advocates claim that the human exploration and exploitation of space is for the benefit of humankind, neoliberalism gives the advantage to corporations, whose prime directive is to make profits.

While it has much deeper, Western European roots, the ideology of neoliberalism got a major kick-start in the twentieth century from Milton Friedman and the Chicago school of economics (Ebeling 2006). Friedman was not only a neoliberal economist but also an avowed libertarian. *The New York Times* described him as "the grandmaster of free-market economic theory in the postwar era and a prime force in the movement of nations toward less government and greater reliance on individual responsibility" (Noble 2006). His thinking continues to inspire today's neoliberals and their more extreme libertarian cousins.

Libertarianism

As to libertarianism, David Boaz (1997, 3), executive vice president of the Cato Institute, a libertarian think tank, says the framework for this theory of political economy is "liberty under law and economic progress." Key concepts of libertarianism include individualism, individual rights, the rule of law, limited government, and free markets. "We need a limited government to usher in an unlimited future," he asserts (5). "More reliance on markets and individual enterprise would mean more wealth for all of us" (4).

Manifest Destiny

And now, on to the ideology of manifest destiny. The belief in manifest destiny is deeply Christian, brought to North America by Puritan immigrants in the seventeenth century and adopted by American politicians in the nineteenth century to serve expansionist goals. As historian Anders Stephanson (1995, 5) writes, the origins of this belief "lay directly in the old biblical notions . . . of the predestined, redemptive role of God's chosen people in the Promised Land: providential destiny revealed." This belief remains at the core of the American ideology of the human exploration of space. It is a dangerous belief, as it propels efforts to extend Western colonial expansion and exploitation into outer space, which is contrary to international law, that is, the 1967 United Nations Outer Space Treaty.[2]

The Neoliberal Rhetoric of Space Exploitation and Colonization

From the end of the Apollo era to the present, the ideology of human space flight, and the rhetoric of human space flight advocacy, has been sustained in public discourse in large part by so-called grassroots space advocacy groups, such as the aforementioned NSS; the Space Studies Institute; the Space Frontier Foundation (SFF); and the Mars Society.

The NSS says its rationale for promoting space settlement is "survival of the human species." Among the values and beliefs articulated in the Society's "vision" for space exploration and development are "prosperity-unlimited resources," "growth-unlimited room for expansion," "individual rights," "unrestricted access to space," "personal property rights," "free market economics," and "democratic values" (National Space Society, n.d.). Key words here are "unlimited" and "unrestricted"—top priorities in neoliberal/libertarian ideology.

In 1977, physicist Gerard K. O'Neill formed his own advocacy group, the Space Studies Institute (SSI), to promote his space colonization agenda (McCray 2013, 3). In 1988, other free-market colonization advocates created the SFF to promote "opening the space frontier to human settlement as rapidly as possible" (Space Frontier Foundation, n.d.). The Mars Society, established in 1998, also advocates space colonization in its "founding declaration":

> Civilizations, like people, thrive on challenge and decay without it. . . . The settling of the Martian New World is an opportunity for a noble experiment in which humanity has another chance to shed old baggage and begin the world anew; carrying forward as much of the best of our heritage as possible and leaving the worst behind . . . No nobler cause has ever been. (Mars Society 1998)

The rhetoric of the Mars Society, to single out one of these groups, is, to be succinct, absurd. All of these groups claim human exploration, exploitation, and colonization of space will benefit everyone, "humanity." As of this writing, Earth's human population totaled close to 8 billion people (Worldometers.inf, n.d.). Close to three and a half billion people are living in poverty—defined by the World Bank as living on less than $3.20 per day in lower-middle-income countries, and $5.50 a day in upper-middle-income countries (The World Bank 2018). There is no good reason to believe that humans will transform themselves, their values, their beliefs, if living beyond Earth.

And who is "everyone," by the way? So far, the so-called "space frontier" is predominantly open to the few nations with human space flight programs and some (very) wealthy Western white men and their special guests (who generate lots of publicity).

US Space Policy, Reagan to Biden, in a Nutshell

US space policy has always been about American "leadership" (read: dominance) in space. "Opening the space frontier" is a persistent theme. Though certain efforts to "privatize" space flight began in the Carter administration, the administration of US President Ronald Reagan kick-started the drive to develop and exploit outer space,[3] which continues to pick up speed.

Policy documents of the George H. W. Bush administration reaffirmed that the "prime objective" of the US space program was "to open the space frontier" (National Space Council 1990, 17). NASA declared in a report on Bush's Space Exploration Initiative—a proposal to send people back to the Moon and on to Mars—that "the imperative to explore" is embedded in our history . . . traditions, and national character" (1989, 1–1, 1–4).

President Bill Clinton did not overtly promote the neoliberal space development agenda that his predecessor and successor did. But his administration did not break with policy or rhetorical tradition. A proclamation from the Clinton White House on the occasion of an Apollo anniversary stated, "Space exploration has become an integral part of our national character, capturing the spirit of optimism and adventure that has defined this country from its beginnings" (National Apollo Anniversary Observance 1994).

In the George W. Bush administration (2001–2009), White House Office of Science and Technology Policy Director John Marburger took up promoting a neoliberal space agenda on behalf of the President. Bush Jr. had recapitulated Bush Sr.'s call to send people back to the Moon and on to Mars in his so-called "vision for space exploration," announced in 2004. At a space symposium in 2006, Marburger said:

> Our national policy . . . affirms that, 'The fundamental goal of this vision is to advance U.S. scientific, security, and economic interests through a robust space exploration program.' The idea is to begin preparing now for a future in which the material trapped in the Sun's vicinity is available for incorporation into our way of life. (Marburger 2006)

The idea of bounding "our economic sphere" to include the entire Solar System has continued to resonate with human exploration advocates inside and outside the White House.

NASA Administrator Michael Griffin, a George W. Bush appointee, claimed that the aim of space exploration is "to make the expansion and development of the space frontier an integral part of what it is that human societies do."[4] He has also stated that "I believe that Western thought, civilization, and ideals represent a superior set of values, which are irretrievably linked to expansion."[5]

By the time of the George W. Bush administration, the pro-space-colonization crowd had branded itself as a "NewSpace" movement. Figureheads of the NewSpace crowd include billionaire capitalists Sergei Brin and Larry Page, Elon Musk, Jeff Bezos, and Richard Branson. NewSpace people were, and are, overwhelmingly college-educated white males.

In his book, *The Visioneers*, Patrick McCray (2013, 263) observed, "The NewSpace ethos of the early 21st century [is] shot through with . . . libertarian ideology. . . . NewSpace is relentlessly capitalistic." Journalist Paulina Borsook explored the culture of Silicon Valley high-tech entrepreneurs, which includes many of the leading figureheads and financiers of the NewSpace "movement." This community, Borsook wrote, embraces "the attitude, mind-set and philosophy [of] libertarianism" (2000, 3). She found this mindset "confusing, for this passionate libertarian population has for the most part only experienced good things, and not bad, from government" (5). And she noted that what she calls "technolibertarianism . . . is morbidly hypermale" (138).

In an April 15, 2010, speech at NASA's Kennedy Space Center, President Barack Obama advocated for establishing a permanent human presence in space. "In fulfilling this task, we will not only extend humanity's reach in space—we will strengthen America's leadership here on Earth . . . Leading the world to space helped America achieve new heights of prosperity here on Earth, while demonstrating the power of a free and open society to harness the ingenuity of its people" (Obama 2010).

NASA Administrator Charlie Bolden, Obama's appointee, told the President's Council of Advisors on Science and Technology (PCAST) on July 14, 2015, that the United States is all about exploration and expansion through the colonization of new places. And space will be next. Said Bolden: "It is the story of the journey West, you know, of the early pilgrims and other people landing on the shores of the United States, but then just not being satisfied and continually moving west and exploring, and so, we're now trying to get off this planet and farther out" (Henry 2015).

The Obama administration embraced and turbocharged the neoliberal heart of national space policy, advancing hundreds of millions of dollars in direct and indirect subsidies and billions more in contracts to so-called "commercial" space businesses, both established and new, many of which were already well-heeled. The Trump administration traveled even further down this road. As of this writing, the Biden administration has not shown any indication that it intends to change this trajectory.

None of these officials seemed to be concerned about how many people, societies, cultures around the world, and also in the States, could be—and often are, as I am—offended by their statements.

Welfare for Space Libertarians

Since President Obama took office, NASA stepped up its use of so-called Space Act agreements (SAAs) to provide financial and in-kind support to space projects.[6] These agreements are not subject to Federal Acquisition Regulations and are intended "to enhance NASA's ability to advance cutting-edge science and technology and to stimulate industry to start new endeavors" (NASA Office of Inspector General 2014, 3). NASA's Office of Inspector General (OIG) reported in 2014 that the number of NASA SAAs rose by more than 29% between fiscal years 2008 and 2012. The OIG found that

> NASA's use of funded SAAs is a relatively recent occurrence and to date most funded SAAs have related to the Agency's efforts to develop commercial spacecraft capable of transporting cargo and crew to the International Space Station (ISS or Station). Since 2006, NASA . . . entered into 15 funded SAAs with eight private companies ranging in value from $1.4 million to $480 million, with a total value of more than $2.2 billion. (2014, 2)

Though many companies have benefited financially from SAAs, I will single out one in particular: SpaceX, a privately owned launch company founded by

Elon Musk, who at the time of this writing (November 2022) was the world's richest man, with a net worth approaching $200 billion. NASA awarded SpaceX a funded SAA valued at $396 million in 2006 to work on cargo and crew space transportation capabilities. SpaceX got two more funded SAAs from NASA worth $75 million (2011) and $460 million (2012). While Musk may not have been the world's richest man when SpaceX received these funds, he was already a billionaire.

In addition to the federal government, state and local governments in Florida, Texas, and New Mexico have been subsidizing "commercial space" development: for example, Virgin Galactic (New Mexico), headed by Richard Branson, whose net worth at the time of this writing (November 2022) was close to billion. Another example is Blue Origin, the rocket company owned by Jeff Bezos (the world's second-richest person at the time of this writing, with a net worth close to $113 billion), who landed around $40 million worth of state, regional, and local incentives to build his launch facility in Florida (Klotz 2015).

While Bezos, Musk, and other space libertarians argue for letting the private sector take over the development of space, they are more than willing to seek government handouts to ensure that their space businesses are profitable. "NewSpace" is nothing new—it is business as usual. Journalists and taxpayers would do well to pay closer attention to how public monies are spent toward advancing the space dreams of the ultra-wealthy.

Conclusion

Space libertarians are, in my opinion, the black flies of the space community—annoying creatures who sometimes bite via ad hominem/ad feminem attacks against anyone who disputes the validity of the conquest-and-exploitation model (I have been on the receiving end of such attacks). They swarm every year to NewSpace, Mars Society, and other similarly oriented conferences, propagating the neoliberal/libertarian ideology of space exploitation and colonization.[7] They are promoting the belief that humanity is destined to propagate throughout the universe. (And please keep in mind that destiny is a religious belief.)

The advocacy groups discussed here are promoting a dangerous ideology as a foundation for national space policy. Why is it dangerous? Because it's elitist, exclusive, and very, very Western. Advocates for space colonization and other forms of exploiting extraterrestrial resources are overwhelmingly Western, white, and male. We females constitute

51 percent of the world's population, and a majority of people on Earth are not white.

How will all of humanity benefit, in the short and long run, from the exploitation of extraterrestrial resources and the colonization of other planetary bodies?

We've had no meaningful national or international dialogue on goals and objectives for space exploration that could benefit all of humanity (I don't care what the advocates say, asteroid mining would benefit mining companies, not humanity). Pretty much any "dialogues" that the pro-space community may point to have been organized by space advocates.

There's a dissonance between nineteenth-century Western/Christian-centric and post-postmodern humanistic, and certainly non-Western and non-Christian, world views. Adherents of the former implicitly believe that humanity—specifically, Western white Christian male humanity—is the pinnacle of creation and has dominion over all of creation. Adherents of the latter, including myself, believe that humans are one of many intelligent terrestrial species, that all life on Earth exists in an interdependent web, and that humanity is responsible for stewardship of Earth, and beyond.

Yes, the US political economy is decidedly neoliberal. However, we do not have a libertarian-style free-market economy in the United States. Not only do we have government regulations and trade protections, but we also have government contracts with bonus awards, government research and development subsidies and tax credits, government bond financing of construction projects, and government promotion (financially and otherwise) of new business development (at the federal, state, and local levels). All of these benefits accrue primarily to business. Corporate executives who espouse the libertarian value of limited government are quick to benefit from government support, in whatever form it's offered.

Science and technology—purportedly the business of NASA—are not value-free. They are value-laden. The science and technology that will, according to human exploration advocates, enable the colonization of other planetary bodies and the exploitation of extraterrestrial resources, will be deeply laden with the values of neoliberalism. What is the public value of human exploration of space? What is the value of human missions to the Moon and Mars—should they ever occur—to the tens of millions of people in the world who are living in extreme poverty (Grameen Foundation, n.d.)?

People who care about the future of life on Earth—not just human life but all life, and even nonliving things, as Daniel Capper writes in Chapter 16 of this volume—need to resist this neoliberal bent toward the colonization and exploitation approach to space exploration. As colonization and exploitation have demonstrated here on Earth, it's a dangerous belief system, and it certainly does not serve the interests of humankind, no matter what the advocates of human exploration say about how space exploration is "for the benefit of all mankind [sic]."

The argument that Earth is dying—resources depleted, environment ruined—is an argument often used to justify the expansion of human society (I do not want to say "civilization") into space. (And which "society" are we talking about? As Evie Kendal writes in Chapter 24 of this volume, choosing a sufficiently diverse/representative group of humans to establish a space colony would be an extremely challenging task.) As Ingrid LaFleur writes in Chapter 13 of this volume, humans are planning to extend their presence into space without first healing themselves on Earth. Humanity has trashed its nest, and we need to clean it up—our natural, sociopolitical, and cultural environments—before we get serious about extending human presence into off-world environments.

Notes

1. See, for example National Space Society (n.d.); also see Space Frontier Foundation (n.d.).
2. For a brief summary of the history of the idea of manifest destiny, see Billings (2015). For a brief history of the ideology of manifest destiny in American culture, see Billings (2007).
3. See, for example, a Reagan policy statement of 1988 emphasizing the goal of stimulating the commercial development and uses of space ("Presidential Directive on National Space Policy" 1988).
4. Griffin made these remarks at a conference sponsored by the Center for Strategic and International Studies, November 1, 2005, Washington, DC. The author attended this event.
5. Griffin continued to expound on these beliefs at a meeting of the NASA Advisory Council's science subcommittees in Washington, DC, on July 6, 2006. The author attended this event.
6. The National Aeronautics and Space Act of 1958 gives NASA the authority to enter into a variety of agreements with organizations to advance its objectives. These agreements range from traditional contracts to "other transactions," which NASA refers to as Space Act Agreements.
7. The author has listened in on these conferences as time has permitted.

References

Alliance for Space Development. n.d. "Alliance for Space Development—About Us." Accessed January 6, 2022. https://allianceforspacedevelopment.org/about-us/.

Billings, Linda. 2007. "Ideology, Advocacy, and Space Flight: Evolution of a Cultural Narrative." In *Societal Impact of Space Flight*, edited by Steven J. Dick and Roger D. Launius, 483–500. SP-200-4801. Washington, DC: NASA History Division.

Billings, Linda. 2015. "More on Manifest Destiny." *Doctorlinda* (blog). March 27, 2015. https://doctorlinda.wordpress.com/2015/03/27/more-on-manifest-destiny/.

Boaz, David. 1997. *Libertarianism: A Primer*. New York: Free Press.

Borsook, Paulina. 2000. *Cyberselfish: A Critical Romp through the Terribly Libertarian Culture of High-tech*. New York: Public Affairs.

Grameen Foundation. n.d. "Grameen Foundation: Our Impact." Accessed November 26, 2021. https://grameenfoundation.org/solving-poverty/our-impact.

Ebeling, Richard M. 2006. "Milton Friedman and the Chicago School of Economics." Foundation for Economic Education. December 1, 2006. http://fee.org/freeman/milton-friedman-and-the-chicago-school-of-economics/.

Harvey, David. 2005. *A Brief History of Neoliberalism*. New York, NY: Oxford University Press.

Henry, Michael. 2015. "PCAST Discusses New Frontiers in Human Space Exploration." American Institute of Physics. July 24, 2015. https://www.aip.org/fyi/2015/pcast-discusses-new-frontiers-human-space-exploration.

Klotz, Irene. 2015. "Florida County Sweetens Bid for Jeff Bezos' Rocket Company." *Reuters*, September 2, 2015. https://www.reuters.com/article/idUSKCN0R14S120150901.

Lipset, Seymour Martin. 1996. *American Exceptionalism: A Double-Edged Sword*. New York: W.W. Norton.

Marburger, John. 2006. "Keynote Address." 44th Robert H. Goddard Symposium. Greenbelt, MD. March 20, 2006.

Mars Society. 1998. "The Mars Society: Founding Declaration." Accessed January 6, 2022. http://www.marssociety.org/home/about/founding-declaration/.

McCray, Patrick. 2013. *The Visioneers: How a Group of Elite Scientists Pursued Space Colonies, Nanotechnologies, and a Limitless Future*. Princeton, NJ: Princeton University Press.

NASA. 1989. *Report of the 90-Day Study on the Human Exploration of the Moon and Mars*. Houston: NASA Johnson Space Center. November 1989.

NASA Office of Inspector General. 2014. "NASA's Use of Space Act Agreements." Report No. IG-14-020, 5 June 2014, https://oig.nasa.gov/audits/reports/FY14/IG-14-020.pdf.

National Apollo Anniversary Observance. 1994. A Proclamation by the President of the United States of America. July 19, 1994. Washington, DC: Office of the President.

National Space Council. 1990. *Report to the President*. Washington, DC: Office of the President.

National Space Society. n.d. "NSS Statement of Philosophy." Accessed January 6, 2022. https://space.nss.org/nss-statement-of-philosophy.

Nelson, Bill. 2021. "Interview with Bill Nelson." By Larry Sabato. UVA Center for Politics. October 19, 2021. YouTube video, 1:01:04. https://youtu.be/9hH1XEqKlTs.

Noble, Holcomb B. 2006. "Milton Friedman, Free Markets Theorist, Dies at 94." *The New York Times*, November 16, 2006. http://www.nytimes.com/2006/11/16/business/17friedmancnd.html?pagewanted=all&_r=0.

Obama, Barack. 2010. "President Obama's Speech at Kennedy." April 15, 2010. https://www.nasa.gov/about/obama_ksc_pod.html.

"Presidential Directive on National Space Policy." 1988. Published February 11, 1988. Accessed January 6, 2022. http://www.hq.nasa.gov/office/pao/History/policy88.html.

Space Frontier Foundation. n.d. "About the Space Frontier Foundation: Who We Are: The Space Foundation Credo." Accessed January 6, 2022. http://newspace.spacefrontier.org/about/.

Stephanson. 1995. *Manifest Destiny: American Expansion and the Empire of Right*. New York: Hill and Wang.

The World Bank. 2018. "Nearly Half the World Lives on Less Than $5.50 a Day." Press release, October 17, 2018. https://www.worldbank.org/en/news/press-release/2018/10/17/nearly-half-the-world-lives-on-less-than-550-a-day.

Worldometers.info. n.d. "Current World Population." Accessed January 6, 2022. https://www.worldometers.info/world-population/.

3

Space from Āfār

From Africa across the Indian Ocean to the Pacific

Mukesh Chiman Bhatt

Orientation

Cultural mythologies from around the world are often based upon the stars seen in the local heavens. China and India with their broad diasporas and wide-ranging cultural hegemonies have a tradition of thinking broadly about space. These epistemologies are generally ignored in the treatment of space and space exploration. Should these cultures and traditions have no place in the future of space exploration? Do we want to recreate unjust and unethical institutions in space that already exist on Earth? Is it possible to create better institutions on Earth, which we can then export outwards? Or should space be seen as a crucible for systems and institutions which can then be brought back to Earth? Perhaps new, innovative, or existing perspectives from these marginalized regions are relevant to the settlement and exploration of space, whether for multiple uses.

Zheng He (1371–1433 or 1435) likely did not visit Afar, the Ethiopian region that is the presumed origin of the global human diaspora. As a mariner and diplomat for the Chinese Ming Dynasty (1368–1644), he sailed around the Indian Ocean confirming China's knowledge and hegemony of the known world. It also confirmed the Mandate of Heaven conferred upon the Yongle Ming emperor Zhu Di (r. 1402–1424; Duyvendak 1939) upon receipt of a mythical creature (a giraffe) as a gift at Malindi in Kenya. Almost sixty years later, Vasco da Gama (following Bartolomeo Diaz, another Portuguese navigator) rounded the southern tip of Africa and made his way to India relying on a local pilot from Mombasa in what is now modern Kenya, becoming the first European to travel to India by sea. The European voyages led to exploitation, imperialism and extractivism. With frequent tributes, the Xuande Ming emperor Zhu Zhanji (r.1425 to 1435) stopped all voyages on the grounds of impoverishing the tributary regions (Chang

Mukesh Chiman Bhatt, *Space from Āfār* In: *Reclaiming Space*. Edited by: James S. J. Schwartz, Linda Billings, and Erika Nesvold, Oxford University Press. © Oxford University Press 2023. DOI: 10.1093/oso/9780197604793.003.0003

1974; Sen 2009). Zheng He is now better known in the West through a TV series, novels, video games, and in science fiction: for example, an interstellar trading society in the novel *A Deepness in the Sky* (Vinge 1999), and a starship in the TV series *Star Trek: Picard* (Goldsman et al. 2020). Despite the isolationism, technological progress in the region was slowed but not hindered (see discussion below).

The first space lawyer, Andrew Haley, proposed (1956) a meta-golden rule—"treat them as they wish to be treated," riffing on the better known "do as you want to be done by." Instead, Europeans meeting others chose the third of three freedoms proposed by anthropologist David Graeber and archaeologist David Wengrow (2021): move away, disobey, or create and transform social relationships. They transformed the tribal and non-urban cultures of the Pacific and Africa into savages, noble or otherwise; by the late nineteenth century the Asian civilizations were deemed despotic and degenerate through nineteenth-century law professor James Lorimer's (and others') hierarchy of civilizations (Koskenniemi 2016), echoed by the USA at the Nuremberg trials (Wilke 2009). European conceptions of the East somehow changed sometime after philosophers Denis Diderot and Jean le Rond d'Alembert's *L'Encyclopédie* was published in France between 1751 and 1772 (Bhatt 2011). Accepted even today through political scientist Samuel Huntington's myth of the clash of civilizations (Bottici and Challand 2013), there continues the denial of any useful knowledge produced by these cultures (Ambrogio 2020, especially chapter 3). Not much has changed since. A seminar in 2008 at the University of London in the UK considered racist a cartoon showing tribal Africans stabbing with spears at a cinema screen: African tribes are therefore primitive and ignorant. Booing, yelling, and throwing things at a cinema screen by educated Europeans is instead civilized and normal. Surely nothing is wrong with a Western perspective and scholarship? Unlike the seminar leaders, I intend to treat each culture on its own terms. In contrast, American historian Donald F. Lach and coauthors (1994–1998) describe a cross-cultural exchange of technology and ideas across borders and other differences.

In Space

Observations in nature and the physical world coincide across and link many cultures. Despite the coincidence of such observations, each culture may interpret these phenomena in different ways that are consistent with their own internal and coherent frameworks. Various cultural frameworks and their interpretations cannot be treated identical to those within the Western science paradigm. The Western scientific method is itself not coherent or consistent, being largely a methodology incorporating various tools derived from numerous global cultural inputs and sources. Differing cultural interpretations of natural and physical phenomena have their own internal logics despite similarities to the modern scientific method but can introduce contradictions and inconsistencies. Similarities can also arise from the common human experience of embodiment. This secular Western scientific methodology can only be equated with caveats to any single or multiple indigenous or other traditional frameworks involving theology, spirituality, or alternative forms of what may be termed "science." An example would be the possible correspondence between Western science and the Indian and Buddhist cultures. Although Western science is seen as secular, it is the end result of many centuries of Christian and Eurocentric (Greco-Roman) ideas with input from the Arabs and elsewhere. Christian and Arab influences would, of necessity, have been constrained at some stage by theological considerations. In contrast, some Indian cultures, particularly Jainism, Sāṃkhya, Buddhism, and Cārvāka, are atheist; the school of Hindu philosophy called Mīmāṃsā considers all deities to be mental constructs (Haribhadrasūri in Jain ed. 1981) while the school of Hindu philosophy called Vedānta views the universe as monist. Extending this to the spiritual beliefs of African, Australian Aboriginal, or Polynesian cultures thus merits an extreme caution in assuming the equivalence of any seeming homology.

Indian philosophers (Bahadur 1978; Rao 1987) and the Buddhist cultures of India and the Far East (Pánikar 2010) share with Taoism a common concept of the origin of the universe, a "void or vacuity" similar to the quantum vacuum postulated by modern physics (Boi 2011; Bhatt 2021). Given interactions between China, India, and Southeast Asia (Bagchi 2011) through Buddhist travels and the universities established as a result in India, important scientific and philosophical concepts and theories would have been exchanged between India and China. Buddhism continues to be a dominant force in China, the Koreas, and Japan, as well as in the Southeast Asian nations. The Japanese TV series *Monkey* (Wu 1978; Arikawa et al. 1978) follows a pilgrimage by the monk Tripitika to the West, here India, to obtain various texts. The first instance of the circle or dot used as a symbol for the number equivalent of zero was probably the result of trade between India and China in Southeast Asia. Hindu and Buddhist cultural dominance and hegemony for most of the last two millennia resulted in Angkor Wat in Cambodia, a single Hindu temple the size of a small city, and Zen Buddhism amongst other such legacies.

Navigation in the Pacific

Stellar positions helped guide Polynesian and Micronesian voyages, changing only their time of rising with the seasons. Each star provides a bearing for navigation at its rising and setting for a specific destination (Gatty 1958). A sequence of stars could be memorized as a previous one rose too high or set and the bearing redirected toward a newly risen star. Recognizing that stellar movement and elevation over different islands followed the same pattern provided navigators (Holmes 1955) with a sense of latitude (although not the latitude as we can recognize it today) and that allowed sailing along a line of latitude before turning to reach the desired destination. Such knowledge was often codified in star compasses with systems comprising, in general, a few dozen and up to 150 stars with their bearings. That such star compasses (Halpern 1985; Thompson, n.d.) exist suggests that tangible technologies are well integrated into the mythos and cultural framework of the Polynesians. It denies the notion of a primitive culture as postulated in the 19th and 20th centuries.

Australasia and the Aboriginal People

Aboriginal people in Australia also navigate by the stars. An informative website that lists several references useful for further reading is Australian Indigenous Astronomy,[1] run by Aboriginal Astronomy, a group of researchers who acknowledge through their anonymity (and academic references on the site) the importance of their Aboriginal informants. There are many Aboriginal peoples of Australia and given this diversity, this implies a diversity of astronomical traditions, each with their own cosmology expressed in a particular and unique manner. This expression is passed on orally, in ceremonies, and in the many artworks that can be found on the Australian continent. Navigation amongst such aboriginal peoples includes following particular stars or correlating various landscape features to the distribution of stars on a map. Stars, constellations, and other celestial features such as nebulae or clouds in the Milky Way are often named after or signified by major figures or events in the corresponding mythologies for local cultures. In addition, star maps had often been developed to teach navigation outside the limits of the known local country (Australian Indigenous Astronomy 2021). Certain patterns of stars indicated waypoints, which were usually waterholes or turning places on the landscape. These maps were passed on through creating and teaching songlines—paths across the land or the sky—to those who had not traveled the route so described. These songlines helped the oral transmission of local knowledge. Many of these routes follow the same routes as modern roads. This is the result of directions given by the local aboriginal guides and interpreters who used their own songlines to direct European explorers from 1845 to 1846.

Stars amongst the Aboriginal peoples are the homes of ancestors, animals, plants, and spirits. They also served as calendars, as a lawbook, and as a guide to all aspects of daily life and culture. The most subtle changes in brightness, color, and position of the stars correlated with ancestor spirits and served as a mnemonic for the lessons of life and society. These lessons were often in the form of stories associated with the stars or the patterns of stars recognized in the sky. They also correlated with natural phenomena for the finding of food and water during different seasons. Aboriginal astronomers have also used star brightness to predict seasonal change and have been known to recognize the relative variability of numerous stars including Betelgeuse and Aldebaran. More recently (Hamacher 2018), the International Astronomical Union (IAU) has recognized the many stories behind the star names used by different Australian Aboriginal peoples. The Pleiades appears in Australian aboriginal cultures as it does in many other cultures in the world, as do Orion and the Milky Way—other constellations such as the Emu in the Sky are more peculiar and singular in their naming. Current star names are mostly derived from the Arabic (Allen 1963). The IAU has also recognized new star names drawn from Chinese, Coptic, Hindu, Mayan, Polynesian, and South African cultures.

India, China, and the Arabs

The zero and number placement concepts were developed in response to the needs of Indian cosmology. Hindu-Arabic numerals, although introduced to Europe earlier, were finally adopted by Europeans in the fifteenth century CE and led to the development of Newtonian mechanics (Bhatt 2021) and later the space programs of Nazi Germany, the United States, and the Soviet Union. Āryabhaṭa (Indian mathematician and astronomer, sixth century CE) in his *SuryaSiddhānta* and later works laid out the mechanics of eclipses while his successor Brahmagupta (Indian mathematician-astronomer, seventh century CE) proposed a force of universal attraction identical to that of Newtonian gravitational attraction (Subbarayappa and Sarma 1985; Subbarayappa 2013). Given the texts by Arab traveler Al-Birūni and the existence of the Silk Road (a network of trade routes connecting East to West, 2 c. BCE–18 c. CE), such concepts likely traveled and were further developed by Arab intellectuals in Baghdad and later passed to German astronomer Johannes Kepler and other European scientists (Al-Khalili 2010; 2011). Of note is the construction in the early seventeenth century CE of nineteen observational instruments by the then-ruler of Jaipur in India, ranging from observatories to the world's largest sundial, in various cities, and known collectively as *Jantar-Mantar* ("calculating machine"). These were used for measuring time, predicting eclipses, tracking locations of major stars as Earth orbits around the sun, ascertaining the declinations of planets, and determining the celestial altitudes and related astronomical data.

Planetary scientist Jonathan Lunine (2005, 12) mentions how Chinese philosopher Qi Meng (c. 1st–3rd centuries CE) viewed the cosmos as an infinite three-dimensional universe. The philosopher Deng Mu (1247–1306) later elaborated on this and realized the plurality of worlds (Lunine 2005, 13). Given the Chinese view at the time that historical events are presaged by events in the heavens, any changes in the celestial skies were carefully noted—the best example is the 1065 CE supernova that created the Crab Nebula, noted by Chinese astronomers of the time and by modern astronomy, but not by European astronomers of that time. The Chinese view of history is in contrast to the stereotypically linear view prevalent in the West even though either of these is probably better viewed as secondary nonlinear phenomena arising from related and multiple human causes. The science and technology of China are discussed in the massive multipart, multivolume works by British historian Joseph Needham and coauthors (1954–present). A similar attempt for India is made by Indian philosopher D. P. Chattopadhyaya and coauthors in their *History of Science, Philosophy and Culture in Indian Civilization* series (1990 onwards). A detailed discussion of intellectual and other imports into Europe from Asia after 1500 CE is presented in the three-volume work by Lach and coauthors (1994–1998). An extended discussion of these publications and their content is beyond the scope of this work (but see Alvares 1991 for a decolonial perspective).

Islam as a culture extended at one time from Spain through North Africa, into large parts of Africa and through Asia to Indonesia and beyond. It is, however, centered on Arabia and Arab culture. As with Indian, Daoist, and Shinto characters, Islamic entities have been transformed through modern technology and interpreted into novel conceptions: *djinn* (English *jinn* or *genie*) and ghouls (from Arabic *ghul*) may be viewed as extraterrestrial life or manifestations of advanced technologies living in this or other dimensions. Islamic activities in astronomy, mathematics (Harwood 2012), and translation have become well known through recent scholarship (Al-Khalili 2010, 2011; Determann 2015, 2018, 2020).

Indian mathematician-astronomer Āryabhaṭa's (476–550 CE) explanation of eclipses, the astronomical instruments of Jantar-Mantar, and technology exported from China into Europe, as well as Muslim-Arab mathematician Ibn-Haytham's experiments in optics, presage those of Sir Isaac Newton and emphasize the fact that tangible technologies continue, as in previous millennia, to be part of so-called degenerate and primitive cultures and that these technologies are well-integrated into the cultural milieux of the civilizations.

Africa on the Indian Ocean

Star Trek: The Original Series (*TOS*) was first shown in Kenya from 1966 to 1968 before it was shown in the United Kingdom, which is approximately the same time it premiered in the United States. Lt. Nyota Uhura, the *USS Enterprise* communications officer played by Black American actor Nichelle Nichols in *Star Trek*, is named after a speculative feminized word in Swahili: *Uhuru* is a gender-neutral Swahili word meaning "independence"; *nyota* means "star". The character was born in Nairobi and later lived in Mombasa. Later, in the *TOS* novel *The Face of the Unknown*, Uhura was reminded of the long-rains season in Kenya. *Star Trek* fandom[2] notes that in novelizations and related works, *Kenya* appears as a mining freighter,[3] a division of the United States of Africa,[4] and a Federation mining colony. Regrettably, Induna, a twenty-third-century president of the Keep Earth Human League, was also from Kenya.[5]

Star Trek is an important influence in this region because Swahili is the official language of Kenya and Tanzania (extending inland and further south to Congo and Zimbabwe). However, it also draws from and feeds into the postcolonial independence of the African countries in the 1960s, as well indicating the new status of women and of Africans. The Broglio ground station run by the Italians on behalf of ESA near Malindi in Kenya is the site of a geological feature known as Hell's Kitchen: the 1981 *Rough Guide to Kenya* (now out of print) describes it as reminiscent of a *Star Trek* landscape. The ground station previously acted as a command center for launches from the offshore San Marco launch platform, in use until 1988, and from which the 1970 *Uhuru*, carrying the first ever satellite-mounted x-ray telescope, was launched. Wikipedia provides a very informative timeline of telescopes, observatories, and observing technologies (Wikipedia, n.d.) to which the interested reader is referred—even a cursory discussion is beyond the scope of this work. China and Peru stand out in the post-Egyptian and post-Babylonian period before the Common Era. The earliest such works listed for the Common Era are by the Indians.

The Kenya coast from Somalia down to Mozambique has been open to Roman, Arab, and Egyptian trade and interaction with the native tribes for several millennia, also visited by the Indians and the Chinese. The European route to India turned eastwards from Mombasa in Kenya. The Swahili, resulting from intermarriages between Indigenous Bantu and Arabs of the East African coast, are named after the Arabic (*S(u)ahel*) for coast; this is also the lingua franca. As a trade entrepôt and stopover, it saw many European and other cultures through to South Asia and the Far East. As such the local populations have, for many centuries if not millennia, been exposed to new peoples and novel conceptions and ideas for at least that length of time. The arrival of new technologies after the industrial revolution and after the scramble for Africa in the nineteenth century is not considered a watershed except insofar as these technologies were new in many parts of the world. Note that the interstate highway system in the United States dates from the 1950s, mass jet travel only took off (pun intended) in the 1960s, color TV in the 1960s and 1970s. Telecommunications and quantum mechanics-based transistor technologies also became widespread in the 1950s. Home computers, the mobile (smart)phone, and the internet are even more recent and available in all parts of the world. The Space Age, in comparison, dates to the 1950s, almost seventy years ago.

2001: A Space Odyssey (1968) and *Star Wars* (1977) were just as iconic and influential in the Global South as they have been in the North Atlantic countries. All or most of the regions discussed here have had access to science fiction in literature, television, and film whether in English or translated formats. South, Southeast, and East Asia have, of course, the benefit of the numerous tales derived from Hindu and Buddhist cultures. In India in Puranic tales, in China the Daoist, and in Japan the Kami stories are full of what can only be interpreted as technological models including travel and habitation in space. Witness multiverses, superheroes, and planet-busting weapons, which find their Western expression in series such as the *Mighty Morphin Power Rangers* or in Japanese anime and manga. The influence of science fiction cannot be discounted: it would seem to be more a part of the quotidian normal rather than a cognitive disconnect with reality.

And in Time

In 2018, I was the only student at the International Space University's Space Studies Program who remembered the Apollo Moon landing: the other students were all around twenty or thirty years old. In the 1960s, I traveled as much by jet and car as by bullock cart and bicycle. These transportation modes coexisted in the 1960s: they still do. Various populations still live without electricity or running water - whether on ranches in New Mexico in the USA or in the underdeveloped countries of Africa. Populations without urban amenities are often described as Indigenous but effectively live by complementing the urban population of whatever national region is under consideration.

South, Southeast, and East Asia together comprise a third of the world's population on a significant part of the world's landmass. Add to that Africa, South America, and the Pacific islands. China and India have millennia-long histories of science, philosophy, and technology, and have ongoing and current space capabilities and plans for humans in space: they alone also contribute over 20 percent of personnel to the global research and Big Tech industries (Bhatt 2021). The UAE wants to colonize Mars. The Pacific Rim countries are major engines of economy and industry. Satellite tracking stations were set up at Woomera in Australia and Malindi in Kenya. The Arecibo telescope in Puerto Rico and the Atacama cosmology telescope in Chile helped image a black hole. There are plans to recycle telescopes in Namibia and to build an extensive "Square Kilometer (Antennae) Array" across Australia and South Africa. Satellite technologies are extensively used for earth observation, catastrophe, and disaster management. The list continues. What do the populations of the Global South see as priorities for space exploration? How does space exploration impact the daily activity of people in these countries? Are space-based communications and navigation technologies a sufficient benefit, or do they contribute inspiration and ambition? What mythologies do they hold with respect to space, what histories of space inspire them, and how are their cultures altered (if at all) as result of the exploration of space? The Global South is clearly contributing to the exploration of space and deserves to be written into the discussion.

Whether the Nazi program in rocketry can be traced back to the development of the Congreve rocket in 1805, which itself was the result of technology captured by the British at the Battle of Mysore in the 1780s on the Indian subcontinent, is moot. What is clear, however, is that the period after World War II was a time of tremendous ferment and change where the Western imperialist mode of government and economics was imposed on the newly independent countries of Asia and Africa. It is then in the context of the postwar 1950s that space lawyer Andrew Haley in his seminal *Space Law and Government* (1963) recognized the reality of colonial and imperialist genocide. He thus links it directly to the development of space law as we know it today: in the 1961

Declaration of Principles for the Peaceful Use and Exploration of Outer Space and the Celestial Bodies, which was ratified as a treaty in 1967, and related international instruments. The same considerations had led him to propose the details of "metalaw" (Haley 1956, mentioned above), a principle based on the "golden rule" on how to handle relationships with extraterrestrial non-human cultures. Haley (1963, 67) mentions that it was in Bogota in 1957 that the International Law Association deliberated on the principles that led to the law of outer space.

In 1976, seven equatorial countries (Ecuador, Columbia, Congo-Brazzaville, the then Zaire [now the Democratic Republic of the Congo], Uganda, Kenya, and Indonesia) declared sovereignty over those parts of the geostationary orbit that lie over their sovereign national territories, known as the Bogota declaration. The arguments were simple: that the space above their national territories did not lie in outer space and therefore that that space should be considered a natural resource; that the segments of these orbits that lay above the high seas were the common heritage of all mankind and should therefore be governed collectively. To have accepted these arguments would have led to questions of ownership of sections of outer space, of the geostationary orbit, and of the delimitation between air space and outer space affecting global telecommunications. However, these arguments did not receive widespread support and the declaration was mostly abandoned (Gangale 2006). To have supported such arguments would have had a much larger effect in terms of ownership and the sovereign national appropriation of outer space and the celestial bodies. What it did instead was to center the rights of developing countries, and to then emphasize the relevant articles regarding cooperation and the benefit of all mankind mentioned in the Outer Space Treaty, thus leading to the declaration regarding the use of space for the benefit of developing countries. Further developments in the allocation process of geostationary slots led to the tiny nation of Tonga having access to sixteen geostationary orbital slots; in 1991 Tonga dropped its claim to ten of these, retaining six – otherwise there would have been problems with claim-staking of orbital slots. The compromise struck allowed the view that such orbital slots were in part property and therefore liable to ownership; in part they remained subject to the principle of non-sovereign appropriation in the Outer Space Treaty (Collis 2009). It is ironic that the United States objected to the Bogota declaration on the grounds of property and ownership: it later objected to the Moon Treaty by insisting on property rights on the Moon and celestial bodies and continues to do so in the Artemis Accords.

To Infinity

While many countries claim to be spacefaring, only a few can launch payloads to orbit. Many countries rely on existing space agencies for their achievements and support. The European Space Agency provides facilities and expertise to many countries for the launch of Earth observation satellites and other types of spacecraft. These are discussed in detail in a series of recent (2018–present) publications too numerous to list under the editorship of Annette Froehlich at the European Space Policy Institute, a European space policy think tank. These publications are especially directed toward the improvement of human rights in the Global South, and of the use of space technologies to achieve sustainable development goals by 2030 for South America and Africa in particular. However, the United States, Russia (formerly the Soviet Union), the European countries through ESA and their national space agencies, Canada, Japan, and finally China and India all have extensive space programs. The United Arab Emirates are a recent entrant and major player in space exploration activities. Australia, Kenya, Ethiopia, and several other countries have recently established space agencies that primarily deal with Earth observation. A few countries (Malaysia, Indonesia) along the equatorial latitude are also considering establishing spaceports. Nigeria has a longer-established space program (Abiodun 2017), while of note is the forgotten saga of Zambia's Afronaut training program in the 1960s (Bourland 2020; Marshall 2020).

India has had a space program since 1949, a fact not widely known (Singh 2017). Their space vehicles are named somewhat prosaically after the destination to which they are to be sent, but also after various deities from Hindu mythology or personages from Indian history. This is no different from the countries mentioned above for whom the naming of space vehicles and missions appears to be part of national vanity and pride. The United States names its missions for its Greco-Roman heritage (Apollo, Gemini, Mercury, Artemis, Juno spying on Jove) or after well-known European and American contributors to the field of astronomy. The former Soviet Union and the current Russian Federation operating on its own territory and through Kazakhstan are also prosaic in naming for the destination; there was the additional cachet of naming according to Communist ideology and goals. China uses its own multiple folklores, mythologies, and deities in its naming conventions, such as naming the Change'e 4 relay satellite at the far side of the Moon after a bridge of magpies helping two lovers to reunite. Different cultures see the significance of the Moon and celestial bodies through varied perspectives. In the United States, For All Moonkind—a nonprofit organization dedicated to protecting lunar landing sites and similar sites in outer space as common human heritage—views the Moon as open to all for settlement. The Australian Earth Laws Alliance (2021) applies Earth jurisprudence: the sacred Moon should not be profaned. Both see the Moon as the Common Heritage of Humankind. What non-Australian and non-US cultures think about this is yet unknown. The Moon is male, female, or neither in different cultures. Which aspect should we deify and retain—the male, the female, the neutral, the fecund? The Moon implicates a diverse heritage.

And Beyond!

A decolonial approach to space exploration cannot imply the rejection of all things colonial. Technology of all kinds is not only accepted by but is adapted into local cultural norms without regard for their colonial or neocolonial origins. Technologies that are used for Earth observation from space and for the exploration of outer space have been part of material culture for the last seventy years. As such, they have become normalized into the technological ecosystem that has been developing and supporting human societies since the industrial revolution and before. This technological ecosystem is a consequence of anthropogenic activity. And as humanity is a part of the biosphere, by extension this technological ecosystem is integrated into the biosphere. If accepted into the biosphere, then attitudes toward technology and its adoption would of necessity be those attitudes toward nature. Any ecological philosophy would, of necessity, include the technological ecosystem as part of the terrestrial biosphere. The use of technology becomes the focus of technology, and its incorporation therefore becomes part of the everyday praxis alongside more traditional and older elements without fundamental doctrine (*orthodoxy*) or method of use (*orthopraxis*). If there is any faith involved, it is that the scientific principles that make the technology possible are trustworthy. Ritual structures quotidian human thoughts and actions (Geraci 2018; Gorman 2011). Futurist Anab Jain (2016) describes the worship of the Indian Mars Orbiter. China has a particular relationship with technology (Dasheng and Tsing 1989). Incorporation of technology and outer space into daily life returns ritual to the center of people's lives: in China to the Confucian notion of *Li* ("ritual") (Chan 1957; Wong 2020); in India to the yoga of action (*karma-yoga*). Ritual maintains and sustains the world. Tradition complements but is not confused with and does not replace science, so radar specialist, science fiction and science writer Arthur C. Clarke's (1917–2008) putative Third Law, such that magic and a sufficiently advanced technology are indistinguishable, does not hold (Beech 2012,190). Such incorporation (Zhu and Tong 2019; Baronov 2010) can then lead to a harmony between technology and nature, an eco-philosophy being thus developed (Drengson 1995). A good example of similar integration is neo-Confucianism, which integrates Buddhism, Daoism, and Confucianism into a seamless whole (De Bary 1970; Capra 2010; Angle and Tiwald 2017). This worldview becomes forward-looking rather than dystopian or claiming past superiority with respect to science, space, or technology. There remains a recognition that the past of various personal cultures helps; but these are complementary methods allowing the most to be made of the present and the future. Reclaiming space is a narrative of integration rather than restoration or replacement.

The first person on a planet or an asteroid is no more than cultural dominance through geopolitics. Objecting to decolonial space can be a denial of minority or non-Western cultures. Property and ownership of space moves away from humankind's common heritage and equitable access to space. This should not be raised into the heavens, signposting our differences. Emphasizing difference in diversity rather than unity in diversity, any language and culture has a word for people: but it always translates as "just some of us, the people" and does not include "you, the not-people." Division grows in the human universe. These differences are significant. They stop us from marching together, from working together, from achieving together. They stop us, period. Creating new modes of difference, pushing each other away until the distance between us, is greater than any distance between the stars. Those looking to explore space ignore the need for assistance on Earth. Looking down on Earth, they ignore the aspiration and the achievements of a cooperative humanity seeing only catastrophe and disaster. Collaboration and cooperation disappear in the competition for resources; in not recognizing that technological spinoffs can help humanity—the poor, the distressed, the elderly, or the disabled. Recycling for space habitats is there to help sustain, conserve, and renew the terrestrial environment, enhancing lives. Space technology is a dual-use technology. To claim space is to construct walls and borders excluding others, to reclaim space to confirm boundaries erected by the other, and to continue exclusion. Space exploration and the movement of humanity into space is a cooperative and collective enterprise. Space has already been colonized and is being colonized by all cultures. In the cosmos, there is nothing but an endless diversity immune to the differences born of Earth.

Notes

1. http://www.aboriginalastronomy.com.au/.
2. https://memory-beta.fandom.com/wiki/Kenya.
3. The *USS Kenya* in the *Star Trek: The Next Generation* video game *Armada II*.
4. In the *TOS* novels *The Starless World* and *Uhura's Song*.
5. In the *TOS* novel *Sarek*.

References

Abiodun, Adigun Ade. 2017. *Nigeria's Space Journey: Understanding Its Past, Reshaping Its Future*. Seattle: African Space Foundation.

Al-Khalili, Jim. 2010. *Pathfinders: The Golden Age of Arabic Science*. London: Penguin.

Al-Khalili, Jim. 2011. *The House of Wisdom: How Arabic Science Saved Ancient Knowledge and Gave Us the Renaissance*. London: Penguin.

Allen, Richard H. 1963. *Star Names: Their Lore and Meaning*. New York: Dover Publications.

Alvares, Claude A. 1991. *Decolonizing History: Technology and Culture in India, China and the West 1492 to the Present Day*. New York: Apex Press.

Ambrogio, Selusi. 2020. *Chinese and Indian Ways of Thinking in Early Modern European Philosophy: The Reception and the Exclusion*. London: Bloomsbury Publishing.

Angle, Stephen C. and Justin Tiwald. 2017. *Neo-Confucianism: A Philosophical Introduction*. Hoboken: John Wiley & Sons.

Arikawa, T., T. Hayakawa, Y. Katori, K. Kumagaya, K. Morikawa, T. Nagatomi, and M. Yamada, producers. 1978–1981, 2004 *Monkey*. Tokyo: Kokusai Hoei, Nippon Television Network.

Australian Earth Laws Alliance. 2021. *Declaration of the Rights of the Moon Draft for circulation and discussion*. February 11, 2021. https://www.earthlaws.org.au/moon-declaration/.

Australian Indigenous Astronomy. 2021. "Navigation & Star Maps." Accessed January 13, 2022. http://www.aboriginalastronomy.com.au/content/topics/starmaps/.

Bagchi, Prabodh C. 2011. *India and China: Interactions through Buddhism and Diplomacy; A Collection of Essays*. London: Anthem Press.

Bahadur, Krishna P. 1978. *The Wisdom of Saankhya*. New Delhi: Sterling.

Baronov, David. 2010. *The African Transformation of Western Medicine and the Dynamics of Global Cultural Exchange*. Philadelphia: Temple University Press.

Beech, Martin. 2012. *The Physics of Invisibility: A Story of Light and Deception*. New York: Springer Science + Business Media.

Bhatt, Mukesh C. 2011. "Diderot and d'Alembert's L'Encyclopédie as Watershed: Ferengi perspectives on Gujarat and its culture." Presentation at Rethinking Religion in India III Pardubice. Czech Republic, Oct 11-14, 2011. Available at https://birkbeck.academia.edu/.

Bhatt, Mukesh C., (2021). "The Universe Decentered: Transcultural Perspectives on Astrobiology and Big History." In *Expanding Worldviews: Astrobiology, Big History and Cosmic Perspectives*, edited by Ian Crawford, 239–268. London: Springer.

Boi, Luciano. 2011. *The Quantum Vacuum: A Scientific and Philosophical Concept, from Electrodynamics to String Theory and the Geometry of the Microscopic World*. Baltimore: JHU Press.

Bottici, Chiara and Benoît Challand. 2013. *The Myth of the Clash of Civilizations*. London: Routledge.

Bourland, W. Ian. 2020. "Afronauts: Race in Space." *Third Text* 34, no. 2: 209–229.

Capra, Fritjof. 2010. *The Tao of Physics: An Exploration of the Parallels Between Modern Physics and Eastern Mysticism*. Boulder: Shambhala Publications.

Chan, Wing-Tsit. 1957. "Neo-Confucianism and Chinese Scientific Thought." *Philosophy East and West* 6, no. 4: 309–332.

Chang, Kuei-Sheng. 1974. "The Maritime Scene in China at the Dawn of Great European Discoveries." *Journal of the American Oriental Society* 94, no. 3: 347–359.

Chattopadhyaya, D. P., ed. 1990-present. *History of Science, Philosophy and Culture in Indian Civilization* series. Delhi: Centre for Studies in Civilization.

Collis, Christy. 2009. "The Geostationary Orbit: A Critical Legal Geography of Space's Most Valuable Real Estate." In *Down to Earth*, edited by Lisa Parks and James Schwoch, 61–81. New Brunswick, New Jersey: Rutgers University Press.

Dasheng, Gao and Zou Tsing. 1989. "Philosophy of Technology in China." In *Philosophy of Technology*, edited by Paul T. Durbin, pp. 133–151. Dordrecht: Springer.

De Bary, William T. 1970. *The Unfolding of Neo-Confucianism*. New York: Columbia University Press.

Determann, Jörg M. 2015. *Researching Biology and Evolution in the Gulf states: Networks of Science in the Middle East*. London: Bloomsbury Publishing.

Determann, Jörg M. 2018. *Space Science and the Arab World: Astronauts, Observatories and Nationalism in the Middle East*. London: Bloomsbury Publishing.

Determann, Jörg M. 2020. *Islam, Science Fiction and Extraterrestrial Life: The Culture of Astrobiology in the Muslim World*. London: Bloomsbury Publishing.

Drengson, Alan. 1995. *The Practice of Technology: Exploring Technology, Ecophilosophy, and Spiritual Disciplines for Vital Links*. Albany, New York: SUNY Press.

Duyvendak, J. J. L. 1939. "The True Dates of the Chinese Maritime Expeditions in the Early Fifteenth Century." *T'oung Pao* 34, no. 5: 341–413.

Gangale, Thomas. 2006. "Who Owns the Geostationary Orbit?" *Annals of Air and Space Law* 31: 425.

Gatty, Harold. 1958. *Finding Your Way Without Map or Compass*. Mineola, New York: Dover Publications.

Geraci, Robert M. 2018. *Temples of Modernity: Nationalism, Hinduism, and Transhumanism in South Indian Science*. Maryland: Rowman and Littlefield.

Goldsman, Akiva, Michael Chabon, M., Kirsten Beyer and Alex Kurtzman. 2020. *Star Trek: Picard*. Los Angeles: CBS Television Studios.

Gorman, Alice. 2011. "The Sky Is Falling: How Skylab Became an Australian Icon." *Journal of Australian Studies* 35, no. 4: 529–546.

Graeber, David and David Wengrow. 2021. *The Dawn of Everything: A New History of Humanity*. London: Penguin.

Haley, Andrew G. 1956. *Space Law and Metalaw: A Synoptic View*. Presented at the VIIth Annual Congress of the International Astronautical Federation, Rome, Italy; September 19, 1956 printed in *Harvard Law Record* 23 (November 8, 1956).

Haley, Andrew G. 1963. *Space Law and Government*. New York: Appleton-Century-Crofts.

Halpern, Michael D. 1985. *The Origins of the Carolinian Sidereal Compass*, Master's Thesis, College Station, Texas: Texas A & M University.

Hamacher, Duane W. 2018. "The Stories Behind Aboriginal Star Names Now Recognised by the World Is Astronomical Body." *The Conversation*, January 14, 2018. Accessed November 11, 2022. https://theconversation.com/the-stories-behind-aboriginal-star-names-now-rec ognised-by-the-worlds-astronomical-body-87617.

Haribhadrasūri, Ṣ. and M. K. Jain, eds. 1981. *Ṣaḍdarśanasamuccaya of Haribhadra Sūri. With the Commentaries of [sic] Tarka-rahasya-dipikā of Guṇaratnasūri and Laghuvṛtti of Somatilaka Sūri and an Avacurṇī*. New Delhi: Bharatiya Jnanpith.

Harwood, Richard. 2012. "Islamic Science and Mathematics: New-Found Wealth." In *Learning and Teaching about Islam: Essays in Understanding*, edited by Carline Ellwood, 127–136. Woodbridge: John Catt Educational.

Holmes, Lowell D. 1955. "Island Migrations (3): Navigation Was an Exact Science for Leaders." *Pacific Islands Monthly* 26, no. 2: 113–115.

Jain, Anab. 2016. "Rockets of India." Blog post, October 17, 2016. Accessed November 11, 2022. https://medium.com/@anabjain/rockets-of-india-f043c5b39b34.

Koskenniemi, Martti. 2016. "Race, Hierarchy and International Law: Lorimer's Legal Science." *European Journal of International Law* 27, no. 2: 415–429.

Lach, Donald F., ed. 1994–1998. *Asia in the Making of Europe* series. Chicago: University of Chicago Press.

Lunine, Jonathan I. 2005. *Astrobiology: A Multi-Disciplinary Approach*. San Francisco: Benjamin Cummings.

Marshall, Colin. 2020. "Meet 'The Afronauts': An Introduction to Zambia's Forgotten 1960s Space Program." *Open Culture*, March 4, 2020. Accessed November 11, 2022. https://www.openculture.com/2020/03/meet-the-afronauts.html.

Needham, Joseph, ed. 1954–present. *Science and Civilisation in China* series. Cambridge: Cambridge University Press.

Pánikar, Austín. 2010. *Jainism: History, Society. Philosophy and Practice*. Delhi: Motilal Banarsidass.

Rao, Vetury Ramakrishna. 1987. *Selected Doctrines from Indian Philosophy*. Delhi: Mittal Publications.

Singh, Gurbur. 2017. *The Indian Space Programme: India's Incredible Journey from the Third World Towards the First*. Manchester: Gurbir Singh/Astrotalkuk Publications.

Subbarayappa, B.V. and K.V. Sarma. 1985. *Indian Astronomy: A Source-Book, Based Primarily on Sanskrit Texts*. Bombay: Nehru Centre.

Subbarayappa, B.V. 2013. *Science in India: A Historical Perspective*. Delhi: Rupa Publisher.

Sen, Tan Ta. 2009. *Cheng Ho and Islam in Southeast Asia*. Singapore: Institute of Southeast Asian Studies.

Thompson, Charles N. n.d. "Hawaiian Voyaging Traditions." Accessed January 13, 2022. http://archive.hokulea.com/ike/hookele/star_compasses.html.

Wikipedia. n.d. "Timeline of Telescopes, Observatories, and Observing Technology." Accessed January 13, 2022. https://en.wikipedia.org/wiki/Timeline_of_telescopes,_observatories,_and_observing_technology.

Vinge, Vernor. 1999. *A Deepness in the Sky*. New York: Tom Doherty Associates.

Wilke, Christiane. 2009. "Reconsecrating the Temple of Justice: Invocations of Civilization and Humanity in the Nuremberg Justice Case." *Canadian Journal of Law and Society/La Revue Canadienne Droit et Société* 24, no. 2, 181–201.

Wong, Pak-Hang. 2020. Why Confucianism Matters in Ethics of Technology. In *Oxford Handbook of Philosophy of Technology*, edited by Shannon Vallor, 609–628. Oxford University Press.

Wu, C. 1978–1980, 2004. *Monkey* [Original title: *Saiyūki*]. Tokyo: Nippon TV and International Television Films.

Zhu, Xu and Tong Wu. 2019. *Returning to Scientific Practice: A New Reflection on Philosophy of Science*. London: Routledge.

4

Cold Warrior Magic, Africana Science, and NASA Space Race Religion, Part One

Laura Nader's Contrarian Anthropology for Afrofuturist Times

Edward C. Davis IV

Overview

Years ago, anthropologist Laura Nader became my PhD mentor after she showed her students the 1980 documentary *The Day After Trinity*. Recounting the events before and after July 16, 1945, the film follows the Manhattan project in the New Mexico desert, whereby J. Robert Oppenheimer led a group of acclaimed physicists who questioned whether splitting atoms would destroy the universe. Nader noted that NASA's Apollo 11 lunar landing mission launched July 16, 1969, to which I revealed I was born in Chicago on July 16 in the 1980s. Strangely, on July 16, 2018, HBO began filming the Afrofuturist drama series *Lovecraft Country* (2020) in Illinois. To say that I have an atomic birthdate would be an understatement.

During COVID-19, *Lovecraft Country* aired as a solution to anti-Black racism. In the series, adapted from Matt Ruff's novel of the same name, young Black American Korean War veteran Atticus Freeman defeats whiteness in his maternal bloodline embodied by cousins who descend from a rapist, Massachusetts enslaver. By patrilineal descent, Freeman's ancestors were never enslaved. Similar to my Illinois Trail of Tears ancestors who descend from Angolans enslaved in colonial Virginia, characters in *Lovecraft Country* challenge hegemonic genocides of Native Americans and the Tulsa massacre of 1921, the Korean War, and the Mississippi murder of Chicagoan Emmett Till, which all take place within the series' storyline, set in the 1950s.

Edward C. Davis IV, *Cold Warrior Magic, Africana Science, and NASA Space Race Religion, Part One* In: *Reclaiming Space*. Edited by: James S. J. Schwartz, Linda Billings, and Erika Nesvold, Oxford University Press. © Oxford University Press 2023. DOI: 10.1093/oso/9780197604793.003.0004

On October 1, 1958, the National Aeronautics and Space Administration (NASA) became a branch of the United States government aimed at challenging the Soviet Union's Sputnik satellite launch on October 4, 1957. Within a continuum of Euro-American Manifest Destiny, traveling into outer space became a way of entering into sacred religious space (Weibel 2017). In harmony with the mission of this volume, the present chapter seeks liberation from systems of knowledge rooted in binary oppositions, such as East-West, North-South, or Black-white. The US-Soviet race into outer space separated humanity during deadly wars for mineral wealth to get to the Moon and back; however, researchers ignore neocolonial conflicts on Earth for minerals with Cold War space-race geopolitics. For the first time, moral ethical logic (MEL) will come to the fore to examine space exploration in the tradition of Laura Nader's contrarian anthropology. Author and activist James Baldwin reminded us of the American lie about the mistreatment of people of color in the United States and abroad (Glaude 2020, 8). Maintaining intersecting lies upholds the myth of US exceptionalism and global white supremacy, which the space race extended from Earth into the heavens.

Kinship and Cold War Anthropology

Many social anthropologists detest kinship studies. To the contrary, I find links between economic production and human reproduction, especially how the trans-Atlantic slave trade produced capitalist negation of Black kinship as the foundation of white supremacy, witnessed in the practice of *partus sequitur ventrem,* a legal doctrine of English colonies in North America which established that all children would have the legal status of their mothers—meaning that children born to enslaved women would also be enslaved. If the children of African women and European men have been denied access to generational wealth by becoming property of their paternal kinsmen, then we can observe present-day parallels in how the Silent Generation found themselves dispossessed after World War II. Any hopes the Silent Generation had for equity in the 1950s were up against racism and McCarthyism. Moral ethical logic rationalized the murder of Emmitt Till and Jim Crow status quo as necessary to preventing communism and un-American behavior, whereby denying Black humanity constitutes core American values akin to killing Korean people and denying Black American veterans of the Korean War, like my grandfathers, or the fictional Atticus Freeman, full access to the GI Bill in the 1950s. Thus, we might hear Gil Scott-Heron's 1970 rap "Whitey on the Moon" as a discussion in kinship, where the poet's Black American sister struggles to survive while NASA astronauts walk on the Moon.

Kinship anthropology allows us a lens to view intergenerational conflict beyond class, gender, and race so commonly used in African American studies, which fails to fully analyze Cold War politics. Significant relationships exist between generational enculturation and social problems in capitalist modes of production, which arise as correlations in this study. Racist, sexist, xenophobic Cold Warrior political agendas and government funding for science shaped the NASA Cold War space race and sustained Black suffering. We can allow an inference that abundance for the few accompanies cruelties upon the masses. One could further infer that late-stage Euro-American capitalism thrives on the magic of racism, by negating Black Indigenous science and personhood, in order to worship Euro-American NASA space racist religion of the seen, video-recorded, techno-heavens, absent from the metaphysical realm of MEL.

Charting Nader's Contrarian Anthropology

Revisiting classic anthropological texts allows us to orbit Cold Warrior academia within contrarian anthropology, using new eyes to discover Polish-born physicist and father of anthropology Bronislaw Malinowski's 1925 work *Magic, Science, and Religion* (Malinowski 1948). As a Black American, I am curious as to how German-American anthropologist Franz Boas (1928) engaged with Black American author Zora Neale Hurston, as well as how Malinowski shared knowledge with his mentee Jomo Kenyatta (1938), who became the first president of independent Kenya. These engagements impacted the Harlem Renaissance and Kenyan independence. Rarely has anyone traced pre-Einsteinian physics to anthropology and Africana studies as I do. Within Nader's book *Naked Science*, her chapter "The Three-Cornered Constellation: Magic, Science, and Religion Revisited," provides a deeper analysis of Malinowski's 1925 text, which he articulated within a multilayered, geolinguistic guise. My first readings of *Magic, Science, and Religion* in my late teens and mid-twenties led me to view such words, on the surface, as racist and deeply troubling. However, as a millennial in my thirties, I saw coded language allowing Malinowski to praise Trobriand Islanders' geospatial technical knowledge in an era of white supremacist anti-Blackness. Of course, Malinowski carried racist and sexist views from his day; however, the multilingual father of anthropology, originally trained as a physicist and questioned Eurocentric notions of the scientific method as never exclusive to the Northern Hemisphere.

The work of Laura Nader provides me with a foundation to repair systems of power and knowledge for new generations. Nader's essay "The Phantom Factor: Impact of the Cold War on Anthropology" outlines her days in graduate school at Harvard in the 1950s when McCarthyism silenced intellectuals believed to share connections to the so-called Red Scare of communism (Nader 1997). Cold Warrior academics served US and British efforts against African liberation and anticolonial self-determination. After July 16, 1945, Oppenheimer received royal treatment beyond the Alamogordo Test Range in New Mexico. Euro-American capitalist magicians praised the murders of over 120,000 people on August 6 and August 9 at Hiroshima and Nagasaki, Japan, in 1945. American necropolitical religiosity had launched a new crusade by air, armed with the magic of physics to destroy *others*, and to ignore any equilibrium with the Universe with respect for God. Thus, the genocidal actions of America—against *others*—fulfilled holy Manifest Destiny against favorite sons and heroes, like Black American performing artist and activist Paul Robeson or J. Robert Oppenheimer, rendering them *personae non gratae*. If we question the witchcraft of capitalist greed, then we could expect generational silencing as necessary for perpetuating the system.

The Silent Generation includes people born before World War II ended, raised during the Great Depression. Typically born after 1925 and before 1945, the Silent Generation were not old enough to fight in World War II. The Silent Generation became parents of late Baby Boomers and early Generation X, occupying multiple layers of liminality and in-between spaces. Key to silencing millions who were too young to fight in the war, Great Depression–era scarcity reminded the youth that they should be seen and not heard, even on V-E Day and V-J Day. They fought the Korean War and made up the earliest Americans in Vietnam, with the US government paying little respect for their lives. Both US presidents Jimmy Carter and Joe Biden belong to the Silent Generation, as do Laura Nader, her brother Ralph Nader, and Grand Ati Max Beauvoir, a Haitian biochemist and Vodou priest.

Cold War academic politics played a key role in silencing young minds, as witnessed in the reign of terror produced by Wisconsin Republican Senator Joseph McCarthy. In *Demagogue: the life and long shadow of Senator Joe McCarthy*, Larry Tye (2020) introduces us to future Trump associate Roy Cohn, an attorney who worked for McCarthy. Toxic masculinist white supremacy in 1950s echoed Black misandry and racism for self-hating white republican queer men. The first Red Scare (1917–1920) aided General Dwight Eisenhower's run for president; however, the second Red Scare (1947–1957) during the Korean War paralleled the Lavender Scare against homosexuals in government (mid-20th century), setting the stage for future government funding. One may wonder how closeted homosexuals Roy Cohn and FBI director J. Edgar Hoover may have set the stage for eliminating the possibility for social equity measures that might have prevented the Vietnam War, COINTELPRO (a counterintelligence program conducted by the FBI from 1956 to 1971 to weed out "subversives"), or even the intensity of the HIV/ AIDS pandemic. Sadly, Roy Cohn would die of AIDS in 1986, as Donald Trump purged Cohn from his inner circle. Cohn provides a clear example of Silent Generation political economics at play with the status quo, to the detriment of humanity. Millennial intersectional Black queer feminist paradigms, argued by Black political strategist, community leader, and writer Charlene Carruthers (2018), move away from self-hating hegemonies within interconnected matrixes of oppression that must come down.

One should not interpret opposition as simple politics of the human body, but as greater hegemonic forces and ideologies that manifested in every aspect of what Nader calls Cold Warrior academics (2018, 230). Cold War ontologies of respectability silenced an entire generation into collective obedience, as well as those born to later generations. As generations suffered in silence, a counterculture of activists and scholars hoped to oppose the status quo, especially toxic notions of atomic weaponry. In the 1950s, contrarian anthropologists saw their careers suffer if they did not adhere to a Cold Warrior stance against colonized people. Nader notes that during her years as a graduate student, she became aware of how the Carnegie Corporation, Rockefeller Foundation, Rhodes Trust, and many other funding bodies protected their mineral interests in Africa. The National Science Foundation developed at this time to reward Cold Warrior academics with grants for so-called development research designed to keep Africa and the Southern Hemisphere underdeveloped, but as a source of raw materials for the West. Uranium discoveries in Belgian Congo led to the decimation of Nagasaki and Hiroshima, and mining of the same element found in Brazil, Venezuela, and rural Canada (Saskatchewan) threatened to destroy indigenous ways of life (Nader 2018, 232). Geological theft remained essential to nuclear rocket propulsion.

Individuals who stood up against inequity and tried to use their careers for good found themselves erased. Nader notes how American cultural anthropologist Jack Stauder held degrees from both University of Cambridge and Harvard, but found the latter university fired him despite excellent work on environmental preservation and subsistence in Ethiopia. In South Africa, South African-British social anthropologist Max Gluckman appeared to be too sympathetic to the Zulu nation in his Marxist analysis of economic inequalities. Cold Warrior academics countered legal arguments for equity, whereby E. E. Evans Pritchard received extensive British government funding during and after World War II for research on the Nuer people in Anglo-Egyptian Sudan, which shored up British control of the Nile River and Suez Canal. During World War II, British sociologist Peter Worsley left his studies at University of Cambridge for military service with the Tanganyika Groundnut Scheme, the British government's ultimately failed attempt to produce peanuts in this East African land, a British territory from 1922 to 1946. Whether or not the League of Nations British mandate in East Africa aimed to deliver mass education or simply prevent the former German colonies from falling into Nazi hands, Worsley remained dedicated to his mission, in collaboration with his mentor Gluckman. After World War II, Worsley found himself "forced out of anthropology" (Nader 2018, 229).

For Nader, her unbowed commitment to moral and ethical concerns in law, nuclear energy, and global historiography reached a critical moment around 1970. In her essay "Barriers to Thinking New About Energy," Nader reveals highs and lows of the US Committee on Nuclear and Alternative Energy Systems (CONAES, 1976–1977; Nader 2018).[1] When NASA asked Nader to take part in CONAES, she met other scholars in Monterey, California where only three women took part in a collective of all white academics. Nader notes how backward thinking among scientists recycled irrational ideas about the world among so-called geniuses, who lacked the ability to question the origins of their own knowledge. They also lacked compassion for others and could not properly assess their own conclusions about the future of *solar* energy, because the word was not allowed to come up in meetings. Furthermore, the group of white men hurled disrespectful sexist comments at Nader with a consistent ability to ignore societies that produce zero commercial quad, or what we can call communities that live off of the grid. Notions of knowledge, or the overarching epistemologies that governed CONAES and NASA, prevented true progress within the omnipresence of white supremacy.

Conclusion

This brief chapter aimed to revisit the work of contrarian anthropologist Laura Nader in conversation with space exploration, echoing the continuum of nonconformist anthropologists engaged in the study of magic, science, and religion. Beginning with Bronislaw Malinowski helps to center historical debates; however, mirroring Black studies or Africana studies within the continuum of knowledge formation helps to include emerging scholarship of the last year and last century in democratic debates on human integrity. With a keen eye upon kinship anthropology, intergenerational personality dynamics transposes multiverses of humanity within space and time well beyond fixed dates. Human history will always remember July 16, 1945, as the day the first atomic reaction occurred on Earth with a bomb tested in New Mexico. Likewise, July 16, 1969, remains part of human history as the date Apollo 11 launched for the Moon. The date this anthropologist was born in Chicago in 1982 matters, just as the Black lives fictionalized in *Lovecraft Country* matter. If we do not consider the impact of geological warfare since 1619 or during the Cold War, then humanity stands to miss the teachable moments witnessed today with billionaires traveling into outer space with minerals harvested from the Southern Hemisphere. Building electric cars and flying into outer space both require the exploitation of Black and Brown people in the Southern Hemisphere, who have lived atop mineral deposits for millennia. By rationalizing the nonhuman status of Black Indigenous People of Color around the world, Euro-American societies see the conquest and destruction of Earth as a necessary evil in reaching the cosmos. Yet, if people destroy Earth below, we must ask what will become of notions of heaven above.

Note

1. In Nader (2018), the author notes that she originally published this article in 1981 with the journal *Physics Today,* vol. 34.

References

Boas, Franz. 1928. *Anthropology and Modern Life*. New York: W. W. Norton & Company.

Carruthers, Charlene. 2018. *Unapologetic: A Black, Queer, and Feminist Mandate for Radical Movements*. Boston: Beacon Press.

Glaude, Eddie S., Jr. 2020. *Begin Again: James Baldwin's American and Its Urgent Lessons for Our Own*. New York: Crown.

Kenyatta, Jomo. 1938. *Facing Mount Kenya*. London: Secker and Warburg.

Lovecraft Country. 2020. Developed by Misha Green. Moneypaw Productions, Bad Robot Productions & Warner Brothers Television Studios.

Malinowski, Bronislaw. [1925]1948. *Magic, Science, and Religion*. London: The Free Press.

Nader, Laura. 1997. "The Phantom Factor: Impact of the Cold War on Anthropology." In *The Cold War and the University: Toward an Intellectual History of the Postwar Years*, edited by Noam Chomsky, et al., 107–148. New York: The New Press.

Nader, Laura. 2018. *Contrarian Anthropology: The Unwritten Rules of Academia*. New York & Oxford: Berghahn.

Tye, Larry. 2020. *Demagogue: The Life and Long Shadow of Senator Joe McCarthy*. New York: Houghton Mifflin Harcourt.

Weibel, Deana L. 2017. "'Up in God's Great Cathedral': Evangelism, Astronauts, and the Seductiveness of Outer Space." In *The Seductions of Pilgramage: sacred journeys afar and astray in the western religious tradition*, edited by Michael A. DiGiovine and David Picard, 223–256. London: Routledge.

5

Global Participation in the "Space Frontier"

Alan Marshall

Touted as a move into the "final frontier," human expansion into the extraterrestrial space of the Solar System has been expressed as the next large-scale exploration and settlement project for modern humanity. Through such expansion, it is supposed that vast resources will be opened up for the general benefit of humankind (see, for example, Mankins 1996; Pyle 2019; Zubrin 2021). If this is so, then it is appropriate to inquire about the real and potential participatory mechanisms whereby the global public could be involved in such a grand project.

In July 2021, while the world was suffering through record-breaking extreme weather events as well as a global pandemic, two of the world's richest men, Jeff Bezos of Amazon/Blue Origin and Richard Branson of Virgin Galactic, separately launched two groups of space tourists high into suborbital skies—each of which included themselves. In addition, Elon Musk's SpaceX launched his space tourism business with a four-day orbital mission whose tickets per passenger were estimated to have cost in excess of fifty million dollars. As Kelvey (2021) explains, none of these companies have divulged the exact ticket price of their space tourist jaunts but for sub-orbital flights the estimates vary between the Virgin space plane flight priced at about $US500,000 to around 25 million dollars per passenger for an eleven-minute Blue Origin rocket flight. For orbital flights the price jumps to around 200 million dollars or more for a crew of four. While these flights helped space journalists get a slot on the front pages across the global media for a few days, there was also a backlash of dismissive commentary variously suggesting the missions were "wasteful," "depressing," "childish" "joyrides" jaunting into "astronomical inequity." Furthermore, the flights served as "monuments to tax evasion" by "exploitative," "billionaut," "fantasist," "space cowboys" whose phallic space vehicles certainly signified something but which, actually, never even made it into space at all.

Alan Marshall, *Global Participation in the "Space Frontier"* In: *Reclaiming Space*. Edited by: James S. J. Schwartz, Linda Billings, and Erika Nesvold, Oxford University Press. © Oxford University Press 2023. DOI: 10.1093/oso/9780197604793.003.0005

Despite the backlash, Bezos, Branson, and Musk believe that such spectac-ular aerospace stunts foster both public interest and commercial support for future space development. Those of us who cannot afford the extremely ex-pensive ticket to travel into space can still ride along by watching our screens back on Earth. This is often how many space projects are sold, whether they be short-term or long-term, or privately or publicly funded: all people participate in space exploration because its pursuit can be seen by all (see, for instance, media reports discussing this in Brown 2021; Brandon 2021; Sharp and Smith 2021; Zahn 2021; Jackson 2021; Framke 2021; Nolan 2021; Rivera 2021).

Such participation is quite shallow, of course. It is nothing but the one-way dispersal of the results of already predetermined plans. And these plans are decided upon and set up by a powerful few. Indeed, with regards to the space exploits of Bezos, if public participation in spaceflight is what he craves, it might be better for him to pay his fair share of taxes so that democratically elected representatives can decide and plan together about how to spend excess money on space development (or other scientific and technological pursuits) in a more inclusive manner.

For Bezos, Branson, and Musk, space tourism is only the beginning. It is a way to "build a road to space," as Jeff Bezos puts it (as reported by Glen 2021). The ultimate goal of space development this century is space coloni-zation: the setting up of permanent settlements in orbit and on other planets. After space tourism takes off, colonization advocates say, the next step is to launch human missions to the Moon and Mars and to set up bigger and better space stations in orbit. As it stands, NASA is joining forces with Musk's com-pany SpaceX to enact the first step via Project Artemis, which plans to put a new set of American footprints on the Moon by the end of the 2020s (as noted in Carlson 2021). After these developments, Bezos and Musk plan to start exploiting the resources of space by setting up mining facilities on the Moon and other planetary bodies and by industrializing "the space fron-tier." (For a foray into the ideas and plans of space frontierists over the years, see von Braun 1967; Heppenheimer 1977; National Commission on Space 1986; Oberg and Oberg 1987; Michaud 1987; Zubrin 1996; Seedhouse 2013; Fernholz 2018; Weinzearl 2018; Gregg 2021; Reanau 2021. As an example of the frontierist zeal of these—and many other—writers, witness how Robert Zubrin, in one short paragraph, attempts to neatly tie space frontierism in with social freedom, universal human happiness, the discovery of America, European expansionism, United States history, and the rationalism and hu-manitarian progress that underlies western humanism: "Free societies are the exception in human history, they have only existed in the four centuries of frontier expansion of the West. That history is now over, the frontier that was opened by the voyage of Christopher Columbus is now closed. If the era of western humanist society is not to be seen by future historians as some kind of transitory golden age, a brief shining moment in an otherwise endless chronicle of human misery, then a new frontier must be opened" (Zubrin 1996, 38). The imperialist and colonialist framework of this frontierist framework of space development is explored in the other chapters of this book and in Marshall 1995.)

If space resource exploitation does proceed as they desire, prospective space industrialists claim there is at least an indirect avenue for tangible global participation, since the benefits of their exploitation would trickle down to all of humanity, including to the poor and needy of the world. However, the dream of opening up the space frontier is perhaps more likely to foster space-baron capitalism rather than global participation. (For an explanation of the evocative term "Space Baron," see Davenport 2019, and for links between Bezos' space development program and capitalism, see Jackson 2021.) This is evident in the way that Bezos, Musk, Branson, and other space-exploitation advocates have tried to reinterpret international laws of space exploration and exploitation.

Presently, the fundamental international law applying to our Solar System is the 1967 United Nations Outer Space Treaty. This treaty has been signed by all space-capable nations and, to date, governs space activities. In decades past, the Outer Space Treaty was an important piece of international law drafted by the superpowers of the 1960s to enable free and peaceful access to the bodies of the Solar System without fear of land-grabbing annexation or territorial space wars. However, nowadays, this is not all that the Outer Space Treaty represents.

Though it prohibits the appropriation of areas upon extraterrestrial bodies, the treaty remains ambiguous with regard to substrate and materials contained within such areas. To quote the treaty itself, Article II states: "Outer space, including the moon and other celestial bodies, is not subject to national appropriation by claim of sovereignty, by means of use or occupation." Thus, the treaty indicates clearly that no one is allowed to claim any particular place of the Solar System for themselves. However, many prospective space industrialists, Bezos and Musk included, interpret the Outer Space Treaty to mean that while Solar System bodies are prohibited from being claimed, any material removed from such bodies can become the rightful property of the remover. Under such an interpretation, an industrial space settlement cannot own the surface upon which it opens operations, but as soon as it removes any material from that surface, the material becomes the property of the removers.

If one believes that the free market will then fairly disseminate these extra-terrestrial materials throughout the world via supply and demand and under established pricing mechanisms, then there may be no problem with this interpretation of the Outer Space Treaty. However, since the likes of Bezos and Musk can only get into the position of running an industrial settlement on another world through massive state support and investment of public funds into general space infrastructure and into what they call preparatory missions, it seems incredible to class such extraterrestrial endeavors as operating according to free market principles (as is maintained, for example, in Weinzearl 2018 and Zubrin 2019b). As for the supposedly "private" / "commercial" space tourism businesses of Bezos and Musk, American journalist Michael Heltzic reminds us that the real target of these entrepreneurs' efforts isn't the small customer base of individuals rich enough to afford quick space jaunts, but the multibillionaire contracts given out by NASA and the Department of Defense to put either humans or military hardware into orbit, onto the Moon, and into the Solar System. (See Hiltzik 2021, who also notes that Elon Musk has received billions in federal subsidies for his car company Tesla, for his solar power projects, and to a lesser degree, for SpaceX. SpaceX used federal subsidies to move on to huge federal contracts, mostly from the Defense Department—and some with NASA. Bezos' Blue Origin has not received so many subsidies but it has received many NASA contracts which bolster the company's capacity in space. Bezos's Amazon, in turn, has received billions in federal, state, and local subsidies to build up his business. It is from his Amazon profits that Bezos then funds Blue Origin's operations. See, Marshall 2022.)

It seems apparent that the showy space jaunts of Bezos' Blue Origin and Musk's SpaceX act as attempts to seduce the public into supporting government-funded space expansion by dampening down public worries about the use of public money for such space activities.

In some ways, the 2021 space jaunts of Bezos, Musk, and Branson have been successful, since wealthy (or starstruck) investors have started pouring their money into space companies. But it's also been a grand failure given the massive public backlash against those same missions, a backlash that threatens the way space activities are supported by the government.

When discussing participation in Solar System resource use, the issue is not whether you believe in the efficiency of the free market versus the egalitarianism of a mixed economy. The point here is that although we all know—and Bezos and Musk admit it as well—that while getting into space necessarily involves public contributions on a massive scale, the Outer Space Treaty may allow for private appropriation once humans are there. The first or "public" phase, signified by the likes of the Mercury, Gemini, and Apollo projects, is cast as a glorious human pursuit that transcends inter-human and international quarrels. (Though, of course, projects like Mercury and Apollo could only have been put into motion because of Cold War competition between the United States and the Soviet Union). The second or "private" phase is cast as the incurable and ineffable operation of the free market. This "private" phase uses the smoke screen of enterprise rhetoric and the cult of the technopreneur—plus the ambiguity of the Outer Space Treaty—to plan for what may as well be labeled space-baron imperialism, whereby commonly owned resources are appropriated by space corporations. After public resources help Bezos and Musk get into space and set up industries there, it seems that they will be legally entitled to claim all the resources and profits for themselves. For some space fans, this is alright, since humans will at least be exploring space. And, after all, Bezos and Musk will be supposedly required to pay tax on their space profits (except, on this last point, they do not have a good track record when paying taxes on Earth—see Propublica 2021).

Of course, it might be thought that space development does not *have* to occur this way and that provisions can be made so that space mining and space industrialization must benefit all the peoples of the globe. However, legal mechanisms aimed at encouraging the sharing of space resources in a global manner via the medium of the United Nations have all but failed. Of relevance here is the attitude of space-faring nations (like the United States, Russia, China, Japan, and the member countries of the European Space Agency) to the attempted introduction of a new, "fairer," space treaty—and their attitude towards calls by nonspace-faring nations and by developing countries for the Outer Space Treaty to be adjusted to be fairer.

In the late 1970s, to combat the vagaries within the Outer Space Treaty, some nonspace-capable nations, such as Chile and the Philippines, drafted another treaty under the auspices of the UN. This new treaty, the 1979 Moon Treaty (see UN Moon Agreement 1979) proposed common ownership of extraterrestrial bodies saying that they were the "Common Heritage of Mankind." This indicated that no one would be allowed to extract resources without the consent of the global community.

Space-capable nations and the aerospace industry never warmed to the Moon Treaty. Indeed, ever since it was drafted and throughout its lifetime in the twenty-first century, the Moon Treaty has been criticized as deleterious to space development by those who seek to industrialize or commercialize space (for instance, see Finch and More 1984; O'Donnell and Harris 1994; Zubrin 2019a). As a result, the document has not obtained the status of a treaty and is instead referred to as the Moon Agreement by the UN. As far as prospective industrialists like Bezos and Musk are concerned, any regime that implies that resource use must somehow be regulated to ensure its worldwide sharing is a regime that discourages space expansion and exploitation. How is development going to occur, say prospective space developers, if they have to share space resources or share their profits? The Moon Treaty so offended US President Donald Trump (2017–2021), in particular, that he issued an executive order in 2020 clearly stating that the United States does not recognize the Solar System as being the "common heritage of mankind."

Here we can see some familiar ideological rifts opening up on the space frontier. Solar System expansion is held to be possible when capitalism is encouraged and virtually impossible under any regime aimed at distributive justice or the "socialization of space." Because of this, the Moon Treaty could well be resisted in the United States and in other space-capable nations for the rest of the twenty-first century.

Given this failure to convince industrialized and space-capable nations to sign up to the Moon Treaty, the dozen or so nations that supported it tried another tactic in the 1990s: to augment the provisions of the original Outer Space Treaty. The most relevant part of the Outer Space Treaty of concern to developing countries and nonspace-faring nations is Article I which declares the exploration and use of outer space, including the moon and other celestial bodies, shall be carried out for the benefit and in the interests of all countries, irrespective of their degree of economic or scientific development (UN Outer Space Treaty 1967). The main issue of significance here for developing and nonspace-capable nations has been the meaning of space benefit distribution. In order for the sentiments of Article I to be respected, developing nation representatives campaigned for a substantive written agreement to be formulated so that it became clear to the nations of the world exactly how benefits from space exploration and exploitation should be disbursed (see Benkö and Schrogl 1993; Jasentuliyana 1994). Those nations that have at some stage campaigned for augmentation of the Outer Space Treaty include Argentina, Brazil, Chile, Colombia, Mexico, Nigeria, Pakistan, the Philippines, Uruguay, and Venezuela.

Fearing that they may be made to enter into a binding agreement that obligated them to distribute space benefits in a way that they did not like, the space-capable nations rejected any proposal to augment the Outer Space Treaty with another regime aimed at bolstering the meaning of Article I. In this vein, space-capable nations have decided that they themselves should be free to dictate how space benefit distribution should be undertaken. To do otherwise, these nations suggest, is to impose upon the sovereignty of a state to formulate and implement its own international cooperation and aid policies. Through such claims of sovereignty about running their own foreign affairs, these nations have effectively asserted sovereignty over any resources that they may chance upon in outer space. Their options will be to implement aid plans that fairly distribute the resources gained from other planets by dispersing them equally to the signatories of the Treaty or implementing token benefit distribution plans that merely disseminate inspiring photographs of the conquered worlds of the Solar System. Understandably, nonspace-capable nations are worried that space benefit distribution will follow more closely the lines of the latter rather than the former option, thus leaving them devoid of any substantial gain.

The instigation of an authoritative and uniform regime that dictates exactly the manner in which benefits from space use should be distributed might be considered somewhat extreme, since not only would it attract little or no support from space-capable nations, but it may also lock nonspace-capable nations into inappropriate aid plans. The position taken by space-capable nations, namely that they should be free to choose how, and to whom, they distribute space benefits, is just as extreme, however, since it pays no heed to a Treaty whose ideals they professed and willingly signed when the Space Age was young. What is needed is an intermediate approach that stipulates the very real obligations that space-capable nations and private space companies have regarding space benefit distribution—given that the Solar System belongs to all—while allowing individual nations to negotiate their own plans of distribution. In short, there should be a formulation of guiding principles that lay down the focus and depth of space distribution for every nation, whether they will be primarily donors of space resources or recipients, or both.

In procuring this advice, it seems reasonable to be optimistic with regards to the successful negotiation of the focus of space benefit distribution, since this refers to the particular areas of help that space-capable nations are able to deliver and to the particular problems that nonspace-capable nations are facing. However, it seems equally reasonable to be skeptical when it comes to the issue of the depth of distribution as this refers to a quantitative view of space benefit dispersal. It seems unlikely, given their performance in both space and nonspace related matters, that space-capable nations will ever agree to a scheme that places any emphasis on the amount of help that they should commit themselves to, unless that amount is piddlingly small.

To summarize, it is apparent that if you are interested in space development in the Solar System you can participate in it in only indirect ways. Either:

- You get yourself into a position that enables you to formulate space policy,
- You make do with being happy about receiving the audio-visual and scientific results from projects that others plan,
- You campaign for those others to do what you desire, or
- You do it by yourself on the back of riches you made in other industries (as in the case of Bezos, Branson, and Musk).

My contention is that unless you are content to watch space launches on your smartphone, none of the above allows great forms of global public participation in space. Also, the backlash against space tourism to date also indicates that many people are not happy to see extravagant space adventures show up on their screens—let alone pay taxes that contribute to the business success of space adventure companies.

This dearth of real participation in space development may be paralleled with equally deficient participation with regards to the global distribution of space benefits. Though couched in terms of peace and inclusivity, and of technological development, space industrialists are determined to push space development (and the laws that govern it) into a pathway that allows for their domination of the space frontier. Under this pattern, commonly owned resources of the Solar System fall into the hands of a technological elite.

From a global perspective, while developing countries and nonspace-faring nations have in the past been demanding that some real substance be attached to the sentiments expressed in the Outer Space Treaty, the nations of the world that are actually in the position to use space resources would like to see the provisions of the Outer Space Treaty remain as skeletal and ambiguous as possible since it allows them to interpret space benefit distribution in as self-interested and miserly way as they desire. In this manner, current space exploration activities are not on a route toward international equality. Indeed, they may be harbingers of great future economic disparity.

References

Brandon, John. 2021. "Jeff Bezos and Richard Branson Face Pushback About Their Planned Spaceflights." *Forbes*, July 8, 2021. www.forbes.com/sites/johnbbrandon/2021/07/08/jeff-bezos-and-richard-branson-face-pushback-on-social-media-about-their-planned-space-flights/?sh=7af37efb1d98.

Benkö, Marietta, Kai-Uwe Schrogl, and Willem De Graaf, eds. 1993. *International Space Law in the Making*. Gif-sur-Yvette: Editions Fronteires.

Brown, Craig. 2021. "Space Tourism: It Will Never Give Me a Buzz." *Daily Mail*, July 13, 2021. www.dailymail.co.uk/debate/article-9782251/CRAIG-BROWN-Space-tourism-never-Buzz.html.

Carlson, Cajsa. 2021. "SpaceX wins NASA contract to design moon lander." *Dezeen*, April 20, 2021. www.dezeen.com/2021/04/20/spacex-starship-nasa-artemis-moon-landing/.

Davenport, Christian. 2019. *The Space Barons: Elon Musk, Jeff Bezos, and the Quest to Colonize the Cosmos*. PublicAffairs.

Fernholz, Tim. 2018. *Rocket Billionaires: Elon Musk, Jeff Bezos and the New Space Race*. New York: Houghton-Mifflin.

Finch, Edward and Amanda More. 1984. *Astrobusiness: A Guide to Commerce and Law of Outer space*. New York: Praeger Publishers.

Framke, Caroline. 2021. "Watching Jeff Bezos Go to Space Was More Depressing Than Inspiring." *Variety*, July 20, 2021. www.variety.com/2021/tv/columns/jeff-bezos-space-richard-branson-1235023553/.

Glen, Alex. 2021. "World's richest astronaut Jeff Bezos reveals masterplan to 'build a road to space'." *Euro Weekly News*, July 21, 2021. www.euroweeklynews.com/2021/07/21/worlds-richest-astronaut-jeff-bezos-reveals-masterplan-to-build-a-road-to-space/.

Gregg, Jack. 2021. *The Cosmos Economy: The Industrialization of Space*. Springer.

Heppenheimer, T. A. 1977. *Colonies in Space*. Harrisburg, PA: Stackpole Press.

Hiltzik, Michael. 2021. "In Billionaires' Space Race, Tourism Is a Sideshow to Quest for Moon Landing Contracts." *LA Times*, July 13, 2021.

Jackson, Tim. 2021. "Billionaire Space Race: The Ultimate Symbol of Capitalism's Flawed Obsession with Growth." *The Conversation*, July 20, 2021. www.theconversation.com/billionaire-space-race-the-ultimate-symbol-of-capitalisms-flawed-obsession-with-growth-164511.

Jasentuliyana, Nandasiri. 1994. "Ensuring Equal Access to the Benefits of Space Technologies for all Countries." *Space Policy* 10, no. 1: 7–18.

Kelvey, Jon. 2021. "Inspiration 4: How Much Does a Ticket to Space Cost?" *Inverse*, September 9, 2021. www.inverse.com/science/inspiration-4-how-much-is-a-ticket-to-space.

Mankins, John. 1996. "Space Technology in the Coming Century: Where Next?" *Ad Astra* 8, no. 3: 48–51.

Marshall, Alan. 1995. "Development and Imperialism in Space." *Space Policy* 11, no. 1: 41–52.

Marshall, Alan. 2022. "The Public Lament of Jeff Bezos' 2021 Space Jaunts." *PSAKU International Journal of Interdisciplinary Research* 11, no. 2: 43–59.

Michaud, Micheal. 1987. *Reaching for the High Frontier*. New York: Praeger Publishers.

National Commission on Space. 1986. *Pioneering the Space Frontier*. New York: Bantam Books.

Nolan, Hamilton. 2021. "Why is Bezos flying to space? Because billionaires think Earth is a sinking ship." *The Guardian*, July 20, 2021. www.theguardian.com/commentisfree/2021/jul/20/jeff-bezos-space-flight-billionaires-earth-sinking-ship.

Oberg, James and Alcestis Oberg. 1987. *Pioneering Space: Living on the Next Frontier*. New York: McGraw and Hill.

O'Donnell, D. J. and P. R. Harris. 1994. "Is it time to amend or replace the Moon Treaty?" *Air and Space Lawyer* 9, no. 1: 121–143.

Reanau, Allyson. 2021. *Moon First and Mars Second: A Practical Approach to Human Space Exploration*. New York: Springer.

ProPublica. 2021. *The Secret IRS Files: Trove of Never-Before-Seen Records Reveal How the Wealthiest Avoid Income Tax*. Available at: https://www.propublica.org/article/the-secret-irs-files-trove-of-never-before-seen-records-reveal-how-the-wealthiest-avoid-income-tax/.

Pyle, Rod. 2019. *Space 2.0: How Private Spaceflight, a Resurgent NASA, and International Partners are Creating a New Space Age*. New York: BenBella Books.

Rivera, Josh. 2021. "Sorry, Jeff Bezos, You're Still Not an Astronaut, According to the FAA." *USA Today*, July 27, 2021. www.usatoday.com/story/tech/2021/07/25/jeff-bezos-federal-aviation-administration-astronaut-wings/8087596002/.

Seedhouse, Eric. 2013. *SpaceX: Making Commercial Spaceflight a Reality*. New York: Springer.

Sharp, Rachel and Jennifer Smith. 2021. "Amazon Customers Slam Jeff Bezos for 'Tone Deaf' Victory Lap." *Daily Mail*, July 21, 2021. www.dailymail.co.uk/news/article-9808657/Bernie-Sanders-slams-newly-minted-astronaut-Jeff-Bezos-not-paying-tax.html.

UN Moon Agreement. 1979. "Agreement Governing the Activities of States on the Moon and Other Celestial Bodies." *UN Office of Outer Space Affairs*. https://www.unoosa.org/oosa/en/ourwork/spacelaw/treaties/travaux-preparatoires/moon-agreement.html.

UN Outer Space Treaty. 1967. *UN office of Outer Space Affairs*. https://www.unoosa.org/oosa/en/ourwork/spacelaw/treaties/outerspacetreaty.html.

von Braun, Werhner. 1967. *Space Frontier*. NY: Holt, Rinehart and Winston.

Weinzearl, Matthew. 2018. "Space: The Final Economic Frontier." *Journal of Economic Perspectives* 32, no. 2: 173–119.

Zahn, Max. 2021. "Bezos Thanks Amazon Customers who 'Paid' for Space Flight, Sparking Criticism from Sen. Warren and AOC." *Yahoo Finance*, July 20, 2021. www.finance.yahoo.com/news/elizabeth-warren-attacks-amazon-taxes-bezos-space-flight-195411533.html.

Zubrin, Robert. 1996. "The Need for a Space Frontier." *Ad Astra* 8, no. 3: 6–9.

Zubrin, Robert. 2019a. *The Case for Space: How the Revolution in Spaceflight Opens Up a Future of Limitless Possibility*. New York: Prometheus.

Zubrin, Robert. 2019b. "NASA's Next 50 Years." *The New Atlantis* 59: 63–67.

Zubrin. Robert. 2021. *The Case for Mars: The Plan to Settle the Red Planet and Why We Must*. New York: Free Press.

6

Phrenology in Space

Legacies of Scientific Racism in Classifying Extraterrestrial Intelligence

William Lempert

> *The most dangerous ideas are those so embedded in the status quo,*
> *so wrapped in a cloud of inevitability, that we forget they are ideas at all.*
> —**Jacob M. Appel (2014, 76)**

Conjuring up images of snake oil salesmen prophesying personality from head bumps, it is tempting to relegate phrenology's relevance to the dustbin of bygone pseudoscience. Indeed, at first glance this discredited nineteenth-century system of measuring skull shapes to interpret intellectual qualities would seem to lack any relevance to current discourse by scientists committed to the search for extraterrestrial intelligence (SETI). However, phrenology was not simply a deeply flawed framework; it was a professionally respected discipline for decades, with devastating and enduring human impacts. It served as an intellectual alchemy that transmuted desires for colonial conquest into the confidently credible language of science.

In this chapter, I argue that several of the same underlying assumptions that animated phrenology's construction of intelligence are also presumed within current scientific discourses about the existence and nature of extraterrestrial intelligence (ETI). While phrenology was originally formulated as a humanistic system to further self-improvement, its central proposition—that measuring the size and shape of skulls reveals qualities of intelligence—went on to legitimize theories of hierarchical racial classification used to justify imperial violence and dispossession.

William Lempert, *Phrenology in Space* In: *Reclaiming Space*. Edited by: James S. J. Schwartz, Linda Billings, and Erika Nesvold, Oxford University Press. © Oxford University Press 2023. DOI: 10.1093/oso/9780197604793.003.0006

Here, I engage the ways in which parallel assumptions structure the following four key pillars of SETI discourse, as well as the distilled question that each one attempts to answer:

- Fermi's Paradox (Where is everybody?)
- The Drake Equation (How can we calculate their existence?)
- The Great Filter (What prevents their existence?)
- The Kardashev Scale (How can we classify them?)

These discourses represent a narrow set of assumptions and questions for imagining ETI. However, I am not whatsoever equating historical scientific racism with contemporary SETI discourses. Rather, my aim is to illustrate the importance of clarifying how, why, by whom, and for whom the intelligence of others becomes codified so that current frameworks for understanding ETI do not similarly lead to unintentional cruelty in the future.

Phrenology's Flawed Foundations

In *Materials of the Mind: Phrenology, Race, and the Global History of Science, 1815–1920*, historian James Poskett describes phrenology as the "most popular mental science of the Victorian age," which held as its core precept that "the brain was the organ of the mind" (2019,1). This now debunked system claimed that it could ascertain a person's intellectual qualities based on the size and shape of their brain by measuring the skull surrounding it. As with current SETI debates, the origin of phrenology was not characterized by explicit malice nor any desire to do harm. None of the three key developers of phrenology would have supported the race science that it ultimately bolstered. In retrospect, however, the potential for this outcome could have been anticipated, as its seeds were baked into phrenology's presuppositions. Here, I briefly trace the intentions, assumptions, and outcomes of phrenology to consider them in relation to the arc of contemporary SETI discourse.

In the late 1790s, the German anatomist Franz Joseph Gall formulated the "fundamental theses" that would shape his work, namely that intellectual ability and morality are innate qualities based on the brain and its components (Greenblatt 1995, 791). Gall was influenced by philosophical materialism (the view that all phenomena including thought can be explained by interactions of matter) and proposed that the size and shape of the skull could reveal insights into the development of specific brain areas (Van Wyhe 2017). After working with Gall for over a decade, physician Johann Christoph

Spurzheim systematized this idea into a model he described as "phrenology." Unlike the empirically minded Gall, Spurzheim was largely motivated by the desire to improve the qualities of individuals and even of human nature itself. He believed that phrenology could break down European class divisions and lead to a more equitable society. This ideological drive led him to become evangelical about it, helping to increase its popularity in Britain in the early decades of the nineteenth century. Ominously, he visited the United States not only to spread phrenology, but also to pursue his interest in examining the heads of Native Americans and African Americans (2017, 795). This early connection between race and phrenology, ostensibly to increase equity, would in time become inverted to naturalize oppression and European superiority.

After decades of theorization without credible supporting studies, phrenology began to lose favor with doctors and scientists in Britain in the early 1800s—rightly so, as recent in-depth analysis has reconfirmed phrenology's lack of any scientific merit (Jones et al. 2018). However, it nonetheless took hold in the broader public imagination, especially in the United States. Lawyer George Combe became its leading proponent, and his book on the subject was so influential that for decades it was the fourth-most purchased text in the English language. Like Spurzheim, he viewed phrenology as a way of improving society, and selected the United States as the ideal place to promulgate it. Spurzheim and Combe inspired influential phrenological societies in Philadelphia, Boston, and elsewhere. These societies pushed for the field to be given full scientific status, and their members regularly published in the *Boston Medical and Surgical Journal* (now the *New England Journal of Medicine*). Even popular writers including Edgar Allen Poe and Walt Whitman were inspired by phrenology, referencing it throughout their writing (Greenblatt 1995,797).

The founders of phrenology are credited with sowing the seeds of Victorian scientific naturalism,[1] which framed much of the science over the following century (Van Wyhe 2017). However, as Poskett argues, phrenology had impacts that transcended its engagement by doctors, scientists, and intellectual societies. Indeed, its implicit conflation of human skull sizes/shapes with morality and intellectual ability positioned it as an adaptable framework for justifying the race science that became mainstream throughout the second half of the nineteenth century, which was part of a broader scientific "politics of oppression" in this era (2019,16). When incorporated with social evolutionism—a contortion of Darwin's theories—phrenology provided the guise of scientific empiricism to instill institutionalized racism. Its focus on skull size and shape bolstered the mass collection and unethical treatment of human remains in museums. Indeed, many of Combe's followers competed with one another to hoard skulls from around the world.

Phrenology's ultimate influence was vast, destructive, and enduring. It was invoked to justify violent dispossession across the globe, having been influential in colonial policies from South Africa to India. It was embedded within the foundations of eugenics, IQ testing, and early anthropological racial hierarchies. Continuing deep into the twentieth century, it helped to inspire Hitler and other genocidal perpetrators. Furthermore, the United States system of crime and imprisonment itself was shaped by phrenology (Thompson 2021). The seemingly progressive lineage of phrenology continued in parallel, remaining popular with missionaries into the early 1900s, who attempted to save not only souls, but also minds (2021, 168). Its legacy echoes within recent popular publications that argue for race-based intellectual difference, such as *The Bell Curve: Intelligence and Class Structure in American Life* (Herrnstein and Murray 1994). As psychologist William Uttal argues (2001), there is even the danger of a "new phrenology" taking hold in cognitive science as cutting-edge imaging techniques foster tendencies to overstate the correlation between brain area functions and inherited traits.

Measuring ET Intelligence

At this point, a reader might reasonably ask, "how is any of this relevant to SETI?" Here, I aim to demonstrate this connection by relating the foundational assumptions of phrenology to four key frameworks of contemporary SETI discourse: Fermi's Paradox, the Drake Equation, the Great Filter, and the Kardashev Scale. Before engaging them individually, I briefly discuss their shared assumptions. My goal is not to caricature any of these multifaceted literatures nor to suggest any negative motivations by individuals involved. Rather, it is to encourage those engaged in these discussions to further consider the core assumptions of these debates, and why they remain central in scientific and popular considerations of SETI.

While vastly separated by historical time and space, these four intellectual pillars share similar premises—or even a *time-knot*[2]—with phrenology as it was originally conceived. These four emphasize a materialist approach to evaluating intellectual attributes based on greatly constrained electromagnetic data. For phrenologists, the visual was paramount. Skull shapes served as proxies for brain shapes, which were further proxies for brain regions, whose functions were then surmised. For SETI scientists, various forms of imaging serve as proxies for interpreting clues regarding signs of ETI. Both share multiple layers of interpretation necessary to bridge data and analysis, and which invite myriad opportunities for ethnocentric projection.

Despite the largely hypothetical nature of SETI, like phrenology, it is wrapped in the language of science. This carries profound rhetorical power, especially for the general public, who might not understand the limited empirical basis for these discussions. SETI discourse also shares phrenology's tendency toward devising classificatory intelligence hierarchies, which often conflate size with capability and, implicitly, inherent value.

Fermi's Paradox (Where Is Everybody?)

Also known as "the great silence," Fermi's Paradox is often framed as the most fundamental challenge to the existence of intelligent life in the universe (Lingam and Loeb 2021). Although physicist Enrico Fermi never published on the topic, it bears his name in light of the now famous remarks he made in 1950 at a lunch in New Mexico's Los Alamos National Laboratory, when he exasperated the question, "Where is everybody?" in relation to the lack of ETI contact. Astrophysicist Michael Hart later answered this by arguing that "they are not here; therefore they do not exist" (1975, 128). Fermi himself held a more measured view, suggesting that interstellar travel might be too difficult or simply not worth the effort (Gray 2015, 2).

This framework posits that if just one intelligent species began expanding outward at a relatively moderate speed, they could theoretically reach the edge of their solar system, galaxy, and universe in a relatively short period of time. Thus, even if intelligent life occurs very rarely, Hart's argument is that it should by now be omnipresent. Indeed, for many SETI scientists, this is *the* most important question regarding ETI, and its answer would have profound implications for humanity itself (Ćirković 2009). This strong view of Fermi's Paradox was so influential that it was cited in the United States Congress as a reason to defund NASA's SETI program (Gray 2015, 5).

Before summarizing the multitude of responses to Fermi's Paradox, I pose a straightforward, though uncommon question within these debates: why do we assume that ETIs would be compelled to expand ever outward from their home? Debates on Fermi's Paradox take for granted a Western behavior that is exceedingly rare in the history of our own species: the impulse to endlessly colonize. Indeed, nearly all proposed answers to this question assume imperial expansion as a given.

In his book devoted to the topic, astronomer Stephen Webb (2015) summarizes seventy-five proposed solutions to Fermi's Paradox. Having devoted decades to the question, he divides them into three sections. The first, "They Are (or Were) Here," considers ten solutions including contemporary UFO claims, ancient alien visits, and humans-as-aliens. It also includes theories suggesting that we are exhibits in a cosmic zoo, are only visited infrequently, or that our intelligence is so relatively inferior that we are of no interest to ETIs. The second category is titled "They Exist, But We Have Yet to See or Hear from Them." It includes forty solutions, many of which engage limits to exploration including those of time, space, and radiation. Other theories suggest that they are contacting us, but we do not yet understand how to listen. Still others hypothesize that ETIs only communicate on interstellar highways made up of physical probes, have transcended into information itself and live in black holes, are hibernating until the universe cools down, or have migrated into other dimensions. Some suggest that apocalyptic events regularly destroy all regional intelligence, or that they are hiding in a "dark forest" for fear of being destroyed by more powerful ETIs. The third category, "They Don't Exist," describes twenty-five explanations for why ETIs have never developed. The solutions suggest various limiting factors, including the rarity of Earth's stable protective qualities and the sparking of life itself, as well as evolutionary barriers to achieving "high intelligence."

Virtually all of these solutions suggest that ETIs have formerly visited, are visiting, will visit, or would if they existed. There are only a few exceptions, with one suggesting that ETIs simply have no desire to communicate or are so different that we could not possibly understand one another. Astronomer Milan Ćirković asserts that the entire premise of galactic colonization itself is flawed. As he suggests, Fermi's Paradox represents an anthropomorphic philosophical failure by which "we are searching for ourselves." He suggests that ETIs may have transcended biological need, realized that distance travel is too arduous, or developed moral codes against conquering (2018, 337). However, expansionism endures even in this thesis, which suggests that ETIs will have technologically or morally *moved beyond* innate desires to conquer.

What if we were to make a singular change to the premise of Fermi's Paradox and instead assume that ETIs (even those with high levels of technological complexity) inhabit a home area, live relatively sustainably, and have no desire to significantly expand their territory? Such an assumption is congruent with the corpus of anthropological insights (Graeber and Wengrow 2021). Indeed, when considering the totality of human societies past and present, the desire to expand endlessly is incredibly rare; this is also true for other species broadly understood to be highly intelligent. The preponderance of data demonstrates that any species or society that overshoots its environmental carrying capacity will thereafter experience rapid decline, an outcome that technology is more likely to hasten than to stave off (Catton 1982). The anomaly of human expansionism is similarly correlated to the rarity of secular worldviews that understand the universe as spiritually inert, rather than animated—a topic engaged deeply by authors throughout this book.

I suggest that the assumed behavior of endlessly expanding ETIs shares a key quality with lethal malignant cancer; in the relatively rare case in which one develops, the host tends to be quickly extinguished. I propose the metaphor of "civilizational cancer" to describe the expansionist behavior that results in self-destruction due to the colonial impulse itself. As engineer Colin McInnes (2002) suggests, there is an amoral calculus to expansionist disintegration, even if short-term environmental destruction is avoided. He demonstrates that there may be a "light cage" limit to human migration, in which expansion leads inevitably to collapse due to a mismatch between population, resources, and the maximum limit of lightspeed travel.

The civilizational cancer model provides one possible resolution to Fermi's Paradox that aligns with anthropological understandings of the prevalence and outcomes of human expansionism. It is compatible with (and even supports) the existence of untold numbers of ETIs throughout the universe. As is the case on Earth, many of these sentient beings would have incommensurably different forms of intelligence than we could scarcely comprehend, though perhaps a small percentage would be vaguely understandable to us. Regardless, our current and permanent lack of communication with any of them would be expected—rather than confounding—since sending distant signals into the universe would only be pursued by short-lived ETIs stricken with acute civilizational cancer.

The Drake Equation (How Can We Calculate their Existence?)

Shortly before a 1961 conference on "Extraterrestrial Intelligent Life" in Green Bank, West Virginia, astronomer Frank Drake realized that he needed to form an agenda as the host. As Drake recalls:

> I wrote down all the things you needed to know to predict how hard it's going to be to detect extraterrestrial life. And looking at them it became pretty evident that if you multiplied all these together, you got a number, N, which is the number of detectable civilizations in our galaxy. (Drake 2003)

The Drake Equation arose organically and casually, similarly to Fermi's lunch table discussion (Ćirković 2018, 2). However, when Drake was writing this equation onto the conference chalkboard, he may as well have been carving it into stone tablets. Indeed, it has remained virtually unchanged over the past sixty years as the dominant quantitative framework for calculating the existence of ETIs. Science journalist Michael Lemonick even declared it to be the second most important equation of the twentieth century, eclipsed only by Einstein's $E = mc^2$ (1998, 45). The Drake Equation is also considered by some to be the least understood of the grand scientific questions (Ćirković 2018, 1). To estimate the number of technologically advanced civilizations in the Milky Way galaxy, it includes seven variables that gradate left to right from relatively certain to highly speculative (Fig. 6.1).

As with Fermi's Paradox, here I aim to unpack some of the assumptions in this model. Most dubious seem to be the cluster of three f variables denoting the total fractions of planets on which life appears, planets on which intelligent life then emerges, and civilizations that develop detectable signal technologies. As with phrenology, there is a seductive logic here. Just as surely as a total understanding of the brain would lead to psychological illumination, so too would understanding these variables lead to profound insights. As with the Drake Equation, it is not difficult to imagine why so many people during the 1800s were entranced by phrenology, as "it transformed abstract metaphysical conceptions . . . [into] entities amenable to the interests and understandings of the practical minded" (Greenblatt 1995, 794). However, as with phrenology, the equation's variables are hampered by the fact that we have a dearth of empirical evidence to draw conclusions from. Indeed, it is unclear whether the

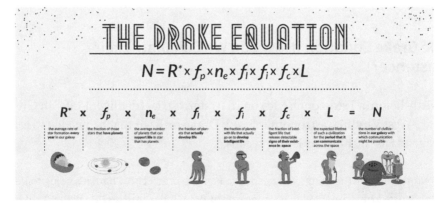

Figure 6.1 An infographic of the Drake Equation. Credit line: Luciano Ingenito ©, used with permission.

three *f* variables could ever be precisely calculated by humans (Gertz 2021). Thus, one might ask the provocative question of whether the Drake Equation might in future centuries be seen similarly to how scientists currently view phrenological maps, such as the one in Fig. 6.2.

While this might at first seem to be an outrageous comparison, consider that both phrenology and the Drake Equation have filled entire books with what appears to be rigorous empirical science, including complex numerical tables and detailed charts. And yet, neither have provided verifiable or precise answers. Unlike phrenology, the Drake Equation may actually have specific answers that scientists could make some progress toward. However, solving

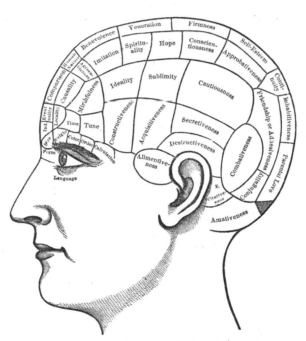

NUMBERING AND DEFINITION OF THE ORGANS.

1. AMATIVENESS, Love between the sexes.
A. CONJUGALITY, Matrimony—love of one. [etc.
2. PARENTAL LOVE, Regard for offspring, pets,
3. FRIENDSHIP, Adhesiveness—sociability.
4 INHABITIVENESS, Love of home.
5. CONTINUITY, One thing at a time.
E. VITATIVENESS, Love of life.
6. COMBATIVENESS, Resistance—defense.
7. DESTRUCTIVENESS, Executiveness—force.
8. ALIMENTIVENESS, Appetite—hunger.
9. ACQUISITIVENESS, Accumulation.
10. SECRETIVENESS, Policy—management.
11. CAUTIOUSNESS, Prudence—provision.
12. APPROBATIVENESS, Ambition—display.
13. SELF-ESTEEM, Self-respect—dignity.
14. FIRMNESS, Decision—perseverance.
15. CONSCIENTIOUSNESS, Justice equity.
16. HOPE, Expectation—enterprise.
17. SPIRITUALITY, Intuition—faith—credulity.
18. VENERATION, Devotion—respect.
19. BENEVOLENCE, Kindness—goodness.

20. CONSTRUCTIVENESS, Mechanical ingenuity.
21. IDEALITY, Refinement—taste—purity.
B. SUBLIMITY, Love of grandeur—infinitude.
22. IMITATION, Copying—patterning.
23. MIRTHFULNESS, Jocoseness—wit—fun.
24. INDIVIDUALITY, Observation.
25. FORM, Recollection of shape.
26. SIZE, Measuring by the eye.
27. WEIGHT, Balancing—climbing.
28. COLOR, Judgment of colors.
29. ORDER, Method - system - arrangement.
30. CALCULATION, Mental arithmetic.
31. LOCALITY, Recollection of places.
32. EVENTUALITY, Memory of facts.
33. TIME, Cognizance of duration.
34. TUNE, Sense of harmony and melody.
35. LANGUAGE, Expression of ideas.
36. CAUSALITY, Applying causes to effect. [tion.
37. COMPARISON, Inductive reasoning—illustra-
C. HUMAN NATURE, Perception of motives.
D. AGREEABLENESS, Pleasantness—suavity.

Figure 6.2 Phrenological chart published in 1857. Credit line: From *The Illustrated Self-Instructor in Phrenology and Physiology* by Orson Squire Fowler, 1857, public domain.

for N with a meaningful degree of precision seems highly improbable for the foreseeable future.

It is important to emphasize here that, like Fermi, Drake meant for his framework to provide a pragmatic way to provoke interesting discussions. However, this too mirrors the goals of phrenology's progenitors, who believed that their work was harmless at worst, while at best it could elevate and serve humanity. There have been substantive technical critiques of the equation (Ćirković 2004), some even by Drake himself, who questioned the built-in assumption that ETIs would develop at a similar rate to human societies on Earth (Dick 2015, 10). Indeed, the variables f_i and f_c imply the incredibly specific human qualities of bipedalism, grasping appendages, three-dimensional (3D) color vision, and a calorically expensive brain featuring language and individualized self-consciousness (Gertz 2021, 264–265).

Perhaps the most relevant critique of the Drake Equation is Walters et al.'s (1980) suggestion that a new parameter of "C" be added to estimate the fraction of civilizations that are motivated to colonize. While this was proposed in the positive sense as a prerequisite for signal dispersal, the "civilizational cancer" model would suggest that this variable operates conversely, extinguishing that society's existence via self-destruction. Indeed, if the framework of civilizational cancer has merit, it suggests that the Drake Equation may be an ultimately futile effort—akin to devising a convoluted and interstellar search for late-stage cancer patients.

The Great Filter (What Prevents Their Existence?)

The Great Filter represents a set of responses to Fermi's Paradox and was proposed by economist Robin Hanson (1998). He takes seriously Hart's assertion that the lack of ETI presence implies its lack of existence. Hanson posits that there must be a powerful limiting factor, or filter, at some point before the end stage of developing intelligence. He argues that since we know that humans with current technology exist, we can reasonably assume that such a level of advancement is relatively common, and that the Great Filter more likely lies in humanity's future than in its past. He surmises that few if any societies survive long enough to communicate or travel interstellar distances before collapsing due to conflict or environmental destruction. Hanson is particularly adamant that colonization is not only a human quality, but a natural inclination of life itself. He asserts that making it through the Great Filter before self-destruction would all but guarantee immortality, as humans would reach an "explosive point" of outward expansion approaching the speed of light. This mission to make it through the Great Filter has been taken up by a generation of scientists (Jiang et al. 2021), futurists (Kaku 2019), and even celebrity space billionaires such as Elon Musk.

However, from the perspective of the civilizational cancer model, the Great Filter as described reverses the cause and effect relationship with danger. Thus, trying to ascend through the Great Filter may operate similarly to what historian Edward Tenner (1996) describes as a "revenge effect," in which the consequences of a technological goal are in opposition to its creators' intentions. There may in fact be numerous ETIs throughout the universe, with the actual filter being the sustained *avoidance* of the self-destructive expansionist behavior that Hanson celebrates. Furthermore, it is worth considering who the Great Filter excludes: namely, every nonexpansionist human society past and present, and every one that has not produced radio telescopes or attempted space travel. This includes every Indigenous society, not to mention nearly all others. By this logic, Aboriginal Australians who have lived relatively sustainably for at least 80,000 years have failed to pass through the Great Filter; meanwhile, Western societies that threaten to render Earth uninhabitable in the near future are positioned at the precipice of transcendence. A similar logic was wielded by later phrenologists attempting to create a framework for naturalizing the superiority of Western societies. When taken seriously, the Great Filter implies that the structural mistreatment of non-Western people would be a small price to pay for the ends of immortalizing humanity and consciousness itself.

The Kardashev Scale (How Can We Classify Them?)

In 1964, astrophysicist Nikolai Semonovich Kardashev developed a scale that classified civilizations into types based on their level of energy usage. While the Kardashev Scale has been expanded over the decades (Gray 2020), his original model captures its enduring core concepts. Type 0 is understood as harnessing limited power from organic-based raw materials: type 1, the full solar power of a planet; type 2, the total power of a sun through a Dyson sphere[3]; and type 3, the power of every star in a galaxy (Kardashev 1964). This scale (much like phrenology, Fermi's Paradox, and the Drake Equation) was created as a tentative heuristic to provoke rich debates. Kardashev was specifically interested in furthering discussions on the detectability of hypothetical civilizations, and it has been refined for more practical use in recent decades (Ćirković 2016). However, as happened with phrenology, the creation of a value system for relating

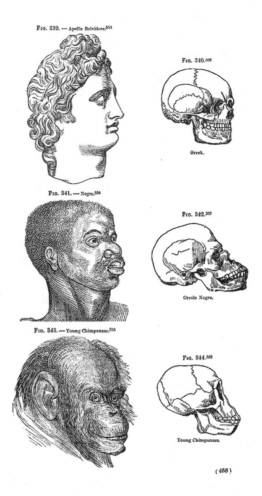

FIG. 339. — Apollo Belvidere.[553]

FIG. 340.[556]

Greek.

FIG. 341. — Negro.[554]

FIG. 342.[557]

Creole Negro.

FIG. 343. — Young Chimpanzee.[555]

FIG. 344.[558]

Young Chimpanzee.

(458)

Figure 6.3 Illustration of racial hierarchy. Credit line: From *Types of Mankind* by George Gliddon and Josiah Nott, 1854, public domain.

spatial qualities with intelligence is easily adapted into frameworks with hierarchical scales. While phrenology was absorbed by various race science typologies that placed Europeans at their apex, the Kardashev Scale has been taken up by physicist Michio Kaku (2019) and other prominent futurists as a model for elevating particular aspects of humanity in order to pass through the immortalizing type 1 threshold. As with discredited race science typologies that drew on phrenology, the Kardashev Scale describes a typology implying existential value, with Western nations having achieved the highest level thus far, which astronomer Carl Sagan himself estimated at 0.7 (Adler 2014, 327).

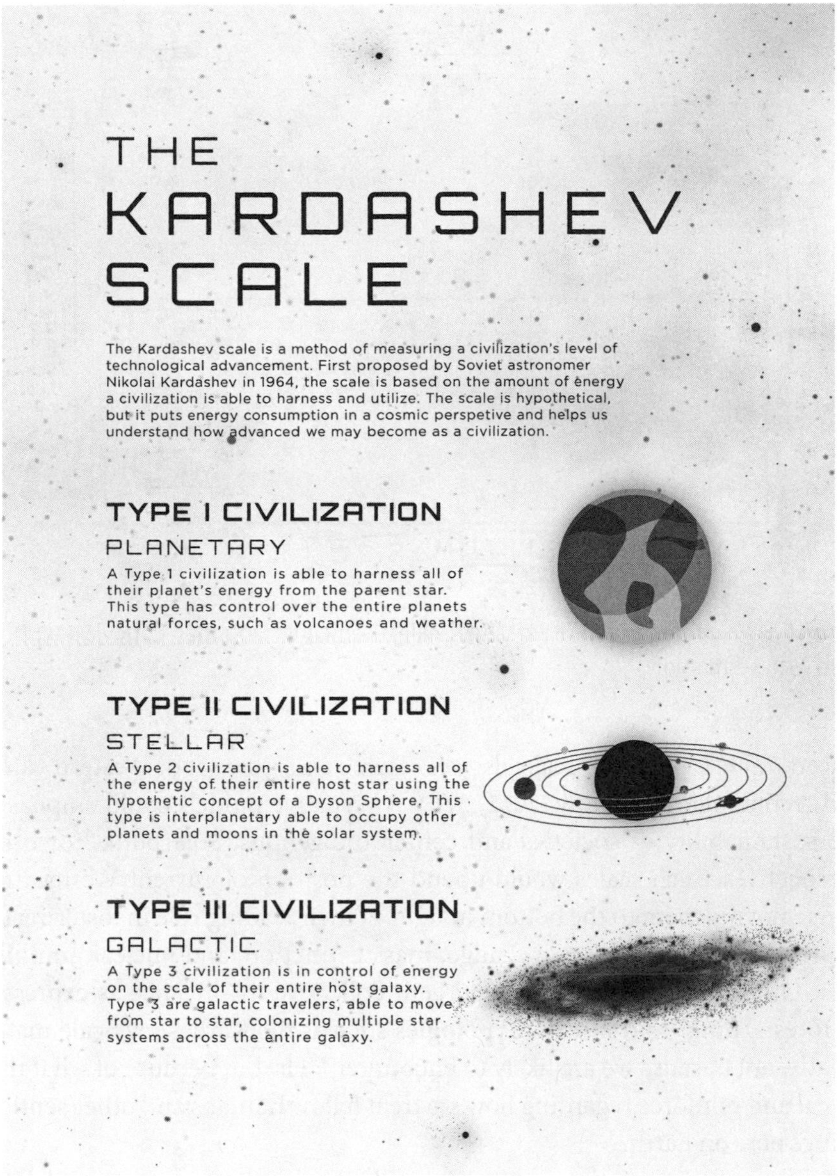

THE
KARDASHEV
SCALE

The Kardashev scale is a method of measuring a civilization's level of technological advancement. First proposed by Soviet astronomer Nikolai Kardashev in 1964, the scale is based on the amount of energy a civilization is able to harness and utilize. The scale is hypothetical, but it puts energy consumption in a cosmic perspetive and helps us understand how advanced we may become as a civilization.

TYPE I CIVILIZATION
PLANETARY

A Type 1 civilization is able to harness all of their planet's energy from the parent star. This type has control over the entire planets natural forces, such as volcanoes and weather.

TYPE II CIVILIZATION
STELLAR

A Type 2 civilization is able to harness all of the energy of their entire host star using the hypothetic concept of a Dyson Sphere, This type is interplanetary able to occupy other planets and moons in the solar system.

TYPE III CIVILIZATION
GALACTIC

A Type 3 civilization is in control of energy on the scale of their entire host galaxy. Type 3 are galactic travelers, able to move from star to star, colonizing multiple star systems across the entire galaxy.

Figure 6.4 Visual description of the Kardashev Scale. Credit line: Brian Smith ©, used with permission.

This model implies that the vast majority of societies and nations are destined for type 0 stagnation (Fig. 6.4). The inclusion of great apes in race science typologies is relevant here (as seen in Fig. 6.3); in the Kardashev Scale, every nonhuman species—including bonobos, dolphins, and octopuses—are

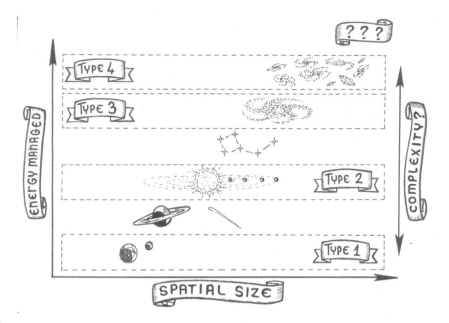

Figure 6.5 Visual chart of the Kardashev Scale. Credit line: Slobodan Popović Bagi ©, used with permission.

placed *below* type 0. The scale also represents a narrow set of Western values and technologies (Denning 2011, 376). Alternative models could emphasize the sustainability of societies and center Indigenous, Solarpunk,[4] or other perspectives. Such scales would upend the position of current Western nations, moving them to the bottom in light of their leading role in fostering the existential threats of climate change, mass extinction, and nuclear annihilation. Other scales could value emergent collective intelligence as expressed by forests, fungi, and insects. Typologies such as the Kardashev Scale matter greatly, not because we are likely to encounter ETIs, but because of what they reveal and reinforce regarding how we treat fellow humans and other sentient beings here on Earth.

Furthermore, from the perspective of the civilizational cancer model, the relentless capturing of all of the energy in the galaxy would seem to court disaster at cosmic scales. As Fig. 6.5 suggests, the only limit to limitless expansion is that of the universe itself. Considering such an endless appetite for exponential growth, one might reasonably ask, "how could any amount of space ever be enough for such an interstellar society?" A cancerous civilization that surpasses Hanson's Great Filter might be akin to opening Pandora's Box or releasing what Kurt Vonnegut (1998) imagined as Ice-nine, a substance that would endlessly and relentlessly spread. Indeed, Types 3, 4, or beyond would seem more likely to resemble destruction metastasizing across the universe than the establishment of enviable societies.

Conclusion

I was thinking this globe enough, till there sprang out so noiseless around me myriads of other globes.

Now, while the great thoughts of space and eternity fill me, I will measure myself by them;

And now, touch'd with the lives of other globes, arrived as far along as those of the earth,

Or waiting to arrive, or pass'd on farther than those of the earth,

I henceforth no more ignore them, than I ignore my own life.

Or the lives of the earth arrived as far as mine, or waiting to arrive.
                                                                    —**Walt Whitman**

Walt Whitman's interest in phrenology began in 1846, when he stumbled upon it in a bookstore. He then sought out members of the Fowler family, who were well known phrenologists at the time, to have his head measured. Their book press, Fowler and Wells, even published his 1856 edition of *Leaves of Grass,* which includes several references to phrenology. The epigraph above is taken from "Night on the Prairie," a poem in his book's 1860 edition. In it, one can almost feel the influence of phrenology as he considers the vastness of the universe. In addition to language on measurement and globes, there is an optimism about self-improvement and wanting to further his understanding of others on and beyond Earth. Whitman's poem is a reminder that SETI scientists tend to be largely motivated by the expansion of human perspectives and empathy.

This chapter is not meant to denigrate SETI. Rather, it aims to constructively interrogate the assumptions embedded within its frameworks. As historian Courtney Thompson reminds us, the fundamental issue of phrenology is that it "has never really left us; or rather that we have never truly abandoned phrenology . . . [since] modern-day cultural assumptions and narratives were created within this epistemology" (2021, 163). Dismissing phrenological thinking as a relic of the past leaves us ever vulnerable to its renewal. Thus, it is essential that we consider the ways in which well-motivated theories and typologies of intelligence can ultimately result in disastrous consequences for humans and other sentient beings.

Acknowledgments

I am grateful to the editors of this volume, who not only organized and framed the book itself, but also hosted a rich and generative virtual workshop with authors. My thoughts regarding this chapter have also greatly benefited from generous feedback from and discussions with David Valentine, Deborah Battaglia, Valerie Olson, Lisa Messeri, Kathryn Denning, Kim TallBear, David Shorter, Sonya Atalay, Michael Oman-Reagan, Claire Webb, Taylor Genovese, Tamara Alvarez, Alexander Taylor, and Jennifer Shannon.

Notes

1. Victorian scientific naturalism aggregates an eclectic group of Victorian scholars in the mid- to late 1800s who shared a common scientific goal of redefining humanity through the careful observation of the natural world. It includes Darwin, Spencer, Huxley, and many of the most canonical scientists of that era.
2. I have previously drawn on Dipesh Chakrabarty's (2008) conception of the "time-knot" to articulate ongoing parallels between contemporary outer space projects and historical colonization (Lempert 2021).
3. Theorized by physicist Freeman Dyson, the Dyson sphere is a megastructure around a star that would capture the vast majority of its energy output.
4. Solarpunk is a speculative genre which imagines future societies that have solved contemporary environmental challenges including climate change, pollution, and mass extinction by returning to a balance with the natural world. Imagined as a worldmaking response to environmental apocalypticism, this hopeful and aesthetically driven genre includes the writing of Ursula Le Guin, Kim Stanley Robinson, and many other science fiction authors.

References

Adler, Charles L. 2014. *Wizards, Aliens, and Starships*. Princeton, NJ: Princeton University Press.
Appel, Jacob M. 2014. *Phoning Home: Essays*. Columbia, SC: University of South Carolina Press.
Catton, William R. 1982. *Overshoot: The Ecological Basis of Revolutionary Change*. Champaign, IL: University of Illinois Press.

Chakrabarty, Dipesh. 2008. *Provincializing Europe: Postcolonial Thought and Historical Difference*. Princeton, NJ: Princeton University Press.

Ćirković, Milan M. 2004. "The Temporal Aspect of the Drake Equation and SETI." *Astrobiology* 4, no. 2: 225–231.

Ćirković, Milan M. 2009. "Fermi's Paradox: The Last Challenge for Copernicanism?" *Serbian Astronomical Journal* 178: 1–20.

Ćirković, Milan M. 2016. "Kardashev's Classification at 50+: A Fine Vehicle with Room for Improvement." *Serbian Astronomical Journal* 191: 1–15.

Ćirković, Milan M. 2018. *The Great Silence: Science and Philosophy of Fermi's Paradox*. New York: Oxford University Press.

Denning, Kathryn. 2011. "Being Technological." *Acta Astronautica* 68: 372–380

Dick, Stephen. 2015. "Introduction: The Drake Equation in Context." In *The Drake Equation: Estimating the Prevalence of Extraterrestrial Life Through the Ages*, edited by Douglas A. Vakoch and Matthew F., 1–20. Dowd. Cambridge: Cambridge University Press.

Drake, Frank. 2003. "The Drake Equation Revisited: Part I." *Astrobiology Magazine*. September 29, 2003.

Gertz, John. 2021. "The Drake Equation at 60: Reconsidered and Abandoned." *Journal of the British Interplanetary Society* 74, no. 7: 258–268.

Graeber, David, and David Wengrow. 2021. *The Dawn of Everything: A New History of Humanity*. New York: Farrar, Straus and Giroux.

Gray, Robert H. 2015. "The Fermi Paradox is Neither Fermi's Nor a Paradox." *Astrobiology* 15, no. 3: 195–199.

Gray, Robert H. 2020. "The Extended Kardashev Scale." *The Astronomical Journal* 159, no. 5: 228.

Greenblatt, Samuel H. 1995. "Phrenology in the Science and Culture of the 19th Century." *Neurosurgery* 37, no. 4: 790–805.

Hanson, Robin. 1998. "The Great Filter: Are We Almost Past It?" Published online at http://hanson.gmu.edu/greatfilter.html.

Hart, Michael. 1975. "An Explanation for the Absence of Extraterrestrials on Earth." *The Quarterly Journal of the Royal Astronomical Society* 16: 128–135.

Herrnstein, Richard, and Charles Murray. 1994. *The Bell Curve: Intelligence and Class Structure in American Life*. New York: Free Press.

Jiang, Jonathan H., Philip E. Rosen, and Kristen A. Fahy. 2021. "Avoiding the 'Great Filter': A Projected Timeframe for Human Expansion Off-World." *Galaxies* 9, no. 3: 53.

Jones, O. Parker, Fidel Alfaro-Almagro, and Saad Jbabdi. 2018. "An Empirical, 21st Century Evaluation Of Phrenology." *Cortex* 106, no. 1: 26–35.

Kaku, Michio. 2019. *The Future of Humanity: Our Destiny in the Universe*. Hamburg: Anchor.

Kardashev, Nikolai S. 1964. "Transmission of Information by Extraterrestrial Civilizations." *Soviet Astronomy* 8: 217–221.

Lempert, William. 2021. "From Interstellar Imperialism to Celestial Wayfinding: Prime Directives and Colonial Time-Knots in SETI." *American Indian Culture and Research Journal* 45, no. 1: 45–70.

Lemonick, Michael D. 1998. *Other Worlds: The Search for Life in the Universe*. New York: Simon & Schuster.

Lingam, Manasvi, and Abraham Loeb. 2021. *Life in the Cosmos: From Biosignatures to Technosignatures*. Cambridge, MA: Harvard University Press.

McInnes, Colin R. 2002. "The Light Cage Limit to Interstellar Expansion." *Journal of the British Interplanetary Society* 55: 279–284.

Poskett, James. 2019. *Materials of the Mind: Phrenology, Race, and the Global History of Science, 1815–1920*. Chicago: University of Chicago Press.

Tenner, Edward. 1996. *Why Things Bite Back: Technology and the Revenge of Unintended Consequences.* New York: Knopf.

Thompson, Courtney E. 2021. *An Organ of Murder: Crime, Violence, and Phrenology in Nineteenth-Century America.* New Brunswick, NJ: Rutgers University Press.

Uttal, William R. 2001. *The New Phrenology: The Limits of Localizing Cognitive Processes in the Brain.* Cambridge, MA: The MIT Press.

Van Wyhe, John. 2017. *Phrenology and the Origins of Victorian Scientific Naturalism.* London: Routledge.

Vonnegut, Kurt. 1998. *Cat's Cradle.* New York: Dial Press.

Walters, Clifford, Raymond A. Hoover, and Rama K. Kotra. 1980. "Interstellar colonization: a new parameter for the Drake Equation?" *Icarus* 41, no. 2: 193–197.

Webb, Stephen. 2015. *If the Universe Is Teeming with Aliens . . . Where Is Everybody? Seventy-Five Solutions to the Fermi Paradox and the Problem of Extraterrestrial Life.* New York: Springer.

PART 2
THE ART OF ENVISIONING SPACE

7

The Language of Space

Mary Robinette Kowal

I am writing this essay in English. This might seem like an obvious thing, but we live on a planet with roughly 6,500 languages, and I live in a country that has citizens who are native speakers of around 400 languages. So, why English? The United States doesn't have an official language, after all.

How often do you think about your language? If I had not called attention to it, would you have noticed that I was writing this in English? Likely not. The choice to write this in English, rather than in Arabic, French, or Portuguese is deliberate because of where the book is being sold. Even though the United States doesn't have an official language, English dominates.

I write science fiction, which means that I have to imagine futures in space and make guesses about how people will communicate. Did you know that Gene Rodenberry's original vision for *Star Trek* was to film it in Esperanto (introduced in the late 1800s as a global language)? The science fiction novels and TV series *The Expanse* imagines a patois for the inhabitants of the asteroid belt, but the common language is still English. The science fiction TV series *Firefly* imagined a future peppered with Mandarin. While it's true that these shows were ultimately filmed in English because that's what their target audiences spoke, it's not that far-fetched to imagine a future with English as the dominant language of space.

After all, it is now.

In space, as I write this, we have native speakers of Russian, Japanese, English, and French on the International Space Station. The Chinese have their own station as well. So there are at least five languages in space right now.

But labels on the ISS are either in English or Russian. To go there, you must know both. So I cannot stop thinking about the languages that may never leave the planet.

As the words form in your mouth and as you listen to a conversation, you are shaping connections and reinforcing cultures. English is the language that my family speaks at home, but we borrow words from French, Spanish,

Mary Robinette Kowal, *The Language of Space* In: *Reclaiming Space.* Edited by: James S. J. Schwartz, Linda Billings, and Erika Nesvold, Oxford University Press. © Oxford University Press 2023. DOI: 10.1093/oso/9780197604793.003.0007

Yiddish, German, and Japanese depending on what we are expressing in a given moment.

The beautiful thing about languages is that they are more than just a collection of words and syntax—they contain culture. If we imagine a future in which humanity spreads out beyond Earth, the languages we speak when we arrive on other planets will be the ones of convenience.

For instance, at one point I lived in Iceland. I tried to use Icelandic, but people would often address me in English before I had a chance to speak. It was frustrating, wondering what gave me away as a foreigner before I even spoke. Once, when a clerk greeted me in English, I finally asked, "How did you know that I needed to speak English?"

"I didn't, but I needed to." He shrugged. "Unless you speak Polish?"

Later, I heard an Icelander and an Italian family speaking together. In English.

They gravitated to the one that they had in common. Rather than going through the effort of acquiring a new language, it was more convenient to meet in the middle with a language that belonged to neither. The menus were in Icelandic and English—so why English and not Italian? Because most of the tourists are English speakers and it makes monetary sense to cater to those tourists by providing things in their language. Which means that if you're traveling abroad and your language isn't spoken outside your home country, it behooves you to at least know English.

With the space programs that we have now, it's easy to imagine a future in which only Russian, Chinese, and English are spoken on another planet and that 6,497 languages die. Oh, some might arrive on the new planet with their native speakers, but to have a living language, it must be spoken in the home. Not just in the home, but intergenerationally. Children must be taught that language.

Let's look at how this might play out.

The International Space Station (ISS) consists of Canada, Japan, Russia, the United States, and eleven member states of the European Space Agency (Belgium, Denmark, France, Germany, Italy, The Netherlands, Norway, Spain, Sweden, Switzerland, and the United Kingdom). Three of those are already English dominant countries, and the rest commonly teach it in schools.

So French, Danish, Italian, Norwegian, Spanish, Dutch, and German have been into space but are not spoken in conversation there.

There's never been a point where there were two French speakers or two Danes or two Germans.

America and Russia hold the two spaceports that launch crewed missions to the ISS.

It is easiest if everyone has common languages, but more particularly, the specific language of reference manuals is so highly specialized that it is safer to train a handful of astronauts to a common tongue than to risk a disaster because they use different words for the same piece of critical space hardware.

English and Russian are used on a regular daily basis both for work and social life because they are the languages that everyone has in common. There's no deliberate effort made to induce people to abandon their native languages, and yet they do for months at a time.

In the larger picture, when we look at why people abandon their first language on Earth, the deciding factors can be cultural, political, or economic marginalization. Arguably, there is political and economic marginalization happening in space simply because the cost of getting into space and the people who control access mean that if you want to go, you must learn English and Russian. Unintentionally, this is an expression of the politics of language in space.

The question of which languages are spoken on a spacecraft is rooted in who has the power to dictate who got to be on board in the first place.

To imagine a future in which thousands of languages survive the move off planet, we also have to imagine one where space travel is cheap and accessible to people from the most marginalized communities in the most disadvantaged nations.

By the time we reach that point, how many languages will have been subsumed as their children move to other countries and learn the languages of space? If they must learn English, or Russian, or Chinese to understand a technical manual, then when they get into space, won't it be easier if they just speak to each other in one of those common tongues?

So I find myself imagining a future in which there are not thousands of languages in space, because all the pieces that would need to be in place for those languages to survive their terrestrial origins seem more far-fetched than the act of traveling between planets.

This essay is written in English. Does the future of space have to be?

8

Spacefaring for Kinship

Vandana Singh

Some years ago, I attended a conference about the International Space Station (ISS), at which were present representatives of government agencies, corporations large and small, and scientists and educators. The ISS is an artificial satellite that has been orbiting Earth since 1998, continuously occupied by astronauts from the United States, Canada, Europe, Russia, and Japan, who conduct experiments of various kinds in low Earth orbit. At the conference, one of the guest speakers was Elon Musk, who spoke of his dream for a manned mission to Mars, and the need to sell the idea of Mars exploration to the public. His suggestion was to involve the film industry and set "Westerns on Mars." I winced and looked around me at the crowd of at least a thousand people. Among those whose faces I could see, I could count perhaps three others whose expressions reflected my own emotions. *Westerns* on Mars?

Later, I overheard a conversation in the hallway about how difficult it was for NASA to get Native Americans interested in space science.

At one point, I asked a question during a session on planetary exploration. I asked whether there was due consideration given to ethical questions and quandaries with regard to the exploration of other planets. The scientists on the panel looked at each other; then one of them said, "Well, that's the UN's department."

At which point, someone in the audience spoke up and reminded us that NASA's Jet Propulsion Laboratory (JPL) had a planetary protection group. In fact, NASA Headquarters has a Planetary Protection Office that is in charge of policy with a twofold aim: protecting Earth from extraterrestrial invaders, and also protecting other worlds from us, such as preventing contamination with lifeforms from our planet.

It was interesting to me that the Planetary Protection Office did not immediately come to mind when the scientists on the panel were talking about space exploration.

Vandana Singh, *Spacefaring for Kinship* In: *Reclaiming Space*. Edited by: James S. J. Schwartz, Linda Billings, and Erika Nesvold, Oxford University Press. © Oxford University Press 2023. DOI: 10.1093/oso/9780197604793.003.0008

Another "this takes the cake" moment was when a senior staffer at NASA informed us that exploration on Earth had really been for the benefit of mankind (*sic*) so, naturally, the same would be true of space exploration.

Although the conference was several years ago, I'm still reeling from that one.

Exploration on Earth benefited everyone? Really?

Tell that to the Native Americans of North America, the Aboriginal peoples of Australia, the various Native peoples of Africa after European conquest.

How could an American scientist possibly say that in the twenty-teens?

As a writer of speculative fiction, and a physicist, I am intensely interested in space exploration. When I was a kid growing up in India, my siblings and cousins and I would visit my grandparents and sometimes be permitted to sleep on the roof of the house. In the days before Indian cities earned the dubious distinction of being among the most polluted in the world, the night skies used to be clear, and filled with stars. Looking up into the sky, I would be filled with a vertiginous sense of falling upward into an endlessly fascinating universe. I would wonder if anyone was looking back at us from those possible other worlds. This was well before scientists discovered multiple planets orbiting other stars; now the number of exoplanets is in the 4,000 range. But although those experiences of wonder predated any knowledge of statistics, my child-self was convinced that some of those millions of suns must surely be orbited by life-bearing planets. And I couldn't have articulated it then, but that childhood communion with the cosmos filled me with a wonder and desire to know more about this vast home of ours, which led me to seek a career in science and a vocation as a science fiction writer.

Nearly every semester, I take my general physics students to the university's planetarium to give them some idea of the vastness of the universe. The planetarium staff take us on a dizzying journey around our solar system and the galaxy and beyond. Although this is necessarily artificially generated, students generally experience a sense of wonder rather like the experiences of my childhood. Yet, when I ask them how their planetarium experience makes them feel, the most common response is "it makes me feel like I'm so small and unimportant."

What's really interesting to me is that those nights under the open sky under the stars, I never felt small or unimportant. I felt a sense of belonging to something vast and wondrous, and it engendered in me a desire to know more, to explore further, at least in my imagination. Decades later, as I have begun to educate myself about Indigenous epistemologies, I discovered that in many of their cosmologies the night sky is no stranger, and the primary feeling it evokes is indeed of belonging, not estrangement or smallness.

Is there a difference between a sense of wonder accompanied by a feeling of smallness and insignificance, and a sense of wonder accompanied by a feeling of belonging? What could possibly be the psychosocial impact of being cut off from the night sky due to light pollution and particulate pollution in cities, when humans have been able to commune with the night sky since we emerged as a species between 200,000 and 300,000 years ago? I am not sure, having no expertise in this area. But I do wonder if that first response—insignificance and smallness—comes from an estrangement with the night sky, and with the rest of nature in general. This seems to be a feature of modern industrial globalized culture—does our atomized, frantic, disconnected existence lead us to a sense of insignificance as soon as we are confronted with the reality of the universe?

I pose this question as a provocation; I have no answers. But a story comes to mind—a teaching story from the Inupiaq peoples of the North Shore of Alaska. There was once a mouse who lived underground, and one day he decided to dig a tunnel up to the surface of the tundra to see what the world was like. So up he went, digging diligently until he broke through the surface. When he did, he found that he could touch the ceiling of the world with his paws. When he stretched his paws sideways, he could touch the walls of the world. He concluded that he was the biggest thing in all the world.

The punchline of the story is that the mouse had come up into an upside-down boot. The top of the world was the sole of the boot, and the walls of the world were the sides of the boot.

When European explorers came to the Americas, they carried with them a certain perspective of the world and their place in it. This is summed up very nicely in the infamous painting by John Gast in 1880, called *American Progress*. A white woman representing Manifest Destiny strides, giant-like, over American soil, driving into the darkness before her the Native peoples. She bears the signs of enlightenment—a book, technology represented by telegraph wires, and behind her come the wagons of the settlers. Colonialism, simply stated, is the domination of one set of peoples by another. Settler colonialism is when the colonizers displace the Natives and take their land. The European invasion led to a massive reduction in the population of Native Americans, due to diseases and/or direct extermination, an act of genocide that is still whitewashed out of history books today. Terms like Manifest Destiny, White Man's Burden, along with justifications like bringing progress, enlightenment, and religion to uplift peoples perceived as less developed and less civilized have been used for the conquest of places as far apart as the Americas, Africa, Australia, and the Indian subcontinent. Nor is colonialism a thing of the past—it has morphed into other forms, such as the neocolonialist dominance of some nations by others for the plunder of their resources.

Unfortunately, space exploration today retains the same old colonialist tropes as exploration on Earth by Europeans, as my experience at the ISS conference demonstrates. We talk about "colonizing" other planets. We speak of resource exploitation on other worlds. You might say, well, places like the Moon and Mars are (as far as we know) lifeless worlds, so how does it matter if we colonize those worlds? That's not colonialism!

However, colonialism manifests even before the rockets have been launched. As space scholar and speculative fiction writer Haris Durrani observes in a 2019 essay titled "Is Spaceflight Colonialism?" the mining and extraction of minerals necessary for space exploration often violently displaces Indigenous communities (Durrani 2019). The fact that rich nations have dominated space exploration and that the new space race is led by techno-billionaires tells us who is (still) in charge. After all, those who have a strategic hold on space also have a militaristic advantage over other nations. And all this before we even leave low Earth orbit, or Earth itself!

Then, of course, there is the question of life on other worlds, and the very real possibility that we might destroy it, either deliberately or due to ignorance. Yet, apart from the little-known planetary protection group at NASA, who is asking these questions?

Space philosopher Frank White, with whom I have had many interesting conversations, coined the term "Overview Effect" to indicate the emotional reaction of astronauts who viewed Earth from the Moon—the photograph of the tiny blue dot swimming in the void of space brought strong men to tears. They had gone away from Earth to explore, and yet some of their most significant moments had to do with the view of the planet as a whole, from space. Astronauts speak of this experience as a profound one that dissolves national and ethnic boundaries and brings forth a vision of one planet and one humanity. But we must be cautious even here. We live in a Eurocentric globalized culture, which means that a one-world conception arising from those who are part of the dominant group may gloss over and erase concepts, dreams, and imaginings of other humans. It is also important to remember that patriarchs, feudal lords, and colonizers have always had a rather different interpretation of a vantage point from which they can see all they own, lords of all they survey. Here we may well see, not so much a sincere acknowledgment of the unity of life on Earth, but instead, a triumphalist validation of the power of the powerful, and perhaps an exacerbated desire for control and dominion.

All this is, of course, also true of space exploration and the potential of human settlement elsewhere. What's the point of traveling to strange and wonderful new vistas if we bring our destructive baggage with us—racism, colonialism, environmental destruction? Why not learn from our experience here on Earth, and ask ourselves—what would be a way of thinking about space and space exploration in an anti-colonial way?

I remember, as a child growing up in India, being fascinated by voyages of exploration. I read about British naturalist Charles Darwin on the *HMS Beagle*, and British explorer Captain James Cook in Oceania. In those accounts, there was no mention of the fact that these great adventures and voyages of discovery were part of European colonial ventures. It was much later, decades after my PhD in theoretical particle physics, that I realized that the science that I love has its roots in colonialism. Botany and zoology, for instance, had developed as a result of European dominion over other nations, which were plundered for their natural resources. It started to become clear to me that science had been shaped and appropriated by vested interests to ram a destructive model of development on subjugated populations. Now we are being sold exciting visions of the same old story played out in space. The movie *Interstellar* is a brilliant example of this kind of propaganda. When I first saw it, I was so shocked that I started writing quotes from the movie on the palm of my hand with a pen in the darkness of the theater. Set on a ruined Earth, the movie's message is that humanity's future and destiny lie in space. There is no sense of acknowledgment that Earth's ruination has been brought about by the very humans who have the gall to complain that "The Earth has turned against us." The hero whines that once they were explorers, and now they are caretakers, as though there is something wrong with caretaking. At a time when our planet's biogeochemical systems are being systematically destroyed by the forces of greed and profit, caretaking needs to be seen as a sacred duty, as so many Indigenous cultures do. The escapist, irresponsible attitude of the hero is echoed in the arrogance of techno-billionaires like Elon Musk and Jeff Bezos, who have appointed themselves directors of "humanity's" glorious venture into space. The US Space Act of 2015, which allows corporations unprecedented leeway with regard to resource extraction from space, became law entirely under the radar as far as the US public was concerned. While the billionaires are at the receiving end of public funds through public-private partnerships, the public also shoulders the risk of accidents or malfunctions. And the public is barely aware of all this!

Which is why planetary scientist Lindy Elkins-Tanton is so concerned about the need for democratization of space policy. In a *Slate* essay, she writes with coauthor Jessy Kate Schingler, that grandiose declarations of techno-billionaires like Jeff Bezos "betray the same paternalistic attitudes and hero narratives that led to centuries of political and economic subjugation of those with less power. That's alarming to those of us who want to decolonize visions of the future. Going into space isn't just about where we're going but how we get there" (Elkins-Tanton and Schingler 2019). Where are the voices, ideas, and concerns of "ordinary people?" The authors go on to talk about the importance of narratives that shape how we feel about space exploration. While there are bound to be disagreements about what decolonization of space entails, these are necessary conversations that cannot be consigned to the margins of public debate and policy.

As a speculative fiction writer and physicist, I am very interested in the stories we tell—explicit and implicit—about space exploration. I believe that while much of science fiction reproduces the familiar tropes of domination and colonialism disguised as adventure, it has the potential to do something far more revolutionary. The driving questions of science fiction often begin with "what-ifs," and many of these are the "what-ifs" of classic "golden age" colonialist science fiction. "What if we terraformed Mars?" is one such example. But another category of "what-if" questions might, instead, ask: "What if we went into space with an entirely different set of values and purposes, such that from start to end, the journey was inspired and informed by the best lessons we have learned from our troubled history on Earth?" "What if there was a reason to go into space that had nothing to do with making a few rich men richer?" "What if our journeys into space did not reproduce the colonial paradigm?"

We live within, and are shaped by, paradigms—frameworks and concepts through which we co-construct and make sense of the world. The dominant paradigm of modern globalized Industrial civilization may arguably be considered the Newtonian, or mechanistic, paradigm, which views the world and the universe as a machine. While we may not *literally* think this, the fact is that such socioscientific-cultural paradigms represent our defaults, our unquestioned assumptions, our invisible guide-ropes. The clockwork metaphor that illustrates this paradigm sees the world as reducible to parts, where each part can be considered separate from the whole. This characteristic separateness and reductionism can be seen in the classical medical model of the human body as a machine, the division of knowledge into disciplinary silos, the separation of ethical concerns from technological ones. Alienation and separateness are features of this way of seeing the world, and the desire for control—after all, that's what a machine is for, isn't it?—becomes primary. Our billionaires are exemplars of this paradigm. Most of us who inhabit modern industrial cultures are caught in this way of thinking. Is it any surprise, then, that our complex, deeply interconnected world, as far removed from the clock metaphor as possible, is in so much trouble?

What speculative fiction can do, I believe (although it mostly hasn't risen to the challenge because of its own colonialist roots), is to make the invisible gantries and ramparts and superstructures of the Newtonian paradigm visible, so that we become aware of our default, unquestioned assumptions about the world. By immersing us in entirely different ways of being, thinking, and exploring, we can stretch the boundaries of the imagination past the mechanistic, Eurocentric, colonialist ways of thinking about Earth and Space. We can listen to the histories and stories that have been pushed aside by colonialism and think about reasons to explore both Earth and space that are just, inclusive, and motivated by the best that we are capable of as a species.

What might some first steps be, for both storytellers and space policymakers? As astrobiologist Lucianne Walkowicz puts it, we can start by having conversations with people who are generally left out of discussions on space exploration and policy (Mandelbaum 2018). Physicist Chanda Prescod-Weinstein wants us to think about "why our language for developing understandings of environments that are new to us tends to still be colonial: 'colonizing Mars' and 'exploring' and 'developing,' for example. These are deeply fraught terms that have traditionally referred to problematic behaviors by imperialists with those that we would call 'indigenous' and 'people of color' often on the receiving end of violent activities" (Mandelbaum 2018).

Conversations are already under way about what it might mean to decolonize space exploration—in real, material terms—so that it doesn't reproduce the violence that has been done to communities and species on Earth. And, of course, decolonizing space cannot happen without decolonizing Earth. Speculative fiction's potential role is to enable us to imagine these possibilities through story, the oldest means of communicating across generations and peoples. For my own part, I have speculated through story about the possibility of going into space as an extension of our biophilia, a search for a wider kinship (Singh 2017). I have imagined spacefarers like the Polynesians on their ships, intimately aware of their environs, able to read the signs and signals of the stars. I have even imagined people who chose not to go to the stars but stayed on Earth to mend what was broken, and thereby healed themselves. The fact is, Earth is part of the universe, too. Space is right here, all around us. I used to think of Earth as a spaceship, orbiting our star, the sun. But that's just the mechanistic paradigm pushing beyond its domain of validity. Earth is a planet rich with life, Earth is home, Earth is, for many cultures, Mother. To be at home in the universe, surely we need to know what it might be like to be at home on Earth.

References

Durrani, Harris. 2019. "Is Spaceflight Colonialism?" *The Nation*, July 19, 2019. Accessed November 12, 2022. https://www.thenation.com/article/world/apollo-space-lunar-rockets-colonialism/.

Elkins-Tanton, Lindy and Jessy Kate Schingler. 2019. "Who Gets to Decide What Our Space Settlements Look Like?" *Slate*, July 8, 2019. Accessed November 12, 2022. https://slate.com/technology/2019/07/space-settlement-decolonizing-forum.html.

Mandelbaum, Ryan F. 2018. "Decolonizing Mars: Are We Thinking About Space Exploration All Wrong?" *Gizmodo*, November 20, 2018. Accessed November 12, 2022. https://gizmodo.com/decolonizing-mars-are-we-thinking-about-space-explorat-1830348568.

Singh, Vandana. 2017. "Shikasta." In *Visions, Ventures, Escape Velocities: A Collection of Space Futures*, edited by Ed Finn and Joey Eschrich, 207–240. Tempe, AZ: Arizona State University.

9

Opportunities to Pursue Liberatory, Anticolonial, and Antiracist Designs for Human Societies beyond Earth

Danielle Wood, Prathima Muniyappa, and David Colby Reed

> *Learning to say goodbye*
> *Is finding a new tomorrow*
> *on some cooler planet*
> *barren and unfamiliar*
> *and guiltless.*
>
> —**Audre Lorde** *in Dream/Songs from the Moon of Beulah Land*
> *(Part V)* in *The Collected Poems of Audre Lorde* (2000)

Introduction

Humans collectively face a unique opportunity in the current historical moment. It is the opportunity to reflect on the meaning of increasing human presence in space and to heed a call to action to design our societies on Earth and beyond with commitment to liberation at the core. The practices and insights of Black feminist poets and scholars provide principles that can serve as a guide. There is a strong force of inertia that pushes society to maintain power structures, economic systems and institutional designs that are grounded in white supremacy, colonialism, racial capitalism, ableism, and heteropatriarchy. These are the forces that lead to extreme income inequality, extractive exploitation of nature and human labor, and wasteful patterns of consumption that leave a trail of greenhouse gas emissions in the atmosphere, plastic pollution in the oceans, and space debris in orbit.

Visionaries are calling for action to imagine and remake a society on Earth that fosters a sustainability that encompasses economic, social, and environmental flourishing. This requires building on a foundation of solidarity

Danielle Wood, Prathima Muniyappa, and David Colby Reed, *Opportunities to Pursue Liberatory, Anticolonial, and Antiracist Designs for Human Societies beyond Earth* In: *Reclaiming Space*. Edited by: James S. J. Schwartz, Linda Billings, and Erika Nesvold, Oxford University Press. © Oxford University Press 2023. DOI: 10.1093/oso/9780197604793.003.0009 among people who live in different geographies and hold different identities but seek a common sense of well-being, through a process that historian Paul Ortiz calls "Emancipatory Internationalism" (Ortiz 2018). Such progress builds on the social movements led by generations of leaders from local and regional communities that have traditionally experienced oppression within racialized capitalist structures. These leaders propose approaches to improving the future of their communities through job creation, police abolition, food security, and liberatory education (Carruthers 2018; Davis 2016; Lorde 1984; Blain 2021). To enable such progress, those with formal authority have a responsibility to eliminate existing sources of oppression that continue to create inequity for intersectionally defined identity groups such as women, people of nonbinary genders, low-income people, immigrants, people with disabilities, Indigenous peoples, and People of Color. People who hold these identities have experienced centuries of oppression, but they also hold visions, practices, artistic forms, and creative traditions that provide a pathway to designing liberatory human societies in space and on Earth.

Black feminist scholars, poets, and teachers have curated a powerful collection of lenses or perspectives that together create a motivation, a map, and a method for humans to design healthy space communities. Black feminists motivate us by calling us to feel the satisfaction of working toward the future we believe is needed; we will discuss this drawing from Audre Lorde's essay "Uses of the Erotic." Black feminists provide a map by showing how we can move from poetry to intellectual ideas to action to poetry again, as seen in Lorde's essay "Poetry Is Not a Luxury." Black feminists provide a method for effecting societal progress by allowing the visions formed by poetry to be converted into action that leads to change, a process discussed in Lorde's essay "The Transformation of Silence into Language and Action." This chapter explores the motivation, map, and method and highlights how complementary ideas from other traditions expand the premises set forth by Black feminists (Lorde 1984).

The Motivation

"We on thy pinions can surpass the wind, And leave the rolling universe behind."

—*Phillis Wheatley (1887)*

The lyricism of the enslaved Black poet Phillis Wheatley's eighteenth-century poem "On Imagination," takes the reader on a celestial journey that expresses the power of the human mind to experience inspiration (Wheatley 1887). Black poet and scholar Audre Lorde reminds us of a reason that humans should shake off the stupor of complacency and historical inertia and draw on such inspiration to design the future world that we believe is healthy, sustainable, and flourishing. Lorde calls this sense of Motivation "the erotic," saying, "The erotic is a measure between the beginnings of our sense of self and the chaos of our strongest feelings. It is an internal sense of satisfaction to which, once we have experienced it, we know we can aspire. For having experienced the fullness of this depth of feeling and recognizing its power, in honor and respect we can require no less of ourselves" (Lorde 1984). Lorde provides a recipe for people to stop complacently watching the dominant power structures of human civilization maintain themselves and move to a posture of creative redesign of society. This is the key decision that confronts society as humans stand on the edge of the capability to live routinely on new planetary surfaces, to use human machinery to terraform celestial bodies, and to acquire and exploit resources from locations beyond Earth. Wheatley reminds us to allow our imaginations to expand beyond our experiences. Lorde (1984) calls us to be awakened by an "internal sense of excellence"—the erotic—and to work as if "the aim of each thing which we do is to make our lives and the lives of our children richer and more possible."

If we take Lorde's invitation seriously, many things must change. Viewed through these lenses, the next steps in human advancement in space become an urgent agenda toward liberatory design, because humans will perform many firsts in the coming years and decades. And what if we allow Wheatley to push us to surpass the wind? What if we allow our imaginations to envision a version of human expansion in space that does not reproduce the destructive patterns of racialized capitalism in which specific identity groups are set apart in society for low-wage, dangerous, or dehumanizing labor (Ortiz 2018; Robinson 1983)? What if we allow ourselves to insist that this next step of human engagement with the cosmos—which is not yet predetermined by the military-industrial complex—will be designed in a way that brings the erotic sense of excellence that comes from making the lives of our children more possible? What if we allow ourselves to sojourn within the "chaos of our strongest feelings" as we ask why it matters how humans behave on a Moon that most people will never visit? Lorde calls the erotic a "creative energy empowered."

Lorde further explains that "the principal horror of any system which defines the good in terms of profit rather than in terms of human need ... is that it robs our work of its erotic value, its erotic power and life appeal and fulfillment." This creative energy of the erotic can motivate us to see that it will be fulfilling to do the work to design liberatory ways for humans to live in space. It will be satisfying to create systems to support humans living on another planetary body that do not create any material waste. It will be fulfilling to design methods to harness resources from the Moon or Mars in a manner that is reversible and does not leave permanent degradation. It will be fulfilling to design approaches to compensate people who earn income in space in a manner that reflects the level of risk they face and treats people from different identity groups with equity. It will be fulfilling to design creative approaches to remove space debris that is already orbiting the Earth. It will be fulfilling to ask how residents of a space station traveling together for years can share their cultural differences peacefully. It will be fulfilling to ask how people living in locations with limited resources can address their medical needs. It will be fulfilling to develop a sense of stewardship and concern for the new locations in space where humans will sojourn before the next generation takes their turn. And each of these fulfilling endeavors can also serve as templates for the changes that are needed to work toward a flourishing society on Earth.

In another essay called "Age, Race, Class, and Sex: Women Redefining Difference," Lorde (1984) interconnects the fate of humans with the fate of Earth saying, "The future of our earth may depend upon the ability of all women to identify and develop new definitions of power and new patterns of relating across difference. The old definitions have not served us, nor the earth that supports us." This last sentence calls to mind both the human community on Earth and the physical planet that provides the energy, food, and water that keeps humans alive. What do we learn by extending the concern for Earth to the new planetary locations where humans will make homes? This feminist lens reminds us that treating other humans with dignity and treating the natural environment with respect are two components of a self-consistent creed.[1]

As we follow Phillis Wheatley on a tour beyond the rolling universe, our imaginations are quickened. We start to feel a connection to the physical places in space where humans will one day create homes. It becomes meaningful to feel concern and a sense of creative energy toward designing human practices that bring life and beauty rather than destruction. When we pause to allow the chaos of our strongest feelings to bubble out, we find that space awakens in us a motivation to work toward liberation. Science fiction writer and social critic Octavia Butler (1993) poetically captures this sense of motivation in the cautionary tale presented in the novel *Parable of the Sower*. The novel describes a version of southern California in which the local and national governments have failed to ensure basic human rights, safety, environmental protection, access to water, or a functioning formal economy. In the midst of chaos and danger, a teenage Black woman named Lauren forms a community and tells them that they are the beginning of a new society called Earthseed that will create a healthy way of life beyond Earth. Whether taken literally or figuratively, Butler taps into the erotic attraction toward space as a location where humans can design freshly. In Butler's words, conveyed in the journal kept by Lauren Olamina and shared with the community that follows her in search of a safe place to live, we see a vision of a new life for the community in the stars:

> The Destiny of Earthseed
> Is to take root among the stars.
> It is to live and to thrive
> On new earths.
> It is to become new beings
> And to consider new questions.

The Map

> *"From star to star the mental optics rove, Measure the skies, and range the realms above."*
>
> —*Phillis Wheatley (1887)*

Lorde's (1984) essay "Uses of the Erotic," gives us a reason to care about what happens in the coming era of humans transforming new locations in space. It gives a reason to demand that humans move beyond a continuation of harmful societal norms, institutions, and inequalities, and instead pursue a dramatic redesign that harnesses innovation in technology, economics, the arts, and social sciences. And the prototypes of liberatory modes of working, living, healing, and sharing in space can be transferred to help re-shape entrenched patterns on Earth. Working on problems of income inequality, environmental pollution, and unhealthy social hierarchy in space presents opportunities when compared to the same effort on Earth, for on Earth those who are currently powerful strive to maintain their power. If no redesign is pursued, human activities in space may be automatically shaped to match the institutions and power structures on Earth. In many ways, that is the current reality when considering the roles of countries as they operate spacecraft in orbit around the Earth or in scientific missions on or around other planets. The same nations that hold strong levels of geopolitical power and who wield nuclear weapons are also active in both civil and military space activities.

When considering activities in Earth's orbit, space holds a paradox. In some aspects, it is an illustration of egalitarian cooperation across nations, such as in the context of the International Telecommunication Union and the process to ensure that many nations, including equatorial nations, have access to radio frequencies needed to operate satellites and other systems. In contrast, there are a few nations or regions that are responsible for launching and operating a large number of satellites in constellations for communications and Earth observation. This means that a few countries—with the United States in the lead—are greatly increasing the complexity of managing collision avoidance and effectively using up scarce access to favorable orbits while many other countries do not have the opportunity to choose these orbits. Similarly, a few countries participate in testing of antisatellite weapons, which creates debris that impacts access to space services for countries that have never operated a satellite. The paradox of space being simultaneously open and exclusionary provides a call for action. Can the historical successes of space cooperation across nations serve as the foundation rather than the nationalist patterns that are dominant?

Lorde's (1984) essay "Poetry is not a luxury" presents a map to allow people to move between the emotional, the intellectual, and the practical. Lorde (1984) notes that from poetry, "we predicate our hopes and dreams toward survival and change, first made into language, then into idea, then into more tangible action." This, then, is the map to start the design of human community practices and technology systems in space that are liberatory. The first step is to allow poetic imagination to inspire us and give us the opportunity to dream of new ways of being that can survive the challenges of space. This first step, of feeling and inspiration, is not analyzable or legible using the tools of engineering and science, though many engineers and scientists experience the awe that supports their design work. Lorde's essay argues that humans must allow ourselves to feel, to be influenced by the poetic. As Lorde (1984) explains, "Poetry is not only dream and vision; it is the skeleton architecture of our lives. It lays the foundations for a future of change, a bridge across our fears of what has never been before." Thus, it is artistic vision, emotional discovery, and non-rational exploration that comes first as we move toward designing updated methods for growing food, recycling water, experiencing virtual conversations, building homes, administering health care. The design challenges of space to keep humans alive and improve human relations can inspire both emotional and intellectual satisfaction and creative energy.

The Method

> *"There in one view we grasp the mighty whole, Or with new worlds amaze th' unbounded soul."*
>
> —*Phillis Wheatley (1887)*

Lorde's pathway of poetry to language to idea to action is a practical guide that can be followed to invent the components of human life in space and to reinvent components of human life on Earth. In this pathway, art helps us imagine what has never been seen, then language and action bring it into being. Lorde's (1984) essay "The Transformation of Silence into Language and Action" further discusses the pathway to work to design a liberatory society on Earth. Once again, Lorde connects the fate of the environment with the fate of people. "We can sit in our corners mute forever while our sisters and ourselves are wasted, while our children are distorted and destroyed, while our earth is poisoned" (1984).

Lorde identifies fear as one of the reasons that people keep silent instead of moving through the Map from poetry to ideas to language to action. What are the sources of fear that seek to derail the practice of designing liberatory approaches to live in space? At the international level, different nations do not have full visibility into the intentions, behaviors, and technical capabilities of other nations. This creates a fear and distrust among nations and leads to technical designs and strategic actions such as antisatellite testing that are driven by fear. At the local level, there can be a fear that spending time on designs enabled by space is a waste and not relevant to improving life on Earth. To address the first of the fears at the international level requires action to improve transparency and trust among nations. Many factors have eroded such trust including recent, intentional antisatellite tests and strong public statements about the commitment of military presence in space.

A counterforce to fear-based actions in space can be global work toward a Space Traffic Management System with input from countries with a variety of backgrounds. Technology policy scholarship by Lifson and Wood has shown a consistent interest among emerging space nations to give input regarding the design of a global Space Traffic Management System (Lifson and Wood 2019; Lifson 2020). Can poetic and artistic creative processes inspire new ideas for coordinating with emerging nations on sharing data to improve space situational awareness? What if we imagine a world in which every country operates its own space situational awareness sensors and shares that information with other countries via a trusted interchange managed by a nonprofit or a distributed-ledger database (Zaid et al. 2021)? What if the principles of relational equality as outlined by philosopher Elizabeth Anderson guide the perspectives that spacefarers have toward each other (Reed 2020; Anderson 2018)? What if the design of laws for human activities in space were guided by a vision of the cosmolegal, which would recognize the wealth of ontological and epistemological diversity in human thought and self-perception, as well as agency beyond the human (Cirkovic 2021, 2023a)?[2] This recognition goes beyond the anthropocentric model and allows for a shift in the imagination and understanding of the cosmos to recognize that humans are only one of many actors of the "cosmos," known and unknown (Cirkovic 2023b). Each of these projects is part of the redesign needed for liberatory human existence in space. The creative energy that Lorde spoke of can give us motivation to keep working on undoing established harmful practices and expand to create new practices.

There is an urgency to responding to the call of the erotic to make designs for a liberatory future in space. If no effort is made to redesign technical systems, legal systems, carceral systems, and food systems in space, a window of opportunity will close. If Earth-based institutions become the standard approach in space, the brief period to design something new that is unencumbered by tradition will be lost. Let this motivation shake us, spur us into action.

Here are three methods that draw from the Black feminist tradition that will help the design process to create human systems in space that increase flourishing and sustainability:

The concepts from Table 9.1 can be illustrated by seeking links between insights from Black feminism and from Indigenous thought. Drawing from the Black feminist lens, lessons can be learned for future human space habitation from those who have experienced colonization, especially

Table 9.1 Feminist methods to inform design

A commitment to consider history and contributions from past individuals, groups, scholars, and activists	Carruthers 2018; hooks 1981; Davis 2016
A commitment to speak truthful, uncomfortable concerns of people experiencing oppression	Lorde 1984
A commitment to speak about personal or "private" challenges that have intersection with social norms, policies, and ideas	Frazier, Smith and Smith 1977; Carruthers 2018

Indigenous peoples. A series of public online seminars hosted by the Space Enabled Research Group at MIT during 2021 invited anticolonial and Indigenous scholars to share their thoughts on human actions in space (Tavares et al. 2021). The conversations included comments by leaders such as Kanaka Maoli scholar, activist, and scholar of Indigenous politics David Uahikeaikalei'ohu Maile highlighting concerns of Native Hawaiians for the ways their sacred land is mistreated by the operation of a telescope system.[3] Several speakers explored the motivation for Navajo Nation president Albert Hale to write a letter to NASA expressing concerns about the placement of the remains of astronomer Eugene Shoemaker on the Moon.[4] Indigenous professionals and advocates in the outer space sector, such as Dan Hawk, paint a new picture by calling for Native students to be active participants and researchers in human space endeavors. Indigenous anthropologist Ren Freeman invites people to study Indigenous research methods and apply them to questions such as the impact of satellite-based Earth observation on tribal decision making and resource management.[5] What might we learn by allowing poetry to conjure up a new imagination for the respect given to plural and diverse indigenous peoples' knowledges related to matters of space? As Harvey et al. (2022) point out, there is tension in trying to engage with anti-colonial thought and scholarship, while working within institutions that are intertwined with the very systems being critiqued.

The methods in Table 9.1 show how we can interpret the concerns and interventions of Indigenous peoples regarding space policies. The commitment to the erotic and to poetry allows us to engage with Indigenous peoples in their various ways of knowing, to listen to their generational knowledge of the heavens that are not based on western astronomy. They allow us to journey with African cultural astronomers (Holbrook, Medupe, and Urama 2008) and receive their wisdom; to enter the high deserts of the Atacama desert in Chile, not to see the ALMA Telescope but to gaze into the ancient pools that reveal the passing of the Llama in the dark spots of the night sky.[6] These leaders have already tapped into their source of erotic, creative energy to turn vision into action. Can learning their stories energize new people to work on designing human space endeavors that support justice and sustainability for the Earth, for other planetary environments and for people of oppressed identities? Can people who know a history of oppression on Earth enter a realm of poetry and imagination to design for space, not from a blank slate, but in the richness of the struggle for liberation? Is there a Black feminist, an Indigenous design, a plural and curious approach that can thrive in the as-yet-uninhabited planetary homes of future generations? Can space help us find the creative energy to work on designs for Earth even as we protect new places in space from human destruction?

There is no guarantee that the inspiring novelty of space will push humans to design liberatory pathways, but there is an opportunity to start, as new vistas open before us, and try designing homes, farms, healing, and teaching where humans have not yet left our mark. Can we do this and leave nothing to regret? Can we do this and foster a place of flourishing?

Notes

1. There is much more to explore about the interactions between feminism and environmentalism, including the variety of perspectives among feminists regarding the relationship between liberation of women and care for the environment. To learn more, consider references such as the following: Shiva and Mies (2014), Carlassare (2000).
2. Editor's note: See, for example, Chapter 10 of this volume for Vermeylen's and Njere's discussion of "African Space Art as a New Perspective on Space Law."
3. For more information, see these references: Maile (2019), Space Enabled (2021a).
4. Further information available at this reference: Ellison (1999).
5. Further information available at this reference: Space Enabled (2021b).
6. For more information see these references: Atacameño Ethnoastronomy Project (n.d.), Ibáñez (2002), ALMA Observatory (n.d.).

References

ALMA Observatory. n.d. "Privileged Location." https://www.almaobservatory.org/en/about-alma/privileged-location/.

Anderson, Elizabeth. 2018. "Freedom and Equality." In *The Oxford Handbook of Freedom*, edited by David Schmidtz and Carmen Pavel, 90–105. New York: Oxford University Press.

Atacameño Ethnoastronomy Project. n.d. "The Universe of Our Elders. ALMA Observatory. https://almaobservatory.org/wp-content/uploads/2016/11/alma-etno_2013.pdf.

Blain, Keisha N. 2021. *Until I Am Free: Fannie Lou Hamer's Enduring Message to America*. Boston: Beacon Press.

Butler, Octavia E. 1993 *Parable of the Sower*. New York: Four Walls Eight Windows.

Carlassare, Elizabeth. 2000. "Socialist and Cultural Ecofeminism: Allies in Resistance." *Ethics and the Environment* 5, no. 1: 89–106.

Carruthers, Charlene. 2018. *Unapologetic: A Black, Queer, and Feminist Mandate for Radical Movements*. Boston: Beacon Press.

Cirkovic, Elena. 2021. "The Next Generation of International Law: Space, Ice, and the Cosmolegal Proposal." *German Law Journal* 22, no. 2: 147–167.

Cirkovic, Elena. 2023a. "The Cosmolegal Approach to Human Activities in Outer Space." In *Institutions of Extraterrestrial Liberty*, edited by Charles Cockell. Oxford: Oxford University Press.

Cirkovic, Elena. 2023b. "International Law Beyond the Earth System: Orbital Debris and Interplanetary Pollution." *Journal of Human Rights and the Environment* 13, no. 2: 324–348.

Davis, Angela Y. 2016. *Freedom Is a Constant Struggle: Ferguson, Palestine, and the Foundations of a Movement*. Chicago: Haymarket Books.

Ellison, Michael. 1999. "Dreamer's Ashes End in Moon Dust." *The Guardian*, July 28, 1999. https://www.theguardian.com/science/1999/jul/29/spaceexploration.internationalnews.

Frazier, Demita, Barbara Smith, and Beverly Smith. 1977. "The Combahee River Collective Statement." Reprinted in *How We Get to Free: Black Feminism and the Combahee River Collective*, edited by Keeanga-Yamahtta Taylor, 15–27. Chicago: Haymarket Books.

Harvey, Alvin, Frank Tavares, Pedro Reynolds-Cuellar, and Saemus Lombardo. 2022. "Developing an Anti-Colonial Practice: Moving from Conversation to Structural and Institutional Change within the Space Community." *Proceedings of the International Astronautical Congress*, Paris.

Holbrook, Jarita, R. Thebe Medupe and Johnson O. Urama, eds. 2008. *African Cultural Astronomy: Current Archaeoastronomy and Ethnoastronomy Research in Africa*. New York: Springer Science & Business Media.

hooks, bell. 1981. *Ain't I a Woman? Black Women and Feminism*. Boston: South End Press.

Ibáñez, Francisco Gallardo. 2002. "Yakana, The Constellation of the Llama." Museo Precolombino. Video, 4:01. http://chileprecolombino.cl/en/arte/narraciones-indigenas/aymara/yakana-la-constelacion-de-la-llama/.

Lifson, Miles and Danielle Wood. 2019. "Implication of Emerging Space Nation Stakeholder Preferences for Future Space Traffic Management System Architecture." *Proceeding of the International Astronautical Congress*. Washington, DC. https://iafastro.directory/iac/paper/id/50959/abstract-pdf/.

Lifson, Miles. 2020. "A Study of Emerging Space Nation and Commercial Satellite Operator Stakeholder Preferences for Space Traffic Management." Master's Thesis, Massachusetts Institute of Technology. Cambridge, MA. https://dspace.mit.edu/handle/1721.1/129198.

Lorde, Audre. 2000. *The Collected Poems of Audre Lorde*. New York: W.W. Norton & Company.

Lorde, Audre. 1984. *Sister Outsider: Essays and Speeches*. Trumansburg: Crossing Press.

Maile, David Uahikeaikalei'ohu. 2019. "For Mauna Kea to Live, TMT Must Leave." *The Abusable Past*, August 14, 2019. https://www.radicalhistoryreview.org/abusablepast/forum-2-1-for-mauna-kea-to-live-tmt-must-leave/.

Ortiz, Paul. 2018. *An African American and Latinx History of the United States*. Vol. 4. Beacon Press.

Reed, David Colby. 2020. "Designing for Voice in the Vacuum: Property in Citizenship for Democratic Equality for Future Spacefarers." Master's Thesis, MIT, https://dspace.mit.edu/handle/1721.1/130609.

Robinson, Cedric J. 1983. "Black Marxism: The Making of the Black Radical Tradition, Second Edition." Chapel Hill, NC: The University of North Carolina Press.

Shiva, Vandana, and Maria Mies. 2014. *Ecofeminism*. London: Zed Books Ltd.

Space Enabled. 2021a. "Panel Discussion with Dr. John Herrington: Indigenous & Anticolonial Views of Human Activity in Space." MIT. May 14, 2021. https://www.media.mit.edu/events/panel-discussion-with-dr-john-herrington-indigenous-view-of-human-activity-in-space/.

Space Enabled. 2021b. "Indigenous & Anticolonial Views of Human Activity in Space: Workshop." MIT. December 3, 2021. https://www.media.mit.edu/events/indigenous-anticolonial-views-of-human-activity-in-space-1/.

Tavares, Frank, Alvin Harvey, Seamus Lombardo, Pedro Reynolds-Cuellar, Dava Newman, and Danielle Wood. 2021. "Centering Indigenous Voices and Resisting Colonialism in Space

Exploration: An Overview of the Ongoing Webinar Series by Space Enabled," Proceedings of the International Astronautical Congress, Dubai, UAE, October 2021. https://www.media. mit.edu/events/indigenous-anticolonial-views-of-human-activity-in-space-1/.

Wheatley, Phillis. 1887. *Poems on Various Subjects, Religious and Moral*. Denver: WH Lawrence & Company.

Zaidi, Waqar, Tom Kelecy, Weston Faber, and Moriba Jah. 2021. "Establishing a Chain of Digital Forensics for Space Object Behavior Using Distributed Ledger Technology." AMOS Conference Proceedings. Maui, Hawaii.

10

African Space Art as a New Perspective on Space Law

Saskia Vermeylen and Jacque Njeri

Introduction

This chapter analyzes African space art told through the work of African artists who retell an alternative history of space travel from an African cultural archive that produces a more inclusive history for humanity in space. The stories that are narrated in the artworks share what the art historian Paul Wilson has called an aspect of retrofuturism (Wilson 2019, 139). This is a future that is firmly rooted in the past, celebrating a rich diversity of African cultural practices and knowledge that have been ignored or derided by dominant Euro-American cultures. The artworks display a vision of the future that is free from past and present geopolitical shackles and multiple sources of disadvantages, such as gender and ethnicity. There is an element of utopianism in the projects that is distinctively African. The artworks envision an African space age that, in contradiction to other space programs, is not interested in rekindling the dogma of being the first in space—be it the first woman on the Moon or the first human on Mars. Instead, the African (art) space program imagines how an inclusive space program would look like in the future by placing it firmly in a liberated past. To achieve this, African artists are using "traditional" stories and cultural regalia from the African continent. They repurpose them to show their audience and viewers that Africa is not the forgotten or doomed continent in space travel, but instead the history of space travel becomes anchored in the cultural history of Africa.

The revised African archive redirects current and future space travel elements so that the idea of *reclaiming space* is about reinforcing the ideals and vision in space law that the exploration and use of outer space must be in the interest of all humankind and for peaceful purposes. But by exploring future space travel from an African point of view, it also introduces new perspectives on space law. The essential principles of reciprocity and relationality, that in

Saskia Vermeylen and Jacque Njeri, *African Space Art as a New Perspective on Space Law* In: *Reclaiming Space*.
Edited by: James S. J. Schwartz, Linda Billings, and Erika Nesvold, Oxford University Press. © Oxford University Press 2023.
DOI: 10.1093/oso/9780197604793.003.0010

simple terms point to an ethics of caring and responsibility for each other's well-being and health, define most African societies. Applying these African ideas to space law can revive the Agreement Governing the Activities of States on the Moon and Other Celestial Bodies (the Moon Agreement). The Moon Agreement was adopted by the United Nations General Assembly in 1979 but, unfortunately, failed to get enough signatories and therefore failed to guard the Moon (and other celestial bodies) *and* Earth and its inhabitants from further exploitation.

In this chapter we will first introduce Afrofuturism and its definitional variations as the main methodology we use for an alternative and more progressive textual analysis of space law. This is followed by a brief description of the Africanfuturists' Nuotama Budomo and Jacque Njeri and their space artworks. Njeri, the coauthor of this paper, also comments about her own work and Africanfuturism as a movement in dialogue with the main author. After we have presented the artworks, we will start making links between the artworks and space law in order to be able to discuss, in the last section, the role African storytelling and philosophy can play in strengthening international space law and the Moon Agreement in particular.

Afrofuturism/Africanfuturism

Interplanetary explorations since the initial Apollo programs have continued to preoccupy African and African American artists who expose in their work how the space race (both old and new) is a continuation of terrestrial colonization characterized by the exclusion of African voices, and a silencing of African cultures (Bourland 2020, 31). Particularly interesting are artworks that seek to Africanize space travel by deconstructing the well-known gendered and racialized trope of the 1960s astronaut by layering it with symbols of African art, history, and philosophy.

The artworks discussed in this chapter include elements of Afrofuturism, and although this genre has mainly been used to label African American literary traditions of science fiction, it is now increasingly perceived as a cultural and political aesthetic that includes many different art forms (Kniaź 2020). Afrofuturism has its roots in jazz and funk music of the 1960s, with artists such as Sun Ra and George Clinton staging their performances in future-oriented imagery (Kniaź 2020). But it was Mark Dery who coined the term "Afrofuturism" for the first time in his essay *Black to the Future* (1993). This was after writers such as Octavia E. Butler in *Wild Seed* (1980) and Samuel R. Delaney in *Nova* (1968) had already included elements of Afrofuturism to describe and reflect about their African diaspora experience in the United States. Over the years, Afrofuturism has become a well-known aesthetic form that works speculatively with the tension between "African tradition and technological progress, between the organic and the synthetic, and between the tribal and the futuristic" (Kniaź 2020, 53). While Afrofuturism is associated with African Americans, there is also a growing group of African artists who are aesthetically working with African histories and cultures. They are using a distinctive African canon as a source to criticize how African knowledge systems are portrayed as primitive and backward in comparison to "Western" technoscientific knowledges.

By illuminating Africa's rich knowledge and belief systems in futuristic artworks, the continent's contribution to various technoscientific evolutions is made visible. As such, African artworks embody and display a sense of empowerment. It is fair to believe that the history of the African continent may be differently remembered and experienced by artists in the diaspora versus artists whose daily life is intimately entangled in the continent's memoirs. The Nigerian writer Nnedi Okorafor argues that Africanfuturism is therefore a more appropriate label for the speculative works that have a strong bond with and are rooted in the African continent (Okorafor 2019; Kniáz 2020, 67). As Kniaź sums it up nicely, "whereas [Afrofuturism] dreams of Africa as a mythical continent and an unfulfilled fantasy of sorts, [Africanfuturism] is the creator and the direct descendent of African cultural heritage" (Kniaź 2020, 55).

A third concept that is useful is the term African retrofuturism (coined by Wilson 2019). It expresses the idea of representing the future steeped in a deep connection with African cultures and ways of being. Despite the attempt of the colonial powers to replace Africa's knowledge systems with western techno-sciences, nevertheless the artworks show that African ways of knowing and being are still thriving and carry a message of hope for the future. But through our conversation, it became apparent that for Njeri, the focus on the African continent and its peoples adds a distinctive feature in Africanfuturism. Njeri's point of view aligns with that of Okorafor, but Njeri introduces the term "AfroCentrism" as a useful "tag" for her artwork.

Notwithstanding the label, the most important message is that the artworks carry some elements of the theorist, writer, and filmmaker Kodwo Eshun's idea that Afrofuturism is all about assembling countermemories in order to resist colonial history (Eshun 2003). In other words, history must be rewritten from an African vantage point. The artworks discussed in this chapter overlay future visions of space travel by including African cultural and material symbols, so they are becoming, in Eshun's words, "chronopolitical acts." They draw together histories and countermemories of the anticolonial struggle and project those onto potential futures (Bould 2019). The artworks turn a critical lens onto the concept of romanticized space travel and layer it with an explicit narrative of "the darker registers of territorial conquest in the United States and the former British Empire" (Bourland 2020, 212). Therefore, these works reflect elements of retrofuturism, perceived as a subset of art practices that use archival materials to explore and reinterpret history by including strong dialectical critiques of masculinity and nationalism; two defining characteristics of the colonial project (Wilson 2019).

The Artworks

The first artwork that is discussed is the digital collage of the Kenyan graphic and digital artist and coauthor of this chapter Jacque Njeri who depicts in *MaaSci* (2017) a future of the Maasai people in space. Njeri explains that in her work she seeks to "[portray] . . . [the Maasai's] rich cultural aesthetic through tech, science and fantasy themes" (Mukhtar 2017). Njeri celebrates in this collection the Maasai culture by putting the Maasai in a science fiction environment. The stark juxtaposition of the Maasai dressed in their traditional clothes traveling into a futuristic space or an astronaut adorning their space helmet with the iconic red beadwork jewelry are a visual expression of Njeri's questioning, "What Would the Maasai Look Like in Space?"

Vermeylen interprets Njeri's digital layered collages as a visual illustration of what the anthropologist Arjun Appadurai would call technoscapes, which represent "semantic, cultural, and creative aspects of science and technology" (Appadurai 1996, 31, quoted in Gaskins 2016, 30). The *MaaSci* series are clearly the result of a technovernacular or African-centric technological creativity and storytelling. The symbolic messages that are layered in the digital compositions are building on the methods that have been used in Afrofuturism, but that Njeri places firmly in an African-centric history. One of the main techniques in Africanfuturism and Afrofuturism is the reappropriation of cultural artifacts in order to counter dominant or political systems (Gaskin 2016). Njeri uses the rich cultural aesthetics of the Maasai as a reminder that, despite attempts by Kenyan and Tanzanian government to "civilize" the Maasai, by placing them in outer space, Njeri shows that the Maasai are not willing to give up their rich cultural traditions—even in space (see, e.g., *Shestory: Female Astronaut* in Fig. 10.1).

In the digital collage *Three Wise Men* (Fig. 10.2), we see images of, for example, three Maasai men wrapped in traditional blankets floating on a flying saucer through the galaxy. This collage is a good example of Njeri's message

Figure 10.1 Shestory © Jacque Njeri, Instagram: @jacquenjerii. See insert for color image.

that Africa and its peoples (in this case, the Maasai) have their own history of scientific knowledge and technological advancements. The *Three Wise Men* embody years of experimentation culminating in the advancement of technological textile engineering. While the space blankets may give the impression that they are out of place and somehow naive because they depart from the traditional white space suit, this reaction is precisely what Njeri seeks to provoke. The viewer's colonial gaze that denies Africa's rich history of scientific and technological knowledge is being mocked by the *Three Wise Men* in the space blankets.

Another image that stands out is *Version 12.2*, depicting a young "humxn" as cyborg against the blue background of an unspecified different planet. The "armor" of the cyborg is an improved and enhanced version of their predecessors, Cyborg Version 12.1. Particularly this image of the African

Figure 10.2 Three Wise Men © Jacque Njeri, Instagram: @jacquenjerii. See insert for color image.

cyborg is a reminder about the essence of Afrofuturism as testified by the composer, poet, and philosopher Sun Ra, who makes the point that introducing Africa's history requires the invention of an identity through the creation of alternate personality or avatar (Sun Ra, quoted in Gaskins 2016, 31). Just like other Afrofuturists, Njeri uses the digital avatar of the African cyborg/astronaut as a tool for transcending and reinventing existing worlds in order to move between different worlds and realities (Gaskins 2016, 31). But simultaneously, Njeri also shows that Africa is a continent that is not backward but is experimental and continuous to advance scientific and technological knowledge, as illustrated by *Cyborg Version 12.2*.

In an interview, Njeri reflects that as a Kenyan artist she has a responsibility to include the African voice into science fiction. To do that well, she has to experiment with different designs and techniques so she can respond to her wider calling to shift the mentality of people by exposing the mindset to new ideas, overlooked histories, and more progressive futures for the African continent. As Njeri comments, she wants to "empower young Africans with skills sufficient for them to be in charge of their future and heal negative ideas and notions regarding Africa" (Robles 2019).

Using the African archive to tell a new story about space travel has also been the main mission of Ghanaian-born film director and filmmaker Nuotama Bodomo; her short film *Afronauts* (2014) retells the story of the Zambian space program during the Cold War. The war veteran Nkoloso, who founded the Zambian National Academy of Science, SpaceResearch, and Philosophy, conscripted twelve astronauts, including the seventeen-year-old girl Matha Mwambwa, to lead a mission to Mars. He trained his space cadets on a deserted farm just outside Lusaka, with the ambition to send a Christian evangelist mission to space. In her own words, Bodomo wants to tell the history of space travel from the perspective of "the people whose stories are lost or silenced to an iconic mainstream history that documents facts." She continues that she is "interested in following the characters that have not been able to find a home on earth and are therefore most attracted to the promise of the space race" (Bodomo, quoted in Francis 2013, 216).

Bodomo's work not only offers a speculative perspective on a better future but uses science fiction tropes to unsettle perceived ideas about Africa in both the past and the present (Magal Armillas-Tiseyra 2016). Bodomo uses science fiction narratives but Africanize them in order to reinterpret the past and recreate a new future wherein Africa is fully present and leading in the technologized world (Bisschoff 2020). Bodomo's film resembles what Marxist literary critic Darko Suvin has identified as the two main components of speculative fiction: "novum" (or the plausibility of scientific innovations) and "cognitive estrangements" (an imaginative world that is different from the experienced world; Suvin 1980, 4). These two ideas represent a commentary against current political contexts and are used by the artist as an ethical-liberating force of current economic exploitations on a global scale (Suvin 1980, 82). Escaping into the future offers a vantage point from which to understand the current sociopolitical landscape and to move to a new form of reality (Suvin 1980, 84). In other words, Bodomo's *Afronauts* offers a visual device for historical estrangements.

The point we would like to emphasize is that despite the upfront recognition that space travel should be for the benefit of humankind, both artists show in their work how space travel is silencing the African continent and reinforcing a history of colonization and frontier thinking. The artworks discussed entangle technological advancements of space travel into a complex web of history that runs parallel but in the shadow of scientific and techno-logical developments. Space travel, though, is not linear and universal. Space travel history includes multiple voices, of which some are a testament to the silencing of race, class, ethnicity, and gender in the space race. Africancentric space art can question the political and social boundaries of space travel (Foster 2012, 1). Artworks like the ones discussed in this chapter point out that events, such as the Apollo 11 landing, can be shown and "consumed in ways that complicate and alter our understanding of the original act" (Foster 2012, 2). These artworks expose new ethical questions that can inform new developments in space law and governance, which will be discussed in the next part of this section.

Maybe more so than any political theory, we posit that art represents an embodied and explicit critique that our political and legal organizations that govern current and future space travel need to open themselves for other ways of being and knowing. Our contemporary liberal democratic understanding of the rule of law is tainted by a colonial history where the law has been used to justify dispossession. To envision space travel as an inclusive and just en-deavor, we need a drastic overhauling of our value systems. As the political scientist Alex Zamalin argues, philosopher and sociologist Jürgen Habermas's suggestion that this can be achieved through the simple introduction of better norms that can produce better facts is not going to be enough. Black uto-pian imagination together with its anti-utopian critiques are, according to Zamalin, more productive elements that can expand and/or replace the basic liberal presumptions of our contemporary political and legal organizations (Zamalin 2019, 139). What we need is a profound questioning of the meaning and relationship between values and the law. We posit that African (retro)fu-turism can shake up the political and legal status quo. We address this in the final part of the chapter by stepping away for a moment from the specula-tive and picking up the thread of African storytelling and African philosophy. Reclaiming space also requires reclaiming the rule of law. We do this by using African jurisprudence as a tool to re-read space law critically and inclusively. But first, we explain how the Common Heritage Principle (CHP) is used as a legal tool to achieve the promise that space should be safeguarded for the ben-efit of humanity.

Space and the Common Heritage Principle

In our textual analysis of international space law, we are specifically inter-ested in the concept of CHP. It is a legal and normative principle that is part of various legal texts and, in broad terms, covers the idea that resources or sites "belong" to all nations and their citizens so that everyone can benefit from the shared economic use of it. It offers an alternative approach to *res nullius* (literally, "nobody's thing") conveying the idea that property rights are attributed to the most economically, politically, militarily, and technologically powerful entity so that they can claim ownership over a new territory or resources. CHP thus offers a *res communis* (literally, "belonging to 'mankind'") interpretation of property rights indicating that some resources should be owned by all nations and their citizens as common property.

CHP was introduced in the Treaty on Principles Governing the Activities of States in the Exploration and Use of Outer Space, including the Moon and Other Celestial Bodies (1967) in Article I declaring that "the exploration and use of outer space, including the Moon and other celestial bodies, shall be carried out for the benefit and in the interest of all countries, irrespective of their degree of economic or scientific development, and shall be the province of all mankind." Article II reiterates the prohibition of ownership and proclaims that "outer space, including the Moon and other celestial bodies, is not subject to national appropriation by claim of sovereignty, by means of use or occupation, or by any other means." But it is in the Moon Agreement (1979) that the CHP is most clearly expressed. Article 11 paragraph 1 states clearly that "the Moon and its natural resources are the common heritage of mankind." Interestingly, Article 11 also specifies the provisions to be made in order to comply with the application of the CHP. Article 11, paragraph 5 mentions that "State Parties to the Moon Agreement undertake to establish an international regime, including appropriate procedures, to govern the exploration of the natural resources of the Moon as such exploitation is about to become feasible." Unfortunately, not enough countries signed the Moon Agreement and therefore it is not binding international law. But it still illustrates how the international community seeks to establish a specific international regime based on benefit sharing to protect the CHP if and when exploitation of resources in space becomes available.

But the equitable sharing of benefits through negotiated contracts or treaty provisions seems like an oxymoron in a global context where there is no level playing field because of economic, political, and technological power differentials. This weakens the effectiveness of the CHP in the Outer Space Treaty as a legal tool that can protect space from unbridled commercial exploitation.

This concern about the unequal characteristic of benefit sharing mechanisms in international law as a tool to preserve the CHP is part of a wider unease about the moral value of international law. As the Third World

Approaches to International Law (TWAIL) school argues, international law does not always deliver on its promises. On the contrary, it behaves more like a hegemonic power that seeks to protect the interest of the powerful (Gupta and Sanchez 2012). According to TWAIL scholars, international treaties are steeped in a legal tradition that is Euro-American centric, which limits the option for a genuine change. International law, for them, is a practice that universalizes and institutionalizes inequality and inequity on a global scale by prescribing standard economic solutions to fight against an unequal world (Rajagopal 2006; Gupta and Sanchez 2012). TWAIL scholars call for international lawyers to become more critical of the colonial discourses that are still embedded in and part of international law (Okafor 2005; Chimni 2006).

We therefore argue that for outer space law to live up to the expectation that the exploration and use of outer space should be for the benefit of humankind, we need a more progressive textual analysis of its core legal principles. In order to present a more progressive, inclusive, and just meaning to space law, we argue that African moral principles of benefit sharing add moral credibility to the concept of CHP in outer space. We do this by using African space art as visual prompts to highlight that African moral principles can offer better guidance on how to reclaim space for the benefit of humanity in an era of commercial space travel and exploitation.

African Storytelling and Philosophy

We put Njeri's iconography of the Maasai in conversation with the narrative element in African philosophy (Bell 2002, 33). According to this tradition, the literary interpretations of oral stories reveal expressive and critical reflections about the lived African experience. The short film *Afronauts* also incorporates elements of African philosophy. It represents space travel by placing it in a storyline that reflects about the meaning of *being* in an African lived reality. This is a complex and multi-layered reality world tied, as the philosopher Richard Bell reflects, "to [its] traditional past, its intervening colonial history, its harsh environments, and its internal human struggles" (Bell 2002, 35). But while both artworks and their (re)interpretations of the African lived experience are indeed firmly rooted in Africa's history, they also speak to universal concerns about justice and what it means to be human (Wiredu 1996).

There is an element of dialogue between Njeri's *MaaSci*'s technoartwork and the tradition of storytelling. Alongside art, orality as a form of literary insights is, what Bell labels, aesthetic entryways into the African experiences as they are. Through the use of an iconography of Maasai cultural objects, most notably beadwork jewelry and woven blankets, Njeri engages and reminds the viewer that they are witnessing and are in conversation with Africa's philosophers, who use their storytelling techniques as a critical dialogue and source of what the writer and essayist Wole Soyinka calls "iconic tradition" (quoted in Bell 2002, 119).[1] The iconic tradition that is evoked in Njeri's artwork through the depiction of cultural symbolism relate to the core moral principles of fairness and equality that are woven into the storyline of many African folktales (Biesele 1999). By lifting African folktales from their usual contextual habitat by placing customary regalia in an outer space context, the MaaSci series also offers an alternative and more progressive interpretation of space law and the CHP.

Both artists also weave into the narratives of their artworks an element of historical fiction which functions as a critique of the nation-state and territorial sovereignty. Bodomo's *Afronaut* (2014) particularly stands out in this respect. Bodomo casts albino African American model Diandra Forrest as the space pioneer Matha, whose albinism functions as a critique against the patriarchal and nationalist limitations that curbed postcolonial aspirations of the newly independent Zambian state (Wilson 2019). Bodomo allows space cadet Nkoloso to reflect upon the deeply flawed colonial construct and imposition of the sovereign nation-state when he advises Matha to declare upon her arrival that she is the "mother, mother of the exiles" and "not to impose Christianity . . . [and] the nation-state" (mothertongue 2019).

This strong critique against the colonial imposition of the sovereign nation-state draws the attention to the fact that Indigenous governance systems have been displaced when the African continent was colonized. But the iconography of traditional artifacts in Njeri's MaaSci series is a strong reminder that precolonial governance systems have not died out. Interestingly, Ubuntu, one of the most well-known and famously reinstated African principles that came embedded in South Africa's postapartheid constitution, can offer, in contrast to international law, better guidance on what kind of rules could be used to implement benefit sharing under the CHP in space law. Unlike international law whose normative framing of fair and equitable relates back to Kant's cosmopolitanism (see, e.g., Kleingeld 2012), Ubuntu's normative framing resides in the daily lived experience of communities across the African continent.

Equitable principles are indeed at the heart of the African ethical principle Ubuntu. In very generic terms, Ubuntu is a Zulu-Xhosa word which can be roughly translated as "humanness." Often its interpretation is reduced to the Nguni expression *umuntu ngumuntu ngabantu*, meaning "a person is a person through other people." Ubuntu acknowledges otherness and restores the belief that it is possible for relational ethics to make their entry into the law. As the legal philosopher Cornell clarifies, "u[B]untu is not a contractual ethic. It is up to me. And, in a certain profound sense, humanity is at stake in my ethical action" (Cornell 2014, 112). In other words, Ubuntu incorporates an ethics of duty, sharing, and generosity. This is in contrast to the liberal individualism that usually defines justice in international law. The concept of justice in the African context is characterized by a "more communal compassion-based moral thinking" than individual rights-based justice (Bell 2002, 59). To conclude, staying true to the message of African (retro)futurism, we are proposing that reclaiming space requires a new form of international space law: one that not resides in the negotiations rooms of the United Nations, but one that is based in the authenticity of Africa's lived experience and storytelling, as portrayed in the "Three Wise Men Version 12.2" on their way to the Moon.

Acknowledgments

The authors would like to thank the editors for their helpful comments on earlier drafts of this chapter. Saskia Vermcylen would also like to thank The Leverhulme Trust for the funding that made this study possible.

Note

1. This was developed in a lecture by Soyinka "Icons for Self Retrieval: The African Experience." Given at Oberlin College, Oberlin lecture, 1985.

References

Appadurai, Arjun. 1996. *Modernity at Large: Cultural Dimensions of Globalization.* Minneapolis: University of Minnesota Press.
Armillas-Tiseyra, Magali. 2016. "Afronauts: On Science Fiction and the Crisis of Possibility." *Cambridge Journal of Postcolonial Literary Inquiry* 3: 273–290.
Bell, Richard H. 2002. *Understanding African Philosophy: A Cross-Cultural Approach to Classical and Contemporary Issues.* New York: Routledge.
Biesele, Meghan. 1999. "Ju/'hoan Folktales and Storytelling: Context and Variability." In *Traditional Storytelling Today: An International Sourcebook*, edited by Margaret Read MacDonald, 59–64. London: Routledge.
Bisschoff, Lizelle 2020. "African Cyborgs: Females and Feminists in African Science Fiction." *Interventions: International Journal of Postcolonial Studies* 22: 606–623.

Bould, Mark. 2019. "Afrofuturism and the Archive, Robots of Brixton and Crumbs." *Science Fiction Film and Television* 12: 171–193.

Bourland, W. Ian. 2020. "Afronauts: Race in Space." *Third Text* 34: 209–229.

Butler, Octavia E. 1980. *Wild Seed*. New York: Doubleday.

Chimni, B. S. 2006. "Third World Approaches to International Law: A Manifesto." *International Community Law Review* 8: 3–27.

Cornell, Drucilla. 2014. *Law and Revolution in South Africa: Ubuntu, Dignity, and the Struggle for Constitutional Transformation*. New York: Fordham University Press

Delaney, Samuel R. 1968. *Nova*. Science Fiction Gateway.

Dery, Mark. 1994. "Black to the Future." In *Flame Wars: The Discourse of Cyberculture*, edited by Mark Dery, 180–222. Durham, NC: Duke University Press.

Eshun, Kodwo 2003. "Further Considerations of Afrofuturism." *CR: The New Centennial Review* 3: 287–302.

Foster, Carter. 2012. *Aleksandra Mir: The Seduction of Galileo Galilei*. New York: Whitney Museum of American Art.

Gaskins, Nettrice R. 2016. "Afrofuturism on Web 3.0: Vernacular Cartography and Augmented Space." In *Afrofuturism 2.0: The Rise of Astro-Blackness*, edited by Reynaldo Anderson and Charles E. Jones, 27–44. Lanham: Lexington Books.

Gupta, Joyeeta and Nadia Sanchez. 2012. "Global and Green Governance: Embedding the Green Economy in a Global Green and Equitable Rule of Law Polity." *Review of European Community & International Environmental Law* 21: 12–22.

Kleingeld, Pauline. 2012. *Kant and Cosmopolitanism: The Philosophical Ideal of World Citizenship*. Cambridge: Cambridge University Press.

Kniaź, Lidia. 2020. "Capturing the Future Back in Africa: Afrofuturist Media Ephemera." *Extrapolation* 61: 53–68.

mothertongue. 2019. "Afronauts: short film." YouTube video, 14:05. July 16, 2019. https://www.youtube.com/watch?v=lb3pu5jXWHU.

Mukhtar, Idris. 2017. "The Kenyan artist who is taking the Maasai to space." *CNN*, August 28, 2017, https://edition.cnn.com/2017/08/28/africa/jacque-njeri-maasai-space/index.html.

Robles, Fanny. 2019. "Jacque Njeri: Life after MaaSci." *Africultures*, May 7, 2019. http://africultures.com/jacque-njeri-life-after-maasci/.

Okafor, Obiora Chinedu. 2005. "Newness, Imperialism, and International Legal Reform in our Time: a TWAIL Perspective." *Osgoode Hall Law Journal* 43: 171–191.

Okorafor, Nnedi. 2019. *Blogpost* http://nnedi.blogspot.com/2019/10/africanfuturism-defined.html.

Rajagopal, Balakrishnan. 2006. "Counter-hegemonic International Law: Rethinking Human Rights and Development as a Third World Strategy." *Third World Quarterly* 27: 767–783.

Suvin, Dako. 1980. *Metamorphoses of Science Fiction: On the Poetics and History of a Literary Genre*. New Haven, CT: Yale University Press.

Wilson, Paul. 2019. The Afronaut and Retrofuturism in Africa. *ASAP/Journal* 4: 129–166.

Wiredu, Kwasi. 1996. *Cultural Universals and Particulars: An African Perspective*. Bloomington, IN: Indiana University Press.

Zamalin, Alex. 2019. *Black Utopia: The History of an Idea From Black Nationalism to Afrofuturism*. New York: Colombia University Press.

11

Embodiment in Space Imagery

Beyond the Dominant Narrative

Daniela de Paulis and Chelsea Haramia

In Cicero's *Dream of Scipio*, Roman general Scipio Aemilianus (185–129 BC) dreams of transcending the Earth to be among the stars (Cicero 1984). He looks to the lands and nations of Earth from above and feels the insignificance of his terrestrial conquests when compared to the immensity of the cosmos. That dream delivered a fictional experience of departing Earth and being positioned in outer space. These celestial entities surely seemed unreachable to those humans living millennia ago. Humans were not creatures who could travel into space or be physically present on celestial objects. The ancients could only dream. They could not know what it's really like to see, say, the surface of the planet Mars. But members of humanity have now seen the surface of Mars. We are not so unaware of planetary realities as Scipio was, and what it means to be human in relation to the cosmos has changed since Cicero's time. Humans are now creatures who, in an important sense, travel through the cosmos and position ourselves on and near planetary bodies. We have the privilege of closeness, and we can achieve a remarkably good understanding of the actual details of other objects in our Solar System. Outer space provides us with unique opportunities to better understand ourselves as humans, and it manifests this potential whenever we advance in our ability to represent the space environment.

Our focus here is on photographic representation. Through photography, humans have extended certain sensory abilities much farther into the cosmos than our physical, Earthly bodies otherwise allow. As technology permits the physical extension of certain empirical senses, it at the same time enables many humans to mentally transport themselves to the celestial realm and, in some cases, onto the land of another planet. But photographs are not a neutral medium. This photographic access we have to the planets is heavily mediated. The typical narrative that carries through is one of mastery, strength, and accomplishment. However, these photographs represent much more than that.

Daniela de Paulis and Chelsea Haramia, *Embodiment in Space Imagery* In: *Reclaiming Space*. Edited by: James S. J. Schwartz, Linda Billings, and Erika Nesvold, Oxford University Press. © Oxford University Press 2023. DOI: 10.1093/oso/9780197604793.003.0011

If we look past the dominant paradigms and discourse regarding space exploration, we may know and understand more about ourselves in the process. We may find new knowledge, not only of the land of distant planets, but also of the realities of how humans experience life on Earth.

The history of photographs taken from the surface of another planet is in fact a history of self-portraits. As humankind progresses in the exploration of the skies and outer space, photographs and film recording have been documenting the progressive achievement of the ancient dream of flying and of exploring the cosmos. Here we focus on three iconic photos from the history of space exploration that pose as self-portraits of the remote embodiment on other planets by humans. These photos are examples of a new form of knowledge produced through remote embodiment, mediated by space probes operated from immense distances, and relayed back to Earth by radio transmissions across space. This process of representation is, most of the time, completely inscrutable to the eyes of the public, and only the final photographic result is shared with society at large.[1] The photos analyzed here are also a self-portrait of a rover, inhabiting and documenting eerie and remote landscapes, mimicking the curiosity, the sensory response, the gestures, and the gaze of a sighted human visitor. As a human artifact, the automated actor is calibrated to typical human scale and spatial understanding, and it is designed to extend human movement and action into the inhospitable environment of distant planets.

The first photos ever taken from the surface of another planet are those by the Soviet lander Venera 9 in 1975 of the planet Venus (Fig. 11.1). The photos are framed as landscape pictures taken from a first-person perspective, with parts of the rover visibly embodied in the photographed landscape and symbolically projecting the presence of humans on the landscape, which in turn prompts a sense of embodiment on the landscape for the viewer. In the photos, the landing gear of the lander and its arm are emblematic features

Figure 11.1 *Venus surface panorama from Venera 9.* This 1975 panorama from the Soviet Union's Venera 9 probe includes the first images ever taken from the surface of another planet. *Credits:* Russian Academy of Sciences/Ted Stryk (Bruce Murray Space Image Library, n.d.).

of human footing on the surface of Venus. This was the first time in history that humankind remotely touched the surface of another planet outside the terrestrial boundaries of Earth, the first time in history that the extension of the mind and the extension of the bodily movements that conceived and built the lander materialized almost as a mirage against the landscape of a remote planet. The photos taken by Venera 9 are highly evocative and imbued with metaphors, both for exhibiting the self-affirmative gesture of having set foot on the surface of another planet for the first time in human history, and also for evoking glimpses of the detailed landscape of a celestial body previously unknown. The content of these images represents not only the eerie features of an extraterrestrial world, it also represents human values and aspirations and our eagerness to extend our reach beyond the terrestrial cradle, a yearning deeply rooted in the history of space exploration from its start.

In 2013, a pivotal new image took the concept of remote embodiment even further, when NASA's Curiosity rover sent back to Earth the photo of the first hole drilled in the surface of another planet to extract a rock sample (De Paulis 2021, 23). The photo (Fig. 11.2) shows in clear detail the texture of the Martian soil and the hole drilled into it with extreme precision by the robotic arm, remotely guided by humans. The perfection with which the hole is carved shows the level of sophistication and mastery reached by scientists and engineers in exploring and shaping the matter of another planet, raising questions about the responsibilities these remote actions might bear. The realism with which the scene is portrayed might trigger memories and emotions of familiar environments in a terrestrial viewer, inviting the mind and the body to

Figure 11.2 The very realistic photo of Curiosity's first full drill on Mars. The smaller hole in the lower right of the image is from a partial test drill previously conducted by the rover's drill. Credit: NASA/JPL-Caltech/MSSS (Clark 2013). See insert for color image.

associate the textured and bright Martian landscape with landscapes here on Earth that have a similar terrain and lighting. This type of wandering might take a viewer on an imaginative journey to Mars simply with the mind, to experience the wind, the smell, and the environment of this inaccessible land. Where the memories of terrestrial landscapes might trigger a sensory reaction to the rocky Martian dune portrayed in the photo, the sensory immersion is destined to remain mental for the time being and thus deficient of the actual physical experience (de Paulis 2019, 200).

Those humans who take this imaginative journey are doing so from the relative comfort and safety of Earth. They will experience some realities of outer space but not others. They will know how the particular hole and the soil and dust appear to a sighted person, but they will be detached from other sensations that would come from actually being present on the planet. Whereas the scientists and engineers designing and building the rover engage in a type of remote embodiment on the Martian landscape that allows them to perform concrete actions on it and to change its previously untouched terrain, the rest of us who experience only a fraction of the entire process—the photo of the hole drilled on the surface of Mars—are passively exposed to this new form of geographical history, a history made of remote actions imprinted on the geological features of faraway and otherwise inaccessible planets. Beamed back to Earth across the ether by invisible radio waves, the photos of Mars could be imagined as spells transmitted towards humans by an immaterial yet all powerful entity. Captured on a remote planet by highly specialized machines, maneuvered by a relatively small team of scientists, these photos are becoming increasingly popular and influential in our contemporary culture (Eldridge 2018). The type of very specialized process that leads to the making of these images risks reinforcing a unilateral narrative of space exploration and exploitation, at the same time creating a concerning dynamic in terms of collective impotence versus the determinations of a few specialists (Messeri 2016). That is, the privilege of executing actions on the surface of a remote planet remains in the hands of a small team of individuals, whereas the greater majority of society has no role in the exploration itself and can only exercise the static appreciation of the photos created by a select few.

Besides remote exploration of space, a small number of privileged humans have experienced more direct contact with another celestial body. US astronaut Neil Armstrong and about a dozen others have walked on the Moon. They breached the bounds of Earth, saw this lunar environment, and felt for themselves an un-Earthly gravity. But while their physical bodies were undeniably near this land, they were entirely sealed off from it. They set their boots on the surface, but no human foot touched the Moon. They did not sift lunar dust through bare hands, nor could they. They gazed at this world through the medium of their helmets, their fragile, delicate bodies protected by their high-tech suits. They experienced some sensory realities but not others. And they experienced it as very situated individuals. The first person to walk on the Moon was a nondisabled white American man manifesting a dominant Western culture that prioritizes independence, competition, and achievement. But many humans possess attributes that deviate strongly from that of the paradigmatic astronaut. Likewise, there are humans who did not experience Moon-landing activities according to the prescriptions of the dominant discourse—a narrative that presumed that all of humanity was in awe of and benefited by this particular achievement that signified mastery and conquest.

Just as the photos of Venusian and Martian surfaces subliminally detach the viewer from the harsh physical realities of those planets, the dominant attitudes surrounding the Apollo missions distorted fundamental aspects of human existence. The realities of traveling to and walking on the Moon underscore the fragility and dependence necessarily involved in being a human. But fragility and dependence have historically been characterized as feminine and disabled traits, and dominant values rarely highlight or glorify these aspects of our humanity or of our history in space. Instead, Western notions of strength, competition, and independence have permeated our narratives of Moon landings and many other cosmic projects, and these particular, masculinized traits are presented as universal. Archaeologist Alice Gorman claims that

> [i]t's also noteworthy, although rarely remarked upon, that the lunar sites are gendered . . . In recognising what the astronaut bodies are, we can also think about what they are not, and how both presence and absence shape the sites. (2016, 122)

These bodies were presented as testaments to human accomplishment, and specifically to Western, colonial notions of what counts as accomplishment. Quoting engineer-historian David Mindell (2008, 5), Gorman also writes

> The US astronauts were meant to embody "the age-old American icons of control and mythologies of individuality and autonomy, from cowboys to sea captains." For the story to work, Mindell argues, the astronauts had to be in control, to have mastery. (2016, 122)

The remote presence—and physical absence—of human bodies in photographs of other planets also shapes the narrative in ways similar to the shaping of cultural sites on the Moon. There is the risk that we will reinforce this Western narrative of mastery, strength, achievement, and unexamined awe as we continue to extend ourselves physically, technologically, and mentally into outer space. The robotic bodies of landers and rovers are many things, but there are just as many things that they are not. They mimic human functioning in some ways, but not all. They do not feel discomfort, pain, joy, or fear. They do not need to be sealed off from the elements surrounding them. They are not fragile or dependent in the ways that all human bodies are. This too is a self-portrait, though perhaps more aspirational than realistic. Some humans are so easily transported to these lands via photographs because popular discourse highlights our cerebral and scientific accomplishments and rarely presents the masculinized symbolic body as a metaphor for our species' fragility and dependence. Even our embodied journeys to the Moon were presented to the public as monumental accomplishments for "mankind" instead of as endeavors that magnified our bodies' limitations and our profound physical dependence on the general hospitability of our planet. This is not how we typically project ourselves as humans into the cosmos, even though dependence is the lived reality of all individual members of humanity.

New space imagery stands to tacitly reinforce these overly narrow assumptions about human embodiment and about how we ought to react to new experiences that some may collectively access. Our more advanced technology arguably provides a more accurate and more immersive view of these planetary bodies, but there is a crucial form of knowledge that these images provide that is contingent on individual experience and thereby on individual understandings of the products of this technology. For example, the Mars 2020 mission of the Perseverance rover marks impressive technological advances, and the potential of remote embodiment through visual representation of the surface of Mars is also central to the photos produced by this, our most recent trek to that planet. Equipped with a very high-definition camera, the rover takes center stage in these photos, portraying itself against the surface of the red planet as in a staged setting from an adventure movie. Here the rover becomes the hero explorer, inviting the viewer to empathize with it and to remotely share its adventures across the Martian landscape. In these images, the rover symbolically embodies the self of the viewer, and the viewer becomes the rover. The rover seems to take the viewer onboard for a ride across the Martian dunes, and to explore lands never seen before by human eyes.

The photos taken by Perseverance are even more immersive and detailed than those captured by its predecessors—their realism, enveloping views, dynamic representations, lighting, and overall cinematic qualities are outstanding (Fig. 11.3). These photos seem to be purposefully composed, making

Figure 11.3 A panoramic view of the landscape of Mars taken by the Perseverance rover (NASA, n.d.). See insert for color image.

use of contemporary visual aesthetics borrowed from mainstream movies and video games. They suggest the dawn of a new type of stage photography, set not on our planet but on an extraterrestrial location where no crew or actor might be able to set foot for a very long time. In these photos, the remote embodiment suggested by former images of the surfaces of other planets is enhanced into fictional embodiment, for which the viewer is stimulated in engaging with the scenes shown in the photos through an active imagination and a sense of adventuring alongside a robotic companion. However, the Perseverance photos, with their wide angle and panoramic visions of never-before-seen Martian landscapes, show reality as seen by the mechanical lens, devoid of a perceiving self that is holding and filtering that vision. The technology that gives us 360-degree images of Mars purports to provide a full picture, but there remains an area to be filled. This is where the individual mind resides. These photos are 360-degree landscapes in which the perceiving individual is cut out like the center of a doughnut. When sighted humans transport themselves into this hole and look out to the surrounding planet, they do not look merely through the lens of a mechanical extension of humans' technological selves. They look through the metaphorical lens of their social and cultural assumptions while simultaneously gaining new and unique knowledge.

This is the knowledge of perceptual experience. That is, the rovers' photographs provide an opportunity for what philosophers call "phenomenal" knowledge.[2] This individual, experiential knowledge is the knowledge of what-it's-like to experience some part of the world as a perceiving subject. For example, a person who tastes chocolate for the first time gains some phenomenal knowledge. They gain knowledge of what it's like to taste chocolate. Even if that person knew all about cocoa agriculture, chocolate production, the chemical make-up of a bar of chocolate, etc., they still would not have the unique phenomenal knowledge of what it's like for them to taste choco-late until they try it. Likewise, one could know all the logistical and scientific details of the Mars rover Perseverance. One could know how far it travels and where, and that it is taking samples from the Martian terrain. One could know the dimensions of the holes and the content of the samples. But until the rover returned an image from the planet itself, no human could know what it's like to look at a hole on the surface of Mars. Of course, not all humans can or have visually consumed this imagery. But for those who have, this phenom-enal knowledge allows new understandings that were previously inaccessible. While this experience of the drilled hole involves indirect perception—the hole is viewed through the medium of photography—it nonetheless provides access to a literal understanding of the planet.[3] What-it's-like to look at a pho-tograph of the surface of Mars may not be the same as what-it's-like to look at the surface of Mars while standing on Mars, but even truncated phenomenal knowledge can be new knowledge. While limited in its ability to convey to humans what-it's-like to be on Mars, the rover provides a crucial piece of that experience—the visual representation of an actual alien landscape.

These rovers are physical extensions of typical human bodies, and they grant humanity remarkable opportunities to concretely expand its empir-ical understandings of these planets. But they can also help to reinforce the suppression of deviant reactions to space travel and space imagery. Not all humans react to this new knowledge with the prescribed awe, mastery, or feelings of accomplishment, and these deviant reactions are in fact a valu-able source of knowledge as well. While the contemporary narrative of space imagery and space exploration has been one of mastery, strength, and unex-amined awe, some prominent voices have at the same time challenged this narrative. Astronomer Carl Sagan wrote at length about humans' relationship to the cosmos, and in his discussion of the famous Pale Blue Dot image of Earth, Sagan felt our fragility and insignificance in that cosmic moment as he wrote,

> Our posturings, our imagined self-importance, the delusion that we have some privileged position in the Universe, are challenged by this point of pale light. Our planet is a lonely speck in the great enveloping cosmic dark. In our obscurity, in all this vastness, there is no hint that help will come from elsewhere to save us from ourselves. (1994, 7)

While it is true that the Voyager spacecraft that sent us this image stands as a remarkable achievement of a relatively impressive species, the perspective it bestowed told a different story of fragile and needy beings. We are impressive and fragile and needy and far more, but the suppression of nondominant emotional responses to space imagery skews our ability to recognize ourselves for *all* that we are.

In 1970, not long after Neil Armstrong declared his lunar accomplishment a "giant leap for mankind," the Black poet Gil Scott-Heron released an album containing a track of his famous spoken-word poem "Whitey on the Moon." This poem provides a radically different account of experiencing the Apollo missions. In a country built on racial oppression and systemic subordination of minoritized groups, Scott-Heron's account thoroughly rejected the dominant narrative surrounding the Moon landing. It starts "A rat done bit my sister Nell" and goes on to describe the bleak conditions of those who endure racial oppression, systemic poverty, and an unjust distribution of resources— all while a white man is literally and ceremoniously elevated for helping humanity. The contrast is clear, and this work of art is increasingly referenced in the present-day as we move forward into an era of renewed interest in space exploration. Scott-Heron's vivid descriptions give his audience a sense of what it's like to experience space imagery as a poor, Black American. His deviant emotional reaction to the dominant narrative provides access to another kind of unique and important knowledge regarding space imagery.

That is, beyond phenomenal knowledge from photographs, we gain from Scott-Heron an understanding of an outsider-perspective. Outsider-perspectives stand apart from the dominant paradigm and often reject dominant values and emotions. In Scott-Heron's society, the dominant values and emotions surrounding the Moon landing included things such as competition, satisfaction, mastery, strength, and accomplishment. Scott-Heron instead experienced what philosopher Alison Jaggar calls "outlaw" emotions (1989). Outlaw emotions don't follow the prescriptions of the dominant narrative. For example, a poor person who reacts to the prospect of a free meal with resentment instead of gratitude could be said to be experiencing an outlaw emotion. The interesting thing about outlaw emotions is that they can notify us of persistent injustices that are otherwise ignored in dominant discourse. If the person being offered the free meal knows that the person handing out free meals isn't doing anything to fight the systemic oppression that disproportionately harms the most vulnerable members of society, then the outlaw feeling of resentment is likely tracking that injustice and rejecting the solution offered as insufficient. Scott-Heron's work uncovers the often rampant and persistent injustice on which so many of our cosmic accomplishments have been built. "Whitey on the Moon" takes very seriously not only the fragile and needy nature of human bodies so easily harmed by, say, rodent bites, but also the systems and institutions that succeed at the cost of exploiting and dehumanizing those who are not members of dominant groups.

There is an important sense in which the robotic rovers don't have a race or gender or ethnicity or class. They are materials and programs but not persons. Their photographic self-portraits of humanity, then, have the potential to provide a more neutral representation of humanity as well. But there is an equally important sense in which they will always be accompanied by at least some socially or culturally identifying characteristics. The narrative of current cosmic technology could easily follow historically dominant ideals. But making room for outlaw emotions can help to uncover unjust and exclusionary dynamics in our discourse and attitudes. Indeed, the tale of Roman general Scipio's dream presented at the beginning of this chapter was, in its time, a way of criticizing those in power by demonstrating that their lauded military conquests acquired a much different meaning when viewed from a cosmic perspective. Today, Chanda Prescod-Weinstein, a theoretical physicist and an expert on the ways in which science and oppression intersect, writes critically about her discipline and describes her experiences within it, noting that

> learning about the mathematics of the universe could never be an escape from the earthly phenomena of racism and sexism (and now that humanity is moving deeper into our solar system, racism and sexism are no longer earthbound). (2021, 5)

We do not simply leave behind oppressive attitudes as we technologically extend ourselves into the cosmos. We reliably build dominant social and cultural values and emotions into our discussions of new technology and into the embodiment in and representations of outer space that they provide. But we can move beyond this paradigm. Making room for deviant reactions to space imagery can clue us in to otherwise hidden dimensions of human realities. What it's like to experience space imagery will not be the same for every human. But recognizing that there is no neutral human perspective does not mean that we cannot aim to remove narrow or oppressive attitudes from our mediation and discourse as much as possible. Space imagery has the potential to foster greater connections across humanity, and it presents us with a self-portrait of our human selves in space. We have a historical tendency to suppress, ignore, or simply miss certain aspects of this self-portrait, and in doing so we risk incorrectly presenting dominant ideals as representative of all humanity. One way to combat this is to listen to those who experience deviant emotions in response to the documentation of outer space.

What it is to be human is not static. Humans were once earthbound. Now, we are a space-faring species with vastly new knowledge whose remarkable ingenuity allows us to extend ourselves physically and mentally into places our ancestors could only dream of. We are clearly capable of great change. There is hope, then, that we can continue to change ourselves for the better. Photos of space exploration remind us that the expression of physicality and embodiment in outer space reflects how we experience physical embodiment on Earth, both socially and culturally. By looking from outer space, we can see and understand more clearly the physicality of Earth and realize that Earth is in space, and thus space exploration is not a hiatus from our human societies and cultures. These photos are also expanding human minds and knowledge of the cosmic environment, taking members of humanity to places where true embodiment and direct experience of space is currently unfeasible. When given the opportunity to remotely embody other celestial planes—thanks to the visual documentation of space exploration—we are building our knowledge of these distant landscapes, not through neutral technology, but through mediation embedded in sociocultural dynamics that are often beyond our direct control. What it means to be human is still changing, and a new path emerges as we move away from outmoded assumptions of the past. We can learn from deviant perspectives, too, as we gain new experiences both of outer space and of the diversity of reactions that representations of space evoke.

Notes

1. However, one long-time effort in making planetary image processing public has been spearheaded by Emily Lakdawalla, formerly of the Planetary Society, who built up a considerable group of amateur image processors. These amateurs have at times made processed images publicly available before the science teams could do so.
2. Appeals to phenomenal knowledge became popular in philosophy of mind debates as a result of philosopher Frank Jackson's influential Knowledge-Argument against physicalism (Jackson 1982; Jackson 1986).
3. And, technically, all observation is indirect observation given that we do not observe the physical world in a vacuum.

References

Bruce Murray Space Image Library. n.d. *Venus Surface Panorama from Venera 9*. Accessed October 22, 2021. https://www.planetary.org/space-images/venus-surface-panorama-from-venera-9/.

Cicero. 1984. *De Republica*. Bologna: Zanichelli.

Clark, Stephen. 2013. "Curiosity Collects Powder Sample in First Drill on Mars." *Spaceflight Now*, February 9, 2013. https://spaceflightnow.com/mars/msl/130209drill/.

de Paulis, Daniela. 2019. "COGITO in Space." *Antennae: The Journal of Nature in Visual Culture* 47: 199–211.

de Paulis, Daniela. 2021. "The Metamorphosis of a Periplaneta Americana." *Leonardo* 54, no. 1: 12–22.

Eldridge, Luci. 2018. "Working on Mars: An Inclusive Encounter through the Screen." In *Image—Action—Space: Situating the Screen in Visual Practice*, edited by Luisa Feiersinger, Kathrin Friedrich, and Moritz Queisner, 103–114. Berlin: De Gruyter.

Gorman, Alice. 2016. "Cultures on the Moon: Bodies in Time and Space." *Archaeologies: The Journal of the World Archaeological Congress*, 12, no. 1: 110–128.

Jackson, Frank. 1982. "Epiphenomenal Qualia." *Philosophical Quarterly* 32: 127–136.

Jackson, Frank. 1986. "What Mary Didn't Know." *Journal of Philosophy* 83: 291–295.

Jaggar, Alison. 1989. "Love and Knowledge: Emotion in Feminist Epistemology." *Inquiry* 32, no. 2: 151–176.

Messeri, Lisa. 2016. *Placing Outer Space: An Earthly Ethnography of Other Worlds*. Durham, NC: Duke University Press.

Mindell, David A. 2008. *Digital Apollo: Human and Machine in Spaceflight*. Cambridge, MA: The MIT Press.

NASA. n.d. *Mars 2020 Mission Perseverance Rover*. Accessed October 22, 2021. https://mars.nasa.gov/mars2020/multimedia/images/.

Prescod-Weinstein, Chandra. 2021. *The Disordered Cosmos: A Journey into Dark Matter, Spacetime, & Dreams Deferred*. New York: Bold Type Books.

Sagan, Carl. 1994. *Pale Blue Dot: A Vision of the Human Future in Space*. New York: Ballantine Books.

12

Appreciating What's Beautiful about Space

James S. J. Schwartz

The Kids in the Blue Flight Suits

I can recall being a fifth-grade student in 1996, sitting on the floor of the US Space and Rocket Center in Huntsville, Alabama, near the zero-g underwater training tank. I was with my assigned group of Space Camp attendees, preparing for our mock Space Shuttle mission, while our "crew trainer" (i.e., camp counselor) was doing their best to develop group cohesion among our motley pack of impatient, energetic youngsters. One strategy they used was to have us chant our group name, "TRW," in unison, along with a clap beat: "T! R! Double-U!"::clap-clap; clap-clap-clap::—That chant remains one of my most distinctive memories of attending Space Camp.

TRW (Thompson Ramo Wooldridge, Inc.) was an aerospace and automotive company that played key roles in the development of early US missiles and rockets, including the *Atlas* launch vehicles used in the US *Mercury* program, before it was purchased (and later sold) by Northrop Grumman. The first American to orbit the Earth, John Glenn, launched on an *Atlas*.

The TRW chant was emblematic of the wider space camp experience, where our days were filled with lessons about space history and technology filtered through a US-friendly lens. Rather than feeling like I was training to explore the universe, I felt like I was training to become a NASA cheerleader (or a public relations official). I don't recall learning much from those lessons; the material contained little I hadn't already learned on my own through watching TV or reading books about space. The actual "training experiences" were vanishingly brief and few and far between. I was ultimately assigned to be the "mission specialist" for our group's mock shuttle flight, which was the flagship activity around which the week-long camp program was designed. My role was to put on a simplified space suit and conduct an experiment in the "vacuum" of the open cargo bay of the Shuttle simulator. I had barely donned

James S. J. Schwartz, *Appreciating What's Beautiful about Space* In: *Reclaiming Space*. Edited by: James S. J. Schwartz, Linda Billings, and Erika Nesvold, Oxford University Press. © Oxford University Press 2023. DOI: 10.1093/oso/9780197604793.003.0012 my suit when our "flight director" gave the order for the crew to prepare for reentry. I never even had a chance to attempt my assigned experiments. Perhaps we were getting more of a true-to-life astronaut training than I realized at the time.

So, despite momentary periods of excitement, I remember thinking as I was leaving Huntsville that Space Camp didn't live up to the hype, at least for me. In retrospect, I would say that the experience did little to validate the love and passion I had already developed for space exploration. Although my grandparents offered to pay for me to attend the more advanced program the following year, the prospect of yet another week of relearning the same tired, propagandized, space trivia did not inspire me.

Unlearning Bad Habits

The lesson I draw today from my Space Camp experience is that *loving space exploration* is not the same thing as *endorsing contemporary American spaceflight culture*. But what do I mean by that?

By "contemporary American spaceflight culture" I simply mean the prevailing attitudes about space that are encouraged by NASA and the space industry, that appear in film and television, that are promoted by space exploration social media content creators, and that you can read in the glossy magazines printed by spaceflight advocacy groups (such as the National Space Society's *Ad Astra* magazine). You can consult many of the other chapters of this volume for more perspective on this culture and its impact on the national and international stage, but for the moment just the basics will be enough: Contemporary American spaceflight culture includes apparently *dogmatic* adherence to ideas such as: that the conquering the space frontier will be necessary to avoid societal stagnation; that human expansion into space is only possible by means of free market processes; and that we must do all of this without delay, regardless of the human suffering it ignores or produces along the way. Never mind that we scarcely know how to maintain basic functioning societal institutions on Earth, and show shamefully few signs of openness to learning how to do things any better in space, where it is significantly more challenging for humans to survive.

To be sure, we have been given other reasons to care about space exploration, and at the end of the day some of them withstand critical scrutiny.[1] But many of the ones you have likely heard, even from seemingly trusted sources, are not exactly evidence-based. We have been told that we need to support space exploration to inspire students to study STEM disciplines, but we have not been given clear evidence that supporting space would produce this effect.[2] We have been told that we are explorers who have natural urges to visit space, but our current understanding of human biology and genetics does not validate this claim.[3] We have been told that space mining will grow the economy and raise our standards of living, but only if we have faith in trickle-down economics. And most often we are told these things, not by experts in education science, child development, biology, genetics, sociology, or human development, but by astronauts, politicians, and businesspersons who have been given platforms to speak well outside their areas of expertise. Much of the way Americans talk about space is influenced by these fallacious appeals to "expert" opinion. This is as much the fault of the cultural forces that encourage this behavior as it is the fault of various high-profile individuals who enable it by freely speaking outside their areas of expertise (and also by failing to defer to more qualified experts when asked questions falling outside their expertise).

So, we shouldn't believe everything we are told about space, because not everything we are told about space is responsive to the truth, to the way things really are. And, according to the philosopher Harry Frankfurt, when we speak without bothering to check the evidence or consulting with relevant experts, we are *bullshitting*.[4] So, how can we talk about space, and what is really important about it, without all of the bullshit?

Learning to Love

My suggestion is that we try to notice what *first* caught our attention about space exploration. Do you remember what was in your head the first time you thought "I love space!"? Do you remember what drew you in? Try really hard!

I suspect my answer will be shared by many sighted lovers of space: It was the experience of witnessing detailed photographs of other planets, galaxies, and nebulae for the first time. This brought about an immediate and overpowering recognition of the brooding, pensive beauty of space. Those images represented places I wanted to experience—to learn about, to appreciate, and if possible, to visit in person. However, ignoring the fact that any photograph is an imperfect representation of what it depicts—and ignoring the fact that many of the most famous spaceflight images are processed and enhanced in ways that are not obvious to the casual observer—it was never far from my mind that it was *Mars* I found beautiful, rather than someone else's *plans* for

Mars.[5] Likewise, it was always *the ruddy and creamy gaseous bands of Jupiter* that held my fascination, as opposed to *plans to harvest resources* from Jupiter's atmosphere or its moons.

Unfortunately, when we go searching for information about space exploration, much of what we find is information about *what other people plan to do* with other planets or places in space. It is easy to learn about the things humans have done in space—about where they have visited or sent robotic counterparts, for instance. It is even easier to learn about the things a select few incredibly wealthy humans would like to persuade the rest of us to help pay for *them* to do in space, such as taking joyrides into low-Earth orbit. Which is all to say that it *takes work* to learn about space for what it is in itself, because we have to wade through so many *other's plans* for space, including many plans that do not rise above the bullshit.

Finding Help

Fortunately, there are many who study space studiously and earnestly—with a mind to *learning about space* as opposed to *learning how to exploit space resources*, for instance. Every now and then you will find planetary scientists— the experts of other worlds—describing their engagement with space, and especially their experiences viewing images of other planets. And it is through these works that we can gain insight into and appreciation for what it is like to experience, and, in many cases, find beauty in, other worlds and places. Consider geologist and planetary scientist Sean McMahon's lively and engaging description of Mars:

> To a geologist, the landscapes recall the Canadian High Arctic or the rocky deserts of Libya. The sky, however, is a strange inversion of the Earth's: brown, orange, lilac, yellow or olive green during the day but electric blue in the glow of the setting sun. Sometimes the blue returns before sunrise in high, feathery wisps on the brightening sky—noctilucent clouds in the upper atmosphere. The little moon Phobos rises and sets twice daily, passing in front of its smaller, slower twin, Deimos. In the warmer seasons, low-latitude hillsides and crater-flanks are painted in fine, dark, parallel strokes by trickles of saltwater. On the floors of craters, black and red sand and dust pile into ripples and dunes, the largest in the solar system. Dust devils are ubiquitous, scrawling dark curlicues on the desert floor like the tracks in a cloud chamber. Compared to Earth, Mars is stark and simple. Undisturbed by the messy influence of biology, Martian features have a mathematical regularity; the same patterns recur at many different scales, like those of a fractal: frost polygons,

craters, ripples and dunes, fractures, veins and valley networks. (McMahon 2016, 214)

McMahon describes his overall reaction to this as a recognition that "Mars possesses a great and distinctive beauty" (p. 214). Will we wind up destroying this beauty before we learn how to recognize and appreciate it? Perhaps, if the first Martian settlement does not consist solely of scientists, it should instead consist of poets, musicians, and landscape artists (that is, the kind of people who can make productive use of the Martian environment without exploiting it or despoiling it!). Either way, it is not clear what role on Mars there will be for commercial developers and billionaires and their visions of refashioning Mars into something else (something less).

Of course, planetary science is not our only pathway to a deeper and more authentic engagement with our love for space and what is really beautiful about space. Sometimes *stories* can provide what a purely scientific orientation cannot—at least when what is lacking is a wider sense of perspective. Indeed, the thirst for a deeper engagement with and appreciation of space bears intriguing similarities to the notion of *kinship* that appears in author and physicist Vandana Singh's short story *Sailing the Antarsa*.[6] Singh's story describes the travels and inner thoughts of an interstellar voyager named Mayha, who is flying through space on her own, *seeking kinship* with other places. For Mayha and her culture, seeking kinship is an important driver of space exploration, as we learn from one of her internal reflections:

Why does one venture out so far from home? Generations ago, our planet, Dhara, took my ancestors in from the cold night and gave them warmth. Its living being adjusted and made room, and in turn we changed ourselves to accommodate them. So it was shown to us that a planet far from humanity's original home is kin to us, a brother, a sister, a mother. To seek kinship with all is an ancient maxim of my people, and ever since my ancestors came to this planet we have sought to do that with the smallest, tenderest thing that leaps, swoops, or grows on this verdant world. Some of us have looked up at the night sky and wondered about other worlds that might be kin to us, other hearths and homes that might welcome us, through which we would experience a different becoming. Some of us yearn for those connections waiting for us on other shores. We seek to feed within us the god of wonder, to open within ourselves dusty rooms we didn't know existed and let in the air and light of other worlds. (Singh 2018, 174–175)

What is kinship? According to Mayha's teacher, her aunt, "[k]inship isn't friendship" but instead,

"A kinship is a relationship that is based on the assumption that each person, human or otherwise, has a right to exist, and a right to agency," she intoned. "This means that to live truly in the world we must constantly adjust to other beings, as they adjust to us. We must minimize and repair any harm that we do. Kinship goes all the way from friendship to enmity--and if a particular being does not desire it, why, we must leave it alone, leave the area." (Singh, 182–183)

Mayha's aunt's instructions on how to seek and form kinships sound much like the advice we might give to someone interested in learning how to better respect and care for someone or something they love, value, or find beautiful. Because when we recognize that we love something, or that we find beauty in something, it becomes rational for us to desire to learn more about that thing, to become better acquainted with it.[7] But it also becomes rational for us to desire to preserve and protect what we place value in—especially against destructive forces. And sometimes our desires to learn are themselves destructive, including, for instance, the deliberate crashing of several *Saturn V* third stages into the lunar surface to engage in seismic studies of the Moon. Sometimes, then, it is only safe to express our love at a safe distance, and recognizing this is a sign not of misanthropy but of maturity and respect.

For instance, those who appreciate beauty in artworks that are too delicate or fragile to be observed directly can still express their love in deep and healthy ways through learning—by seeking out and creating knowledge about these artworks. For many, this might involve watching a video, or reading an article or book. Some might use these experiences as inspiration for creative expressions of love (through poetry, artwork, music, etc.). The enhanced perspective and understanding these activities provide sometimes also enhance our senses that preservation and protection are worthy responses to the delicate beauty that we so value.

I would not be the first scholar to suggest that we should apply a similar attitude toward what and how we explore (or what and how we seek kinship with) space. For instance, the environmental philosopher Holmes Rolston III, in one of the very first publications on the ethics of space exploration, had this to say:

We are learning that solar science, too, despite its laws, is full of history. Each planet, moon, place is going to have its own story, a unique world that cannot be predicted in advance, not entirely, not in many interesting details, but which can be enjoyed only upon discovery. (Rolston III 1986, 169)

Here, Rolston III is calling attention to the *particularity* of space. No two planets will ever be exactly alike. No two moons will ever be perfectly similar. No two comets or asteroids will have precisely the same mineral composition, orbital parameters, etc. We can celebrate these differences by trying to learn from them, as Rolston III makes very clear later in his seminal essay:

> Banish soon and forever the bias that only habitable places are good ones (temperature 0–30 degrees C., with soil, water, breathable air), and all uninhabitable places empty wastes, piles of dull stones, dreary, desolate swirls of gases. To ask what these worlds are good for prevents asking whether these worlds are good in deeper senses. The class of habitable places is only a subset of the class of valuable places. To fail as functional for Earth-based life is not to fail on form, beauty, spectacular eventfulness. Even on Earth humans have learned, tardily, to value landscapes and seascapes that have little or nothing to do with human comfort (Antarctica, the Sahara, marine depths). Just as there is appropriate behavior before Earthen places, regardless of their hospitality for human life, so there will be appropriate (and inappropriate) behavior before Martian landscapes and Jovian atmospheric seas. (Rolston III 1986, 171)

Rolston's call for us to recognize "appropriate (and inappropriate behavior) before" places in space recalls Mayha's aunt's instructions on how to seek and form kinship, which resides on a spectrum ranging from friendship to enmity.

But places in space—planets, moons, asteroids, comets, and so forth—are not static. They are ever-changing systems that interact with one another in complicated and difficult-to-predict ways. We cannot appreciate what is really beautiful about *Mars* without appreciating Mars as a dynamic geological system, one that is always in flux, even if the scale of its changes is not always apparent to us. In this respect, our love of Mars and its beauty, and our efforts to protect that beauty, should be directed at protecting the *geological processes* that comprise the real Mars that we wish to come to know better. The same principle holds for our appreciation of the beauty of the bands of Jupiter, which are the result of ongoing atmospheric processes, about which we have much to learn. Protecting Jupiter's beauty requires protecting the integrity of its atmospheric processes from human intervention.

Ultimately, learning how to love requires learning about who or what we love. Those who try to love in ignorance and haste often find that love can be fleeting and even damaging to all parties involved. In science fiction stories, we justly deride violent alien species that come to paradises like Earth seeking conquest and destruction. It would be ironic, at the very least, should we find ourselves endorsing the same impulse when visited upon other places by beings like us. (Are we so vain we probably think this space opera isn't about us?)

So What's the Answer?!

So what is beautiful about space and how do we appreciate it? I could just as easily have asked: What is worth seeking kinship with in space? And the answer is that we need to decide this *for ourselves*.

We do not inhabit a world that makes this easy, because so much of space media is filtered through the perspective of zealous advocates who do not always prioritize the truth but instead prioritize their own *plans* for space. But if you do not start with the truth, or as close to the truth as you can get, you will not be creating *your vision* for space. They will not be *your hopes* and *your dreams* for space that are made manifest. Instead, they will be *someone else's dreams* and ambitions, *someone else's vision*, that get pursued in space.[8] Most likely it will be the vision of an incredibly wealthy individual who demonstrates little to no care or appreciation for the beauty of space; who seeks control and conquest rather than kinship. But we have both the opportunity and ability to craft visions for ourselves. And when people are given the knowledge and resources sufficient for creating their own visions, they often discover their preferences are rather unlike those being forced upon them by the wealthy and powerful. In this respect, the status quo is among the most disabling forces in the universe. We cannot escape the status quo if we put no energy into *imagining* how we can change our collective behavior for the better, on Earth or in space.

We should also plan future spaceflight priorities in ways that validate the wisdom of the youth—especially in what the younger generations find beautiful and fascinating in space. The responsibility of older generations is not to profess or proselytize, but to encourage and enable the visions of the younger generations. After all, the younger among us will always have a key advantage: Their insights are fresher, their moral sensibilities are more compassionate and tolerant and less likely to be constrained by arbitrary or unjust prejudices, their minds are full of more possibilities, and they have more extensive access to the full range of human knowledge and experience compared to the generations which have come before them. Not only do they know more, they are capable of learning more yet. We should listen to the kids more than we do now.

More Kids, More Blue Flight Suits

Several years ago, I visited the US Space and Rocket Center while in Huntsville for an academic conference—my only return trip since I attended Space Camp some twenty years prior. Near the Space Shuttle *Pathfinder* simulator I noticed a "crew trainer" trailing behind her own motley crew, who were keeping themselves occupied with a group-building chant as they walked from the "hab" buildings (the dorms for camp attendees) toward the main building for camp activities. Remembering my own group's reluctance to chant, I asked her how she managed such enthusiasm from them, to which she replied along the lines of "They just come up with chants on their own." Hopefully that rings true going forward—that those who will actually venture into space to live will be doing so according to visions and dreams of their own creation.

We have had enough people try to tell us what is beautiful about space and what we should do when we go there. It is time we started exploring for ourselves.

Acknowledgments

Thanks to Ben Ragan, Deb Michling, Jessica Garner, Vandana Singh, Linda Billings, Erika Nesvold, Zach Weinersmith, and David Hewitt for discussion and comments.

Notes

1. My own position is that *scientific exploration* should be prioritized over other spaceflight objectives (e.g., space mining, space settlement). I defend this position at length in my book *The Value of Science in Space Exploration* (Oxford University Press 2020).
2. See chapter 1 of (Schwartz 2020) for analysis and discussion. See also (Delgado-Lopez 2011).
3. See (Schwartz 2017) for discussion.
4. According to Frankfurt, "[t]he fact about himself that the bullshitter hides, on the other hand, is that the truth-values of his statements are of no central interest to him; what we are not to understand is that his intention is neither to report the truth nor to conceal it. This does not mean that his speech is anarchically impulsive, but that the motive guiding and controlling it is unconcerned with how the things about which he speaks truly are . . . the bullshitter . . . is neither on the side of the true nor on the side of the false . . . He does not care whether the things he says describe reality correctly. He just picks them out, or makes them up, to suit his purpose" (Frankfurt 2005, 55–56).
5. See (Messeri 2016) for a fascinating examination of how space scientists envision places in space.

6. See also Vandana Singh's chapter in this volume (Chapter 8: "Spacefaring for Kinship").
7. Here I borrow from philosopher Linda Zagzebski's conception of *virtue ethics,* of which one tenet is that if we care about something, we ought also care about *getting to the truth* about that thing. See (Zagzebski 2004) for discussion.
8. For more on the need to interrogate the foundations of our visions of space, see (Schwartz 2021).

References

Delgado-Lopez, Laura. 2011. "When Inspiration Fails to Inspire: A Change of Strategy for the US Space Program." *Space Policy* 27: 94–98.

Frankfurt, Harry. 2005. *On Bullshit.* Princeton, NJ: Princeton University Press.

McMahon, Sean. 2016. "The Aesthetic Objection to Terraforming Mars." In *The Ethics of Space Exploration,* edited by James Schwartz and Tony Milligan, 209–218. New York: Springer.

Messeri, Lisa. 2016. *Placing Outer Space: An Earthly Ethnography of Other Worlds.* Durham, NC: Duke University Press.

Rolston III, Holmes. 1986. "The Preservation of Natural Value in the Solar System." In *Beyond Spaceship Earth: Environmental Ethics and the Solar System,* edited by Eugene Hargrove, 140–182. San Francisco, CA: Sierra Club Books.

Schwartz, James. 2017. "Myth-Free Space Advocacy Part I: The Myth of Innate Exploratory and Migratory Urges." *Acta Astronautica* 137: 450–460.

Schwartz, James. 2020. *The Value of Science in Space Exploration.* New York: Oxford University Press.

Schwartz, James. 2021. "Gunnery Sergeant Draper and the Martian Congressional Republic's Vision for Mars." In *The Expanse and Philosophy,* edited by Jeffery Nicholas, 151–160. Hoboken, NJ: Wiley Blackwell.

Singh, Vandana. 2018. "Sailing the Antarsa." In *Ambiguity Machines and Other Stories,* edited by Vandana Singh, 173–211. Easthampton, MA: Small Beer Press.

Zagzebski, Linda. 2004. "Epistemic Value and the Primacy of What We Care About." *Philosophical Papers* 33: 353–377.

PART 3
CULTURAL NARRATIVES AND SPACEFLIGHT

Plate 1 Shestory © Jacque Njeri, Instagram: @jacquenjerii.

Plate 2 Three Wise Men © Jacque Njeri, Instagram: @jacquenjerii.

Plate 3 The very realistic photo of Curiosity's first full drill on Mars. The smaller hole in the lower right of the image is from a partial test drill previously conducted by the rover's drill. Credit: NASA/JPL-Caltech/MSSS (Clark 2013).

Plate 4 A panoramic view of the landscape of Mars taken by the Perseverance rover (NASA, n.d.).

13

Sacred Space

Decolonization through the Afrofuture

Ingrid LaFleur

Introduction

We are at a crossroads. Humans have entered space without healing themselves here on Earth. These humans are carrying trauma that informs their perspective and ultimately their actions, which are affecting the rhythm of the entire universe. Probably most disturbing is that our space exploration is being planned and carried out by white, cis, heterosexual-presenting, wealthy men who seem to be borrowing from the colonial playbook. In this chapter, I focus on only two of them, billionaires Elon Musk, founder of SpaceX, and Jeff Bezos, founder of Blue Origin. The two of them have been actively cultivating the imagination of space in the minds of people in the United States and worldwide. This cultivation is happening while we still grapple with the consequences of colonialism, capitalism, patriarchy, and white supremacy. As an Afrofuturist, I must ask: What does that mean for Black people and marginalized communities? What does that mean for our collective future when space is led by those who are beneficiaries of the course we are currently on?

There is still time to course correct. I offer that the cultural movement Afrofuturism can help carve a path that allows for a more ethical approach to exploring space and all her wondrous worlds. By revisiting African precolonial and ancient past relationships to space we can imagine and investigate new possibilities that will not continue the systems of harm we perpetuate here on Earth. As writer James Baldwin writes:

> You were born where you were born and faced the future that you faced because you were Black and for no other reason. The limits of your ambition were, thus, expected to be set forever. You were born into a society which spelled out with brutal clarity, and in as many ways as possible, that you were a worthless human being.

Ingrid LaFleur, *Sacred Space* In: *Reclaiming Space*. Edited by: James S. J. Schwartz, Linda Billings, and Erika Nesvold, Oxford University Press. © Oxford University Press 2023. DOI: 10.1093/oso/9780197604793.003.0013

You were not expected to aspire to excellence: you were expected to make peace with mediocrity. (Baldwin 1963)

Welcome to the Afrofuture

From 1966–1969, Nichelle Nichols played the role of Lieutenant Uhura on *Star Trek*. Initially, Nichols told the show's creator, Gene Roddenberry, that she would not return after the first season. However, upon meeting the civil rights activist the Reverend Dr. Martin Luther King Jr. at a National Association for the Advancement of Colored People fundraiser, he encouraged her to stay on, emphasizing how Black people of that time, and arguably even today, needed to see her on *Star Trek* (Martin 2011). She represented something very important to her people. He even stated that *Star Trek* was the only show his wife, Coretta Scott King, would let their children watch. Dr. King was explaining to Nichols that the image of a Black female space engineer on television was part

Figure 13.1 Originally displayed in Pittsburgh, PA in 2017, There Are Black People in the Future was created by Alisha B. Wormsley. The billboard shown here was placed in downtown Detroit in 2019 as part of the exhibition Manifest Destiny curated by Ingrid LaFleur.

of the plan to ensure our survival now and into the future (Fig. 13.1). In a time when most representations of Black people were maids, seeing Black people in space was a huge leap, a necessary leap that encouraged what Robin D. G. Kelley, historian and Distinguished Professor, and Gary B. Nash, Endowed Chair in US History, called the Black Radical Imagination:

> By employing the phrase the "Black Radical Imagination" in my book *Freedom Dreams*, I was referring to the ways in which Black Leftists, some nationalists, feminists, surrealists, etc., envisioned collectively, in struggle, what a revolutionary future might look like and how we might bring this new world into being . . . It is not enough to imagine a world without oppression (especially since we don't always recognize the variety of forms or modes in which oppression occurs), but understanding the mechanisms or processes that not only reproduce structural inequality but make them common sense, and render those processes natural or invisible. The Black Radical Imagination is not a thing but a process, the ideas generated from what Gramsci calls a "philosophy of practice." It is about how people in transformative social movements, moved/shifted their ideas, rethought inherited categories, tried to locate and overturn blatant, subtle, and invisible modes of domination. (Red Wedge 2016)

The Black Radical Imagination helps to redirect the destiny of Black mind, bodies, and soul and provides direction. The radical imagination of the 1960s, where Nichols showed us what we could look like in space, is still influencing movements and individual dreams of today.

What Is Afrofuturism?

A working definition of Afrofuturism is as follows:

> Afrofuturism, coined in the 1990s, is a multi-disciplinary cultural movement that visualizes alternate futures for Black bodies. Afrofuturism offers an intersectional, multi-temporal, and inter-disciplinary approach to the future. Also called a liberation movement, Afrofuturism dreams of worlds that empower the mind, body, and soul of Black people, in the past, present and future. Afrofuturism stimulates and encourages imaginations to craft decolonized destinies of pleasure, inclusion, holistic health, joy, and prosperity using speculative modalities such as science fiction, surrealism, magical realism, and horror. Afrofuturists, practitioners of Afrofuturism—artists, architects, choreographers, musicians, urban planners, writers, farmers, etc.—are active in reimagining Black power narratives, warning us

of how a colonized future can control us, but always introducing perspectives and inventing tools to redirect the path. The afrofuture investigates the intersection of race, emerging technology, and science, and is inspired by traditions of ancient and near pasts, mythologies, legends, spiritual sciences, and cosmologies from Africa and the diaspora. (LaFleur, n.d.)

Afrofuturism centers the dreams and imaginations of Black people. It also centers the Black body to protect it from ideas, policies, and conditions that may cause harm, while encouraging the strength and skill of the Black body. By imagining a new destiny, we are able to create a more balanced, peaceful, pleasurable, loving present. Afrofuturism is intent on shifting how we treat humans, Earth and space, and all of their inhabitants. The Afrofuture questions our personal perspective on the world and requests that we reflect on how we engage and communicate with the world and various realities. Afrofuturism gives us permission to dream the impossible and actively work towards making it real.

As Cedric Robinson, political scientist and former Professor of Black Studies and Political Science at University of California, notes, "What concerns us is that we understand that racialism and its permutations persisted, rooted not in a particular era but in the civilization itself" (2021, 27). Power lies in the belief in the imagined. Control what is imagined, and you control the destiny. Afrofuturism wants to break us, all of us, free from all oppressive control. For non-Blacks, Afrofuturism becomes a place of liberation as well. The centering of the Black mind, body, and soul in future visions is a radical act, a mind-altering act that goes against a millennium or more of anti-Blackness. The anti-Blackness battle is an old one that transcends the transatlantic slave trade. This is a global epidemic that has been trained to see Black people as dangerous, and as machines to be used only for labor. There are many lessons to learn from anti-Blackness. Black bodies bear intimate witness to the worst of our humanity. Black people know how any plan can go frighteningly wrong if you are on the vulnerable side of the plan. However, Afrofuturism considers all of this and asks: Where are we going, and is that where we want to go? And how do we create more just ideas and thought patterns to provide the healthiest and most just direction?

Where Are We Now? History Predicts the Future

What *is* the American fetish about highways?
They want to get somewhere, LaBas offers.
Because something is after them, Black Herman adds.

But what is after them?

They are after themselves. They call it destiny. Progress. We call it Haints. Haints of
their victims rising from the soil of Africa, South America, Asia. (Reed 2013, 135)

Just as cities and states constructed highways that cut right through bus-
tling Black communities, thereby destroying them, the United States has
entered space with a familiar arrogance, treating planets like objects instead
of living beings. "Progress" cannot be stopped, and thus, all that is in its way
must be destroyed. And this "progress" was created to serve as convenience
and comfort for a fraction of the population. And here we are at the edge
of "progress" again, centering the needs and comfort of humans as imagined by white men. In a tweet, Elon Musk suggested using thermonuclear
weapons on the poles of Mars to warm up and terraform the planet making
it more livable for humans (Masunaga 2015). Now there are T-shirts reading
"Nuke Mars" that the SpaceX founder and his fans wear. Apparently, Musk
isn't concerned with the ecosystem of Mars, because to him, it's a cold rock
to play with.

As Wole Soyinka, Nigerian novelist, playwright, and poet, told an inter-
viewer, "Colonialism bred an innate arrogance, but when you undertake that
sort of imperial adventure, that arrogance gives way to a feeling of accommo-
dativeness. You take pride in your openness" (Meikle 2010). We are embarking
on settler colonialism on Mars and the Moon, and current modes of racial capitalism and militaristic force are being used to support this mission.
Although we have yet to define what indigenous means in the context of celestial planets, the assumption that the inhabitants of all forms are to conform to
anything we do on their planets is an extension of our colonial past. But first
to assume there are no living beings, however defined, on Mars and the Moon,
is not only arrogant but strategically ignorant. This is where Afrofuturism can
help guide the space industrialists by looking at those ancient pasts that speak
about the cosmos and those planets in particular and create a new process of
engagement. For instance, three Yemeni men sued NASA because of its invasion of their ancestral lands, Mars, and demanded the end of all operations on
the planet. It was bequeathed to them, they say, 3,000 years ago (CNN 1997).
This ancient relationship underscores that Mars is a living place even if it is
only in the imagination.

In 2019, then-President Donald Trump created a new arm of the US military called the Space Force. "Space is the world's newest war-fighting domain," President Trump said during the signing ceremony. "Amid grave
threats to our national security, American superiority in space is absolutely
vital. And we're leading, but we're not leading by enough. But very shortly

we'll be leading by a lot" (Kennedy 2019). Initially the inspiration for many jokes, the creation of the Space Force caused confusion as to why invest in such a military branch. What the public didn't know was that one of the main objectives of the Space Force Guardians is to protect our satellites from attack. This mission will only grow as the industry grows and comes closer to actually creating bases on Mars and the Moon. We are witnessing the beginning of colonialism—a military force to invade the planets, to secure resources and land, and to enforce assimilation and the ultimate erasure of the indigenous and their culture and identity. Will we ever ask: What right do we have to invade and enforce our laws onto celestial lands? Is there another way?

The inequalities we experience on Earth will continue far into outer space. A driving force of this inequality is racial capitalism. US modern racial capitalism is defined by Professor Charisse Burden-Stelly of Carlton College as follows:

> Drawing on the intellectual production of twentieth-century Black anticapitalists, I theorize *modern U.S. racial capitalism* as a racially hierarchical political economy constituting war and militarism, imperialist accumulation, expropriation by domination, and labor superexploitation. The *racial* here specifically refers to Blackness, defined as African descendants' relationship to the capitalist mode of production—their structural location—and the condition, status, and material realities emanating therefrom. It is out of this structural location that the irresolvable contradiction of value minus worth arises. Stated differently, Blackness is a capacious category of surplus value extraction essential to an array of political-economic functions, including accumulation, disaccumulation, debt, planned obsolescence, and absorption of the burdens of economic crises. At the same time, Blackness is the quintessential condition of disposability, expendability, and devalorization. (Burden-Stelly 2020)

"Labor superexploitation" is already moving around in the mind of Musk who, in a tweet, suggested indentured servitude to pay off the debt of traveling to space. He emphasized there will be many jobs on Mars (Musk 2020). By suggesting a form of indentured servitude to access space, Musk is, intentionally or unintentionally, repeating what his colonial predecessors have mapped out for him.

The backbreaking work of the laborer in service to establishing a colony need not be forgotten.

Most slave labor, however, was used in planting, cultivating, and harvesting cotton, hemp, rice, tobacco, or sugar cane. On a typical plantation, slaves worked ten or more hours a day, "from day clean to first dark," six days a week, with only the Sabbath off. At planting or harvesting time, planters required slaves to stay in the fields 15 or 16 hours a day. When they were not raising a cash crop, slaves grew other crops, such as corn or potatoes; cared for livestock; and cleared fields, cut wood, repaired buildings and fences . . . Slave masters extracted labor from virtually the entire slave community, young, old, healthy, and physically impaired. (Digital History 2021)

The settler colonial system relies on the continued subjugation and exploitation of a people. They are to be controlled, completely, and cannot escape, physically, sometimes even if they have financial means. The abuse endured in these moments of subjugation reappears as the continuation of space exploration grows. History tells us this subjugated group that will support the space industry will have no power, recourse, or recognition. They will only serve the visions of a wealthy master.

Racial capitalism is most apparent within the inner workings of Amazon, Blue Origin's funder. The labor practices of Amazon have long been criticized for their mistreatment of employees, but apparently all of this is lost on Bezos. In July 2021, in a statement upon arrival back on Earth after the first public launch of his Blue Origin rocket New Shepard, Bezos commented, "I want to thank every Amazon employee, and every Amazon customer, because you guys paid for all this" (D'Innocenzio 2021).

What did they pay for exactly? Well, both the customer and the employee helped Bezos create the $5.5 billion he initially invested in Blue Origin. This allowed him, his brother and two others to go into suborbital space for four minutes of weightlessness. Bezos extracted these billions by continuing to pay low wages. The median income for an Amazon worker was just $29,007 in 2020. The Amazon CEO-to-worker pay ratio is 58:1 (Dailey 2021). And just months before, Amazon squashed the union efforts created by workers at a warehouse in Alabama (Streitfeld 2021). Poor working conditions prompted the union organizing. For instance, bathroom breaks were so restrictive that it made it difficult to actually take one, which led to many workers urinating in empty bottles just so they wouldn't be penalized for going over time on the break (Picchi 2021). Most recently, a group of workers within Amazon highlighted how the distribution centers were placed in lower income neighborhoods that were majority people of color. These centers were causing pollution in the surrounding areas, resulting in textbook environmental racism (Levin 2021).

Pollution

On May 12, 2021, a robotic arm of the International Space Station was dam-. aged by space debris (Howell 2021). According to the European Space Agency (ESA), as of late 2021, approximately 330 million or more pieces of debris are circulating the orbit (ESA 2021). Objects are sent up into orbit for fun, like Musk's Tesla, without considering the consequence of them floating in space (Ellison 2020). Why are humans intentionally junking up space?

Here on Earth, we have found 150 million metric tons of macro and micro pieces of plastic floating in our waters (Ocean Conservancy, n.d.). Amazon alone produced 465 million pounds of packaging plastic waste in 2019 (Oceana 2020). And 22 million pounds of that waste is in our oceans. The carelessness with which corporations and overall society treat our environments seems to be the modern human's way. So it should not surprise us that our bad habits have extended into space. And why have they extended into space? Because the colonial way is all about the now and the near future. The consequences that our actions today may cause for a far off future does not concern the capitalist. And when the resource is no more, you find another resource to exploit, like the asteroids in space. The same actors polluting Earth are the ones wanting to architect new systems of power in space, and yet, we follow their lead and uphold their cosmic visions.

Light Pollution

According to MIT Technology Review, "83% of the world's population and more than 99% of the U.S. and European populations live under light-polluted skies . . . Due to light pollution, the Milky Way is not visible to more than one-third of humanity, including 60% of Europeans and nearly 80% of North Americans" (Emerging Technology 2016). The cause of the light pollution: streetlamps, digital billboards, lit buildings, highway lighting, and vehicle headlamps, which are all using more often than not LEDs, a light spectrum that makes it less harmonious with animal life.

I would argue that LEDs are not in harmony with human life as well. Urban city dwellers have lost sight of the celestial skies forcing people to focus on that which is here on earth. We have become prisoners of artificial light. The ability to stargaze is one that should not be taken lightly. It can ignite a necessary connectedness to our star ancestry and the cosmos within us (Melina 2010).[1]

However, even if we were able to experience a clear night sky without light pollution, we would still have to contend with the satellites. To date, Musk's

Space X has launched approximately 1,700 satellites into the sky (Thompson 2021) and seeking approval to launch 30,000 more (Wall 2019). The sky is littered so much with satellites, astronomers have complained about the disturbance it is causing in mapping our skies (Resnick 2021). It was the north star that guided American abolitionist Harriet Tubman on thirteen missions to save herself and rescue approximately seventy others from slavery. One hundred years later it has become increasingly difficult to identify the north star because of all of the satellites. We will soon no longer be able to map our liberation in the skies. Are we teaching our children to tell the difference between stars and satellites?

The Afrofuture Approach

We are entering space the same way we treat Black people, engaging without permission, exploiting and sacrificing lives in order to achieve a goal by any means necessary. As if space is to be owned. The American way has a long tradition of honoring aggression over grace.

Afrofuturism can prepare us for a more just space future. Afrofuturism shows us that the dehumanization of the Black body is reflected in how we treat Earth and space. Drexciya is a sacred place within the imaginary that helps to explain and reconcile a horrific event. In 1992, the music duo Drexciya from Detroit, James Stinson and Gerald Donald, created a myth of the same name that served as a foundation for their electronic music (David 2021). Their myth recounts a story where pregnant women were thrown overboard the ships during the transatlantic slave trade. Those women gave birth underwater and created the aquatopia Drexciya. The music Stinson and Donald created were the sounds coming from that aquatopia. Those aquatic sounds helped to keep us tethered to these water souls. The myth ends with Drexciyans, frustrated with the years of pollution of our waters, escaping to planet Grava-4 to create their own colony. Through the myth, we were reconnected with the ancestors who didn't make it to new lands. Here, Drexciya serves as a refuge, a place where Black souls were saved again and again.

In the Afrofuture, space becomes the alternative to human shenanigans. The cosmos breaks through the limitations and categories that are placed upon Black bodies. In the film *Space is the Place* (Fig. 13.2), we see jazz musician Sun Ra giving us permission to dream upon the cosmos. He wants to take Black people back to his homeland, the planet Saturn. It is there where he believes we will be able to create what he calls the alter-destiny. In the film he tells a group of Black youth that they are myths because Blackness is a myth

Figure 13.2 Film still of the film *Space is the Place* (Produced by Jim Newman, 1974). Written by jazz musician Sun Ra in 1974, the film, directed by John Coney, featured Sun Ra and his Arkestra. From Danielle A. D. Howard, (2021) "The (Afro) Future of Henry Box Brown: His-story of Escape(s) through Time and Space," *TDR* 65, 3 (T251). (Screenshot courtesy of *TDR*.)

and reality doesn't exist on this earthly plan. He asks: Why be controlled by something that isn't real? Instead, travel to Saturn, where new experiences and possibilities abound. Through the myth of Saturn, Sun Ra demands that we shift our consciousness and get free.

Afrofuturism reveals how our minds have yet to be decolonized. Science-fiction writer Octavia Butler often projected into the future based on the stories she would read in newspapers. She wrote dystopian futures while showing us how to use our unique skills such as empathy to navigate the dystopia. For instance, in her book *Parable of the Sower*, we are confronted with a United States in complete chaos, an environmental and political disaster together making the country quite dangerous. When the walls that were built to protect homes from fire addicts are breached, we see the main character, seventeen-year-old Lauren Olamina, grab her emergency go-bag. Lauren read the signs, felt the future, and prepared accordingly. After many years of imagining, the star colony Lauren dreamt of is within reach. As the sequel *Parable of the Talents* illustrates, the ship that is used to fly into space was named after the colonial explorer Christopher Columbus, indicating that the spaceship is filled with colonized minds. Butler reminds us that there is no escape from that which is within.

Afrofuture thinking honors our collective celestial legacy and maps the dreams of fugitives and the creation of maroon societies.[2] It follows and records imagined projections and realized imaginings of Black dreams and desire. Since the coining of the term, Afrofuturism has served as an archive of radical dreaming and its practical manifestations. The Afrofuture archive on space exploration and the Black body relationship to space is extensive. There is enough collected knowledge to know what type of interaction to cultivate, where to be cautious, and how to approach with reverence.

Be it a utopia or dystopia, Afrofuture images of space guide us to another consciousness. Because the inner reflects the outer and vice versa, a self-assessment is required before embarking on any imagination exercise. We must be made aware, continuously, of our colonized mind, since all is shaped from it. It is for this reason I begin every Afrofuture workshop with a self-assessment by asking questions like: What is influencing our decisions in the now and how can we imagine differently? What is influencing how we relate to humans, the Earth, and the cosmos and all of their inhabitants? It is more of a point to be honest and less about assessing good vs. bad. Peace and balance is the goal, and to find the rhythm of the universe and be in tune with it, rather than see it as an object to be explored.

After the assessment we must then ask: What makes it our right to travel the skies? Have we asked for permission? Who do we receive permission from? Is it the same for the permission we seek here on Earth? Whose ancestors do we listen to? With tech taking up the cosmic skies, how does that affect our tradition for mapping freedom through the skies and informing how we cultivate land? Simply, how does space exploration affect us spiritually?

Space Is Our Ancestral Home. We Are in Service to Space

In the 1960s and 1970s there was a push to embrace Blackness and the Black body. The embrace included a return to the "roots" of Blackness, Africa. In returning to Africa, Black people were able to establish a decolonized relationship to the self, and thus a new relationship to the skies and land. This relationship was couched in the ancient and revered as such. During that time, the cosmologies of the Dogon reemerged.

The Dogon of Mali have influenced musicians, artists, and Afrofuture thinkers for decades. A favored cosmology, the Dogon tell us of amphibian-like beings called the Nommo, who came onto land to impart a knowledge that still guides Dogon society and spirituality today. The amphibian beings came from the star system Sirius, seen and mapped by the Dogon without the use of a modern telescope. The pair, representing feminine and masculine energy, shared their intelligence to help cultivate a better world. Around the 1960s there was a strong presence of Dogon culture within musical imagery in US American jazz and funk. For example, jazz composer Jymie Merrit wrote "Nommo" for drummer Max Roach in 1966 (Micucci 2020). The Dogon have created a framework in which to think about the cosmos and to reconnect with the ancestral beings (not aliens) that occupy the skies—a friendly reconnection that isn't born out of fear. Instead, the Nommo are caregiving, kind, and parental. However, a millennia later, our world may pollute theirs. How can we make space cleaner? Healthier? How are we showing our respect for all beings in space?

The Dogon's cosmology supports the idea that humans are part of a larger cosmic ecosystem and that outer space can be a resource beyond minerals; it can be a resource of intelligence and healing. How can we find the rhythm and be in harmony? The Afrofuture approach would be to view humanity existing in harmony with all ecosystems. That means no part of the system should be sacrificed to attend to humans' whims.

Another approach is to incorporate cooperative economics into the space industry. Worker and owner could share the profits of this massive emerging industry and thus control labor conditions. It is not too late to create the system now, turning the mega-corporations of SpaceX and Blue Origin into worker-owned companies and viewing all workers as cocreators.

And finally, education. It always comes down to how we are educating our youth, because that determines our future. We must make it mandatory for all youth to learn about the space industry and to learn how to code starting at an early age. It is especially imperative for youth of color who are usually the last to know. Our Black youth cannot be regulated to just worker or consumer; they must be part of the development. Simultaneously, we must teach all youth about Afrofuturism and global indigenous perspectives about Earth and space.

Conclusion

In August 2021, Ghanaian artist Amaofo Boako sent three portraits into space on Blue Origin's New Shepard (Cascone 2021; Fig. 13.3). Only one month later, in September, astronaut Sian Proctor became the first Black woman to pilot a spacecraft (Ottesen 2021). In 2018, artist Tavares Strachan launched "Enoch," an Egyptian funerary vessel, into space (Fig. 13.4). Carved into the canopic jar is the face of the first African American to train as an astronaut, Robert Henry Lawrence Jr. (Meier 2018). Blackness is contributing to space and always has, but now the visibility of our actions is proving useful. All these efforts, as exciting as they may be, humanize the next colonial project, space exploration, while inequalities still persist on Earth.

Imagining space within Afrofuturism is imagining liberation. Afrofuture foresight makes predictions based on our current collective trajectory, which tells us the systems we employ on Earth will be reflected in the interplanetary worlds we invade. However, we have the tools to create a new destiny in space. We can still change course. Afrofuturists can carve another path forward, but we must have a seat at the space industry table and join the conversation and

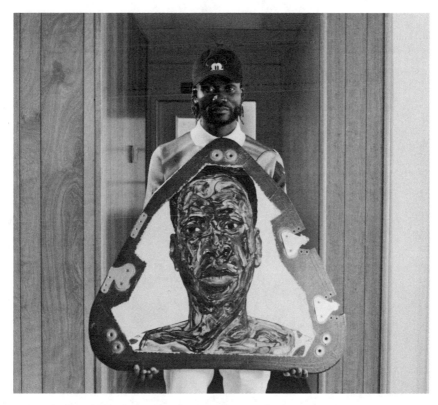

Figure 13.3 Commissioned by Uplift Aerospace, Ghanaian artist Amoako Boafo holds a portrait of his childhood friend Otis Kwame Kye Quaicoe entitled, *Shormeh's Gold Earrings* (2021). Painted on a shuttle launched by Blue Origin, the portrait is part of the series titled "Suborbital Triptych."

Figure 13.4 On December 3, 2018, Tavares Strachan launched Enoch in collaboration with the Los Angeles County Museum of Art.

process. Until then, let's try to treat the cosmos and all its inhabitants as sacred. Imagine the possibilities.

Notes

1. Editor's note: see Chapter 22 of this volume for Tanja Masson-Zwaan's discussion of the legal issues surrounding satellite constellations and the night sky.
2. Editor's note: see Chapter 4 of this volume for Edward C. Davis IV's reflections on Maroon communities.

References

Baldwin, James. 1963. *The Fire Next Time*. London: Penguin Classics.

Burden-Stelly, Charisse. 2020. "Modern U.S. Racial Capitalism: Some Theoretical Insights." *Monthly Review* 70, no. 3. July 1, 2020. https://monthlyreview.org/2020/07/01/modern-u-s-racial-capitalism/.

Cascone, Sarah. 2021. "Three Amoako Boafo Paintings Were Just Launched Into Space Aboard Jeff Bezos's Rocket Ship." *Artnet News*, August 27, 2021. https://news.artnet.com/art-world/amoako-boafo-jeff-bezos-blue-origin-2003190.

CNN. 1997. "3 Yemenis Sue NASA for Trespassing on Mars." *CNN*, July 24, 1997. http://edition.cnn.com/TECH/9707/24/yemen.mars/.

Dailey, Natasha. 2021. "Amazon Reveals How Much It Paid Its Median Employee Last Year: $29,007." *Business Insider*, April 15, 2021. https://www.businessinsider.com/amazon-employee-salary-pay-median-worker-compensation-compared-jeff-bezos-2021-4.

David, S. 2021. "Inside the Stunning Black Mythos of Drexciya and Its Afrofuturist '90s Techno." *Arstechnica*, February 28, 2021. https://arstechnica.com/gaming/2021/02/inside-the-stunning-black-mythos-of-drexciya-and-its-afrofuturist-90s-techno/.

Digital History. 2021. "Slave Labor." *Digital History*. https://www.digitalhistory.uh.edu/disp_textbook.cfm?smtid=2&psid=3041.

D'Innocenzio, Anne. 2021. "Jeff Bezos Thanked Every Amazon Employee and Customer 'Because You Guys Paid for All of This,' Prompting Criticism." *Boston Globe*, July 21, 2021. https://www.bostonglobe.com/2021/07/21/business/jeff-bezos-thanked-every-amazon-employee-customer-because-you-guys-paid-all-this-prompting-criticism/.

Ellison, Stephen. 2020. "Tesla Roadster Launched into Space 2 Years Ago Floats Past Mars: Report." *NBC Bay Area*, October 8, 2020. https://www.nbcbayarea.com/news/local/tesla-roadster-launched-into-space-2-years-ago-floats-past-mars-report/2377411/.

Emerging Technology from the arXiv. 2016. *MIT Technology Review*, September 13, 2016. https://www.technologyreview.com/2016/09/13/157575/light-pollution-atlas-shows-why-80-percent-of-north-americans-cant-see-the-milky-way/.

ESA. 2021. "Space Debris by the Numbers." ESA Website. Accessed November 9, 2021. https://www.esa.int/Safety_Security/Space_Debris/Space_debris_by_the_numbers.

Howell, Elizabeth. 2021. "Space Station Robotic Arm Hit by Orbital Debris in 'Lucky Strike' (Video)." *Space.com*, May 31, 2021. https://www.space.com/space-station-robot-arm-orbital-debris-strike.

Kennedy, Merrit. 2019. "Trump Created the Space Force. Here's What It Will Actually Do." *NPR*, December 21, 2019. https://www.npr.org/2019/12/21/790492010/trump-created-the-space-force-heres-what-it-will-do.

LaFleur, Ingrid. n.d. "What Is Afrofuturism?" *Think Like An Afrofuturist*. https://www.thinklikeanafrofuturist.com/.

Levin, Sam. 2021. "Amazon's Warehouse Boom Linked to Health Hazards in America's Most Polluted Region." *The Guardian*, April 15, 2021. https://www.theguardian.com/technology/2021/apr/15/amazon-warehouse-boom-inland-empire-pollution.

Martin, M. 2011. "Star Trek's Uhura Reflects on MLK Encounter." *NPR*, January 17, 2011. https://www.npr.org/2011/01/17/132942461/Star-Treks-Uhura-Reflects-On-MLK-Encounter.

Masunaga, Samantha. 2015. "What Scientists Say about Elon Musk's Idea to Nuke Mars." *Los Angeles Times*, September 10, 2015. https://www.latimes.com/business/la-fi-elon-musk-mars-20150910-htmlstory.html.

Meier, Allison. 2018. "An Ancient Egyptian Funerary Vessel Heads to Outer Space." *JSTOR Daily*, December 7, 2018. https://daily.jstor.org/an-ancient-egyptian-funerary-vessel-heads-to-outer-space/.

Meikle, James. "England Is 'Cesspit' Breeding Islamists, says Soyinka." *The Guardian*, February 1, 2010. https://www.theguardian.com/books/2010/feb/02/soyinka-england-cesspit-islamists.

Melina, Remy. 2010. "Are We Really All Made of Stars?" *LiveScience*, October 13, 2010. https://www.livescience.com/32828-humans-really-made-stars.html.

Micucci, Matt. 2020. "Remembering Jymie Merritt: Max Roach—'Nommo.'" *Jazziz Magazine*, April 13, 2020. https://www.jazziz.com/remembering-jymie-merritt-max-roach-nommo/.

Musk, Elon. 2020. Twitter post. January 16, 2020, 9:08pm. https://twitter.com/elonmusk/status/1217992175452995584/.

Ocean Conservancy. n.d. "Plastics in the Ocean." *Ocean Conservancy Website*. https://oceanconservancy.org/trash-free-seas/plastics-in-the-ocean/.

Oceana. 2020. "Amazon's Plastic Problem Revealed." *Oceana Website*, December 2020. https://oceana.org/reports/amazons-plastic-problem-revealed/.

Ottesen, K. K. 2021. "Sian Proctor, the first African American Woman to Pilot a Spacecraft: 'The Reality Is that Solving for Space Solves for Earth.'" *The Washington Post*, November 2, 2021. https://www.washingtonpost.com/lifestyle/magazine/space-spacex-nasa-sian-proctor/2021/11/01/52075a50-305b-11ec-a1e5-07223c50280a_story.html.

Picchi, Anne. 2021. "Amazon Apologizes for Denying that Its Drivers Pee in Bottles." *CBS News*, April 5, 2021. https://www.cbsnews.com/news/amazon-drivers-peeing-in-bottles-union-vote-worker-complaints/.

Red Wedge. 2016. "Black Art Matters: A Roundtable On the Black Radical Imagination." Red Wedge, July 26, 2016. http://www.redwedgemagazine.com/online-issue/black-art-matters-roundtable-black-radical-imaginatio.

Reed, Ishmael. 2013. *Mumbo Jumbo*. New York: Open Road Integrated Media.

Resnick, Brian. 2021. "The Night Sky Is Increasingly Dystopian." *Vox*, January 29, 2020. https://www.vox.com/science-and-health/2020/1/7/21003272/space-x-starlink-astronomy-light-pollution.

Robinson, Cedric. 2021. *Black Marxism, Revised and Updated Third Edition: The Making of the Black Radical Tradition*. Chapel Hill, NC: University of North Carolina Press.

Streitfeld, David. 2021. "How Amazon Crushes Unions." *New York Times*, March 16, 2021. https://www.nytimes.com/2021/03/16/technology/amazon-unions-virginia.html.

Thompson, Amy. 2021. "SpaceX Falcon 9 Rocket Launches 60 New Starlink Satellites, Nails Its 2nd Drone Ship Landing at Sea." *Space.com*, May 26, 2021. https://www.space.com/spacex-starlink-28-rocket-launch-and-landing-success.

Wall, Mike. 2019. "SpaceX's Starlink Constellation Could Swell by 30,000 More Satellites." *Space.com*, October 16, 2019. https://www.space.com/spacex-30000-more-starlink-satellites.html.

14

Sherpas on the Moon

The Case for Including "Native Guides" in Space Exploration

Deana Weibel

On May 1, 2009, the satirical news site *The Onion* published a humorous article entitled, "Sherpa Who Led Neil Armstrong To The Moon Dead at 71." The article included such lines as "Following the well-worn path of his ancestors, Dorje and his two yaks hauled dehydrated food, water, space helmets, binders full of flight schematics, and a large silver plaque commemorating the moon landing as he shepherded the astronauts through their historic journey to the moon" (*The Onion* 2009). Despite the farcical nature of the article, it drew an apt comparison between space exploration and other types of exploration with a long history on Earth.

Even though it is carried out in significantly different ways, space exploration is an offshoot of terrestrial exploration and is normally understood as such. The Explorers Club, which was founded in New York City in 1904, now describes itself on its official Facebook page as promoting "the scientific exploration of land, sea, air, and space," essentially ordering the various venues of exploration as they occurred chronologically (The Explorers Club 2022). The verb "explore" is defined (in its second sense) by Merriam-Webster as "to travel over (new territory) for adventure or discovery" ("Explore" 2021) and this human activity is as old as (likely older) than the first hominids. *Homo sapiens* has always sought new land, expanded into new areas, and attempted to learn about (and exploit) nature in this way.

Throughout human history, there have been various periods of intensified exploration. The movement of Asian people into North and South America, the spread of Polynesian peoples throughout the Pacific Ocean and the era of European colonialism and expansion are three well-known instances. These times of exploration often resulted from humans looking for food (such as following herds of prey animals), expanding living space, or, in some cases, pursuing scientific knowledge or increasing a group's standing among other

Deana Weibel, *Sherpas on the Moon* In: *Reclaiming Space*. Edited by: James S. J. Schwartz, Linda Billings, and Erika Nesvold, Oxford University Press. © Oxford University Press 2023. DOI: 10.1093/oso/9780197604793.003.0014 groups (these latter two goals are most evident in European accounts from the 1400s–1800s). Human exploration of outer space is motivated by some of these but not others; the search for food is not typically a motivation for humans to explore space, but according to journalist Lotte Rigter, national pride, scientific knowledge, and territorial expansion do seem to be goals (Rigter 2020).

Unlike other forms of historical exploration, space exploration has been severely limited to a small number of people, and both class and economics have constrained who has been able to participate. However, during the centuries of European exploration, for example, crews on sailing ships tended to include both the rich and poor, and exploration on land frequently included the so-called native guide. There was a recognition that locals to the area being explored could prove helpful because of their connections to other locals and to their knowledge of the region itself.

Perhaps the most well-known "native guide" in the Americas was Sacajawea, the young Lemhi-Shoshone woman who participated in Meriwether Lewis and William Clark's Corps of Discovery Expedition from 1803 to 1806. Gender studies scholar Wanda S. Pillow describes how Sacajawea "has been proliferated, ignored, storied and represented as an Indian woman since the expedition" (2012, 47) and notes her role in American culture as "an emblem of manifest destiny (and white American man) early 1900s to present" (48). Pillow exposes the colonial underpinnings of the way history tends to portray this woman. She is not so much a subject or explorer in her own right as she is a "helper" to Lewis and Clark and a comforting symbol of ethnic and racial cooperation. Pillow writes that "we want Sacajawea whitened, contained and reproduced in images and texts; we want to hear her stories of hardship and struggle, but only enough to support manifest destiny and a multicultural narrative of the beginnings of the USA that denies or downplays any colonial legacy" (53).

Similar stories have been told about Tenzing Norgay (who published as Tenzing Norkey), one of two men who were the first humans to reach the summit of Mount Everest on May 29, 1953. Norkey climbed alongside a New Zealander, the more famous Sir Edmund Hillary. According to Norkey's memoir, he was a Sherpa and born in Nepal (Norkey 1957), although there are indications from other family members that he was actually born in Tibet and later married into a Sherpa family ("Tenzing Norgay" 2021). The initial story of the men's summit of Mt. Everest sometimes omitted or downplayed Norkey, following a pattern where European explorers (or those of European descent) were given credit and glory for some new "first," while the Indigenous explorers who accompanied or led them were often given second billing or seen as mere supporting players to the true, white adventurer.

A newspaper article published in Utah's *Pleasant Grove Review* on June 2, 1953, for instance, begins with the three-line headline, "Be Proud of Britain On This Day, Coronation Day/All This—And Everest, Too!/Briton First On Roof Of World," marking the summit as a clear triumph for Britain and tying it to the ascension of Elizabeth II to the crown. Although Norkey is mentioned in the article (referred to within as "'Tiger' Tensing"), the summit is not described similarly as an achievement for Nepal, and Norkey is mostly praised for his strength as a porter. The unnamed author says Norkey was given the nickname "Tiger" after he "carried a load to 24,000 ft." and that he is "a Sherpa, one of a Nepalese people living in the valleys below Everest famed for their hardiness" (Express Staff Reporter 1953). Although the author notes that the flags of Nepal and the United Nations accompanied the Union Jack to the summit, it is clear that the protagonist of the story is Edmund Hillary, with "Tiger" Tensing along to assist.

While explorers have frequently relied on "native guides," so have members of various disciplines that conduct field research, including my own, cultural anthropology. Cultural anthropologists typically learn about communities by living in them, learning the languages spoken and participating in the rhythms of daily life. In an article aptly titled, "Anthropology's Hidden Colonialism: Assistants and Their Ethnographers," anthropologist Roger Sanjek calls attention to the large amount of ethnographic research performed by members of the communities being studied in the name of helping anthropologists. He writes, "While professional ethnographers—usually white, mostly male—have normally assumed full authorship for their ethnographic products, the remarkable contribution of these assistants—mainly persons of colour—is not widely enough appreciated or understood" (Sanjek 1993, 13). Sanjek goes on to mention, by name, various "anthropologists of diverse non-Western nationality and ancestry" who made "important contributions, as professionals, to the emerging discipline" (16) but whose contributions were erased because they did not meet the stereotypical image of the Western explorer.

Political economist Franklin Obeng-Odoom goes farther, characterizing an "intellectual marginalization" of African academic work resulting from global inequalities writ large. The problem is not a lack of excellent African intellectual contributions or an underappreciation of southern hemispheric cultures, he argues, but with a world where colonialism has created and perpetuated discrimination. He writes, "Although these two perspectives (human capital and southern theory) are quite distinct, they seek to explain, and ultimately address, Africa's intellectual marginalisation apart from, not as part of, Africa's marginalised position in the world system" (Obeng-Odoom 2019, 213). For Obeng-Odoom, African scholars are not properly recognized because their work cannot be properly recognized in a world dependent on an understanding of Africa as inferior and impoverished.

Space exploration is a product of the same world. Perceived inequalities are reified and perpetuated, and, coupled with the sheer amount of capital needed to move machines and people from our planet into our solar system and beyond, this means that only certain populations can afford space exploration to begin with. Moreover, the idea that only some groups have the intellectual capacity to create spacefaring technology is widely held. Finally, even though earlier forms of exploration frequently involved the participation of members of marginalized groups (for practical or symbolic reasons), there is no apparent need for a "native guide" in space, given that the parts of space humanity has explored (so far) have been (as far as we can tell) completely devoid of life. Perhaps someday an indigenous being from the planet Kepler-186f could lead human explorers who have come to study the planet first-hand, but a role for Indigenous Earthlings in space is less clear.

Space exploration, humanity's newest and most exclusive form of exploration, looks very different from older forms. Instead of being a situation where groups of scientists or conquerors strive to explore unfamiliar territories (unfamiliar to them, anyway) by physically going there and perhaps enlisting the help of friendly locals, most exploration of space has involved extremely limited "boots on the ground" presence by human teams or exploration at a distance via robotic exploration. Astronauts (twelve white males so far, all American) have landed on the Moon in pairs and covered small sections of the lunar surface either on foot or via lunar rover. Another twelve (again, all white males) have orbited the Moon. This lack of diversity was very much a product of its time, with astronauts being pulled from the ranks of American military-trained pilots who were all male and overwhelmingly white. Orbital flights, such as those undertaken by the Space Shuttle and on the International Space Station have been less about "exploration" of new territories and more about experimental research in an outer space environment.

Robotic space exploration, on the other hand, has involved learning about and mapping dozens of different celestial bodies in our solar system, including all of our neighboring planets, some dwarf planets like Pluto and Ceres, asteroids, and other objects. According to astronomer Ken Croswell, Voyager 1 and 2 are the most distant human-made objects from Earth, with both having crossed the "heliopause" (which marks the border of our solar system) within the last decade (Voyager 1 in 2012 and Voyager 2 in 2018;

Croswell 2021, 2–3). While diverse populations have been involved in this form of exploration, working as engineers at the Johns Hopkins University Applied Physics Laboratory and Jet Propulsion Laboratory, for instance, space exploration remains an enterprise most closely associated with powerful, rich countries like the United States, Russia, and China, which have the most established programs. Other countries and regions continue to expand and develop their capabilities, with Israel (Clery 2019) and India (Padma 2020) having completed recent high-profile lunar missions.

Space exploration, then, whether involving humans (like the Apollo Moon landings) or robotic craft (like the probes and rovers sent out into space), seems to lack participation by the less technologically focused communities who are so essential for terrestrial exploration. The most obvious argument is that explorers don't need native guides when there is no native population to encounter, but a counterargument may be made when we look at Antarctic exploration. Antarctic exploration has very frequently been compared to space exploration for many reasons, including the distant otherworldliness of Antarctica, the potential deadliness of the environment, and the frequent need for protective equipment on its surface, but also because of its lack of a native population (for more on similarities between Antarctic and space exploration, see Weibel 2020). It would be natural, then, to expect that Indigenous groups were carefully excluded from Antarctic exploration, but in at least one situation this was not the case.

Polar expedition researcher Mary R. Tahan reports that the great explorer Roald Amundsen sought to bring Inuit people with him on an Antarctic survey, for at least part of the journey. The Inuit, of course, live in the northern polar regions, at the farthest possible distance from Antarctica, so their role to play during an Antarctic expedition isn't immediately clear. Amundsen planned to bring Greenlandic huskies with him to pull sleds across the Antarctic ice but, according to Tahan, was not confident in his ability to care for the dogs. She writes, "(F)irst and foremost, he wanted the dogs to be well taken care of—and perhaps trained—by those who were the most knowledgeable and experienced with sled dogs, the Inuit. Because he was staking his South Pole discovery on the Greenland dogs who would take him there, for this reason, he was willing to assume the responsibility of taking two Greenlandic indigenous people with him to take care of these sled dogs" (2019, 35). Amundsen's plans to include Inuit participation didn't come to fruition, but his willingness to entertain the idea that communities with different kinds of experience had different skills and that these skills might well be useful in a place unfamiliar to all, demonstrates the kind of thinking that could inform space exploration.

While it is true that we are aware of no communities in space (at least no communities that we'd identify as "human") and no earthly society with a long history of space travel to draw upon, there are plenty of communities with historic knowledge *about* space. In an article entitled "Why are there Seven Sisters?" astronomers Ray P. Norris and Barnaby R. M. Norris (2021) discuss Greek and Australian Aboriginal myths about the stars that are surprisingly similar to each other, suggesting to the authors that these stories, still readily available in contemporary times, indicate a shared history. They explain:

> These strong similarities suggest a common origin, which appears to predate European contact with Aboriginal Australia. This similarity includes an insistence on there being seven stars, even though only six are visible to most people, together with a story to explain the "lost Pleiad". The evidence presented above shows that, because of the proper motion of Pleione, the Pleiades would indeed have appeared as seven stars to most humans in 100,000 BC. We conclude that the Pleiades/Orion story dates back to about 100,000 BC, before our ancestors left Africa, and was carried by the people who left Africa to become Aboriginal Australians, Europeans, and other nationalities. (Norris and Norris 2021, 5–6)

While mythological stories aren't typically used as sources of astronomical information among scientists, Norris and Norris demonstrate that these stories can be used as evidence for scientific theories about how certain constellations, like the Pleiades, used to look to humans on Earth in ancient times.

It is possible that other ancient stories still told in contemporary societies may offer additional knowledge about the way outer space was once perceived. Astronomer Auke Slotegraaf recounts a story told among a tribal group in Southern Africa: "In /Xam Bushmen star lore, the Milky Way was created by a girl of the ancient race who scooped up a handful of ashes from the fire and flung it into the sky. This made a glowing path along which people could see the route to return home at night. She also threw bits of an edible root into the sky, the old (red) pieces creating red stars and the young (white) pieces creating white stars" (2013, 67). The association in this story between a star's color and its age, while likely coincidence, is tantalizingly close to contemporary scientific understandings of the relative youth of white stars and the relative old age of red stars. Perhaps patterns and unusual phenomena visible in the sky (events that might not be obvious to contemporary astronomers) were noticed through eons of observation and passed along through stories like these.

When space exploration is left to a handful of countries and cultures, many skills, abilities, and experiences relevant for exploration of space are lost to the endeavor. There is an association between space and technology, but information about the history of the solar system can be found in numerous places including Indigenous societies around the globe. Astrophysicist Ray P. Norris and astronomer Barnaby R. M. Norris (2021) identify a record of a celestial change not gleaned via radio telescopes or computer models but through the stories that long-ago people passed down to their contemporary descendants. What other such stories may be found? What other useful information may have been passed down by the grandmothers and grandfathers who watched the sky and told stories that explained the changes and movements they noticed?

When Europeans wanted to climb Mount Everest, they enlisted peoples with a long history of familiarity with that mountain because their historical knowledge and firsthand experience climbing it demonstrated expertise. Expertise in one area can be used in other areas. An Englishman who climbed in the Himalayas, for instance, would not be prohibited from climbing the Andes or the Pyrenees or the Alps. Instead, his experience in one environment would be seen as giving him skills and insights that would be useful in another. Geologist Kathryn Sullivan, who was the first American woman to do a spacewalk, also explored the Mariana Trench in 2020 and described herself as "a hybrid oceanographer and astronaut" (Murphy 2020). Her bravery, insights, and skills were transferable from one locale to another.

There may not have been Sherpas on the Moon in the past, but surely individuals with strong cultural connections to mountains have something to contribute to the exploration of lunar and Martian peaks.[1] Populations with experience crossing deserts and rationing water would certainly have useful wisdom for crossing the plains of Mars and surviving the journey. Given that the widely varying cultures on Earth represent a vast source of experiences and solutions to problems, it would be foolish to limit spacefaring populations to certain groups with certain types of knowledge while excluding other populations with different knowledge. As we see in biology, the more traits that are available in a population, the more easily that population will adapt.

Furthermore, if settling space will be a future endeavor of *homo sapiens*, humans from more than just a few of Earth's populations should have a say as to whether, how, and where we should go. Human rights are a consideration when it comes to climate change, poverty, hunger, and other global issues, and should also influence space exploration. Unheard voices with unheeded wisdom deserve a seat at the table. Expanding recruitment is key. We do need Sherpas (and Maasai, Wai-Wai, Sami, Ojibwe, Mayans, /Xam Bushmen, Nuer, Berber, Bontoc, Inuit, Ainu, Maoris, etc.) in our rockets, on our space stations, in our mission control centers, and on the Moon.

Note

1. Editor's note: See Chapter 16 of this volume for Dan Capper's discussion of the Tibetan cultural relationship with mountains in the context of space.

References

Clery, Daniel. 2019. "Israeli Lander Demonstrates a Cut-rate Route to the Moon." *Science* 363, no. 6434: 1373–1374. doi:10.1126/science.363.6434.1373.

Croswell, Ken. 2021. "News Feature: Voyager Still Breaking Barriers Decades after Launch." Proceedings of the National Academy of Sciences 118, no. 17: 1–4. doi:10.1073/pnas.2106371118.

"Explore." 2021. Merriam-Webster. Accessed September 25, 2021. https://www.merriam-webster.com/dictionary/explore/.

Express Staff Reporter. 1953. "Briton First on Roof of the World." *The Pleasant Grove Review*, June 2, 1953: 1. Accessed September 25, 2021. https://www.newspapers.com/image/291564518/.

Murphy, Heather. 2020. "First American Woman to Walk in Space Reaches Deepest Spot in the Ocean." *The New York Times*, June 09, 2020. Accessed October 15, 2021. https://www.nytimes.com/2020/06/08/science/challenger-deep-kathy-sullivan-astronaut.html.

Norkey, Tenzing. 1957. *Man of Everest*. Dortmund: Lensing.

Norris, Ray P. and Barnaby R. M. Norris. 2021. *Why Are There Seven Sisters?* Cham: Springer.

Obeng-Odoom, Franklin. 2019. "The Intellectual Marginalisation of Africa." *African Identities* 17, no. 3–4: 211–224.

Padma, T. V. 2020. "India Reveals Third Lunar Mission." *Physics World* 33, no. 2: 12. doi:10.1088/2058-7058/33/2/23.

Pillow, Wanda S. 2012. "Sacajawea: Witnessing, Remembrance and Ignorance." *Power and Education* 4, no. 1: 45–56. doi:10.2304/power.2012.4.1.45.

Rigter, Lotte. 2020. "A March to Mars? The Function of Outer Space within US Nationalism in Historical Perspective." Master's thesis, Universiteit Utrecht.

Sanjek, Roger. 1993. "Anthropology's Hidden Colonialism: Assistants and Their Ethnographers." *Anthropology Today* 9, no. 2: 13. doi:10.2307/2783170.

Slotegraaf, Auke. 2013. "African Star-Lore." *MNASSA: Monthly Notes of the Astronomical Society of South Africa* 72, no. 3–4: 62–71.

Tahan, Mary R. 2019. *Roald Amundsen's Sled Dogs: The Sledge Dogs Who Helped Discover the South Pole*. Cham: Springer.

"Tenzing Norgay." n.d. Encyclopædia Britannica. Accessed September 25, 2021. https://www.britannica.com/biography/Tenzing-Norgay/.

The Explorers Club. 2022. Accessed November 8, 2022. https://www.facebook.com/TheExplorersClubNYC/about/.

The Onion. 2009. "Sherpa Who Led Neil Armstrong To Moon Dead At 71." *The Onion*, May 01, 2009. Accessed September 25, 2021. https://www.theonion.com/sherpa-who-led-neil-armstrong-to-moon-dead-at-71-1819570724/.

Weibel, Deana L. 2020. "Following the Path That Heroes Carved into History: Space Tourism, Heritage, and Faith in the Future." *Religions* 11, no. 1: 23. doi:10.3390/rel11010023.

15

Indigeneity, Space Expansion, and the Three-Body Problem

Tony Milligan

I will assume that we need to build Indigeneity and the place of Indigenous peoples into space ethics. We need to build them into our ways of thinking about how we might *belong* on other worlds, into our ways of thinking about what it is that humans want from space exploration, and into the practical day-to-day arrangements of space programs. After all, if we are to think of our expansion into space as fully "our" expansion, as "humanity's expansion," or as an important step in the history of "our entire moral community," then we can hardly ignore the global 5 percent who are Indigenous, and who include around 15 percent of the world's poor, as well as being the main bearers of human knowledge about how to live in extreme environments (World Bank 2021).[1]

For simplicity, let us call this "the inclusion of Indigeneity," but allow that a better way of putting matters may one day be found. My concern is that including Indigeneity within an ethics that stays much the same, fixed in position, may not be good enough. Even if it looks more inclusive, the position that it is fixed into may be oriented a good deal more to the West than to anywhere else. One reason for this is that much of the structure of what we tend to think of as "ethics" bears an imprint of colonial thinking. Not in the sense that there is an open commitment to inequality. Quite the reverse. There is a presupposition of an equality of persons and circumstances that would allow the right set of ethical principles to be devised in one place and then shipped out to everywhere else.

Questioning this presupposition need not mean that we must build an entirely new ethic for a new space age. After all, there are constraints upon how different a practical ethics can be from our current ethics. A practical ethic must be psychologically available in the sense that we must be able to work with it under difficult circumstances. It is no good sending astronauts up into space with a fine set of ethical commitments which then break down whenever

Tony Milligan, *Indigeneity, Space Expansion, and the Three-Body Problem* In: *Reclaiming Space*. Edited by: James S. J. Schwartz, Linda Billings, and Erika Nesvold, Oxford University Press. © Oxford University Press 2023. DOI: 10.1093/oso/9780197604793.003.0015 they have to make difficult choices about continuing missions at the expense of safety standards, or when they have to make terrible decisions about closing hatches to save three people at the expense of someone still outside. We have to be able to stick by the ethics when the going gets tough. Otherwise, ethical commitment will be merely decoration, rather than integral to our expansion into space. Practical ethics also has to be politically viable. It has to be the kind of ethics that can be endorsed by the political institutions through which democracies might operate.

This is not to say that there is no role for more utopian ethics, or ethical deliberation at the margins. But ethics for the routine practice of space agencies should have at least this level of political viability and accountability. Finally, practical ethics must be disciplined in its treatment of problems. "Anything goes" is not a good approach to ethics any more than it is a good approach to science, religion, or gymnastics. However, we may need to draw out some of the overlooked, more dynamic and open-ended features of ethics if the inclusion of Indigeneity, as well as a rethinking of gender, identity, and belonging, are to play a genuine transformational role, enriching space ethics rather than merely reproducing what we already have. In what follows I will try to draw attention to these more dynamic and open-ended features through an analogy between space ethics and the three-body problem: a problem of science which itself remains open-ended, in the sense that it has no simple real-world solution.

I will refer periodically to "we," and have already done so. But this convenient way of putting matters is not intended as an assumption that I can speak for all of us. Rather, it is closer to what the English philosopher and ethicist Bernard Williams has in mind when he says that "It is not a matter of 'I' telling 'you' what I and others think, but of my asking you to consider to what extent you and I think some things and perhaps need to think others" (Williams 1993, 171). It is shorthand, an invitation to consider what commitments we may hold in common and what vulnerabilities we might also share.

Why Ethics Is Like the Three-Body Problem

Let us work with an imperfect analogy and imagine for a moment that space ethics is like celestial mechanics, the science that deals with the motion of the planets. Of course, ethics is not a science, but it is not this aspect of the analogy that interests me. Rather, I am concerned with the broad similarity of science and ethics as techniques for problem solving. Like celestial mechanics, ethics deals with a certain class of problems and tries to find solutions. If we press the analogy a little further, we can ask whether ethics is like a two-body problem or a three-body problem. In the case of two-body problems, we can easily track what is going on by positing a stable center, a dynamic point around which two planets or celestial bodies move in their different ways. When we add a third body, things become chaotic. Readers may be familiar with this idea from an important work of science fiction by the Chinese author Cixin Liu (2015), who uses the three-body problem as a metaphor for a clash of civilizations. In Cixin Liu's novel, *The Three-Body Problem*, the inhabitants of Earth and a group of invaders from elsewhere settle down to an uneasy compromise. They find a balance point around which both can work, one that allows both to calculate risks and opportunities. But everything collapses once a third civilization enters into the picture. It only takes one more body to throw all the calculations off.

An actual three-body problem is like this, as well. If we want a clear solution to it, then we have to start restricting the conditions by assuming that one or more of the bodies has no motion or else that it has negligible mass. In other words, we have to introduce some fairly artificial assumptions. Unless this is done, the required calculations will never end. They will be nonfinite. The simplicity of working around an easily identified, shared, and stable center will be lost.

Pursuing this analogy, I want to suggest that ethics is more like a three-body problem rather than a two-body problem. Yet, when faced with uncontrollable complexity, we are continually tempted to restrict matters in ways that will artificially yield a decision procedure for each and every question. We want ethics to have a kind of determinacy about everything. A determinacy that is a poor fit for how the world is, and for what we are. We want a stable center. Or many of us do, for much of the time. Disagreements then ensue about exactly where this center should be placed. The disagreements can take the form of disputes about which cluster of concepts should be sovereign: those connected to rights, consequences, or virtue. Concepts which all have important roles to play. Or disagreements can take the form of disputes about the kind of "centrism" that is most appropriate. *Anthropocentrism* (Passmore 1974) presupposes a human center. *Biocentrism* (Taylor 1986) presupposes that the flourishing of the biological or of biotic communities should be the measure of all things. *Cosmocentrism* (Lupisella 2020; Dick 2020) is more elusive but tries in some sense to take the universe itself as the fixed point. An attractive but difficult idea given that our human sense of belonging has always involved contrasts: we *belong here* rather than *belonging there*. Belonging everywhere looks hard to square with our understanding of what belonging is for beings like ourselves. Perhaps a god could belong everywhere, but our job description is more constrained. Yet the idea of cosmic belonging remains attractive.

These different theories play upon the idea of a stable center, a stable idea that remains fixed while everything else moves around it or in relation to it. And when we then begin to fold some additional set of issues into the pictures that they set out, issues such as Indigeneity and the place of Indigenous peoples within space expansion, we may still want everything to remain orderly and undisrupted. We may want the cohesiveness of Indigeneity and the cohesiveness of ethical theory to blend nicely together and settle down in readily predictable ways.

I want to suggest that this is a reducing way to think about matters. A way that is at odds with our regular experience of what it is like to be human. In everyday life we care about a multiplicity of things, and they do not always combine well together. They do not revolve around just one thing but around lots of different things. Responsiveness to a plurality of concerns also seems to be an important and stable feature of lives like ours. It forms an important insight of ancient tragedy: the things that we value and desire may not harmonize. They may clash with one another. Philosophers such as Martha Nussbaum (2001) in the US, as well as Bernard Williams (1993), have pointed out that sometimes we have to sacrifice one thing in order to save something else, even when both are required if we are to enjoy a good life. We block our own pathways to well-being and contentment. Valuing many things poses problems that valuing only one thing does not. But rather than evading multiplicity, and the problems that it brings, it strikes me as important that we take the opportunity to do justice to it. Especially when we are thinking about space ethics in the light of various sorts of social critique or as part of an emerging "critical space theory."

Building Indigeneity into the Picture

What this means, in practice, is that we should not expect Indigeneity to slot neatly into place, when we start to build it into our picture of ethics. We should expect it to be disruptive and to frustrate the familiar desire for an ethics that centers upon only one thing or a cluster of things. To gain a sense of how this old longing is likely to be frustrated, let us consider three different but plausible pathways to affirming the importance of Indigenous inclusion in relation to discussions of space. These are three arguments that I have been trying to develop as part of a larger research project of thinking about space in the light of Indigeneity. Superficially, they too appear to be centered upon just one thing, i.e., Indigeneity. But a closer look shows that they draw upon a multiplicity of different "things that matter," and point us in different directions when we appeal to them in order to answer simple questions such as "How rapid should our expansion into space be?" Consider the first of these arguments:

> **The Liberties Argument:** It is common to speak about the conquest of space and to use metaphors of being "at war with" extreme environments. This may work in the short term, but wartime conditions justify authoritarian governance, and authoritarianism ultimately generates its own counterculture. The long-term survival of any settlement will depend upon a process of moving beyond authoritarian options for space settlements, abandoning the idea of being at war with an environment and instead belonging to it. In a sense, long-term survival, even in a comparatively favorable place such as Mars, will depend upon *becoming indigenous*. What this entails can be shaped by thinking about similar processes on Earth among various different Indigenous Peoples, many of whom have their own migration stories.

This argument brings a range of considerations into play. One of which is the extreme vulnerability of humans in space. Any sort of political freedom elsewhere will be what the British astrobiologist Charles Cockell (2015) calls "freedom in a box" with worrying tendencies toward authoritarian control over the basic requirements of life such as air and water. The realization that conditions will be adverse, and far more adverse than on Earth at any point in the near future, has also been an important driver for metaphors of war, colonization, and conquest. The German-born political philosopher Hannah Arendt framed matters in these terms in her classic essay "The Conquest of Space and the Stature of Man" (1963). The German jurist Carl Schmitt assumed much the same in his "Dialogue on New Space" (1954), i.e., there would be continuity of conflict with only a difference in the configurations of power. If this is all that is on offer, there will be little liberty in space—a problem which concerned Arendt (a Holocaust survivor) far more than it did Schmitt (a former official within the Nazi regime). But if we can learn from experiences of *belonging to physically extreme environments,* then something better than our familiar patterns of conflict might be achieved. The theme of belonging also figures in the second argument:

> **The Argument from Belonging:** Human expansion into space is a multigenerational project. More precisely, it involves a series of multigenerational projects: scientific, commercial, political, experiential. Some will be protective, and others extractive.

But almost all of what is done will be done by future generations, people whose lives will not overlap with our own. This poses a problem. Why should *they* continue the kinds of projects that *we* begin? Why should *they* care at all about what *we* want and hope for the future? This would be fine if the actions of future generations were themselves irrelevant to us. But much of what we do presupposes various kinds of continuity, obviously so in the case of space. Otherwise, our activities would be a little like the construction of the feet of a statue, with no body to follow. Our way out of the problem is to try to avoid preoccupation with transitory idiosyncrasies and political fads, and instead to look to what runs deep within human concerns, and in particular what runs deep within our human fascination with space. Indigenous Knowledge is an obvious repository of knowledge which can help us to do this, partly because it is not heavily marked by what one might call "ground bias." Rather, it situates humans within an order of things that is larger than Earth, as beings who belong to something more expansive. Future generations will clearly need a sense of belonging which is of this type.

Some important things are assumed here, such as the idea that Indigenous knowledge really does involve a more expansive sense of belonging than familiar ways of thinking about ourselves. Part of my justification for this assumption is that the knowledge in question tends to be holistic, joined together and entwined with complex cosmologies and origin stories, which are not ground-biased. The other part of my justification is that a larger sense of belonging has tended to erode within non-Indigenous communities during modernity, reducing (at best) to more ecological, Earth-based thinking, and often to something even more local. There are some exceptions within scientific and religious communities, but, overall, notions such as "belonging to God's creation" or "having a place within the Universe," have exercised a decreasing importance within the lives of non-Indigenous agents such as myself. We do not think in the ways that the pioneering astronomers Kepler or Copernicus once did, or in the ways that are spoken of in Dante's *Paradise*, Cicero's *Dream of Scipio*, or in the Qur'an. All are texts in which humans are located as inhabitants of only one region of space out of many such regions. We have demythologized the world, but in the process, we have lost our cosmological bearings, and have come to think only of the ground that we stand upon. **The Argument from Belonging** addresses itself to this problem.

Unlike **The Liberties Argument**, it does not require any actual settlement process on other worlds, or even a human presence on Mars. It will apply just so long as humans regularly use *nearby* space in a more systematic way, so that they begin to see it as included within their world. **The Argument** from **Belonging** also serves as a cautionary note about overreliance upon twentieth-century political debates about the state versus the private sector. These debates may be of no great interest to future generations better able to reflect upon the continuing interplay of both. We need not assume that they will be fervent statists or champions of laissez-faire capitalism. Indeed, it seems unlikely that future generations will be bound by any idea that their economic systems and political arrangements are mere continuations of those we presuppose as basic. Perhaps we may still exercise a disproportionate influence upon what is done, through some kind of "founder's effect" (Elvis, Krolikowski, and Milligan 2022), but the projects which we initiate that have deep roots in our humanity stand a better chance of being continued rather than quickly replaced.

The third and final argument marks a shift away from belonging, although it includes an overlapping appeal to the importance of future generations. Like **The Argument from Belonging**, it will apply even in the absence of a substantial human settlement elsewhere, just so long as extensive activities are conducted in nearby regions of space.

The Right to Knowledge Argument: Do we have a right to knowledge about our remote origins? Here, I mean the kind of knowledge that astrobiology might supply and that might then be cited in international declarations and treaties about space. The knowledge would include knowing whether or not the emergence of life on Earth was simply a chance occurrence, unlikely to have ever occurred elsewhere, or whether we emerged as the result of inbuilt tendencies of the Universe to produce living creatures. The idea that we have a right to such knowledge could work quite well with the two main lines of philosophical thought about rights: i.e., attempts to base them upon autonomous rational agency, and attempts to base them upon an appeal to strong interests. If we take the former route (the classic approach from the philosopher Immanuel Kant), then it will make sense to say that knowledge about our origins could help to shape our decision-making and the ways in which we exercise our autonomous rational agency. Knowledge of this sort could help to optimize conditions for autonomy. But if we appeal instead to rights as grounded in strong interests, in the manner of legal and environmental philosopher Joel Feinberg (1974) and some of the more contemporary animal rights theorists (Cochrane 2012; Garner 2013), then we will have an even more direct route to the idea of *a right to knowledge about our origins*. Narratives about such origins go all the way back and all the way across human cultures. We can even point to a continuity of concern between knowledge-containing Indigenous story-telling about origins and Charles Darwin's *On the Origin of Species* (1859), Soviet biochemist Alexander Oparin's "On the Origins of Life" (1924), and the pivotal discoveries of astrobiology about the origins, extent and future of life. Indigenous traditions of storytelling will count as important evidence for a strong and ongoing human interest in such knowledge.

This is an argument of a very different sort from the previous two. Although, like the others, it also presupposes that Indigenous knowledge is the real thing, i.e., knowledge and not some manner of quaint fantasy or shared delusion. But this is hardly a difficult assumption to support. Indigenous origin stories often connect humans and primates, a connection which we lost sight of in the West for several centuries, as well as connecting the origins of humans and other creatures, humans, and plant life (Bender 2019). Such stories can also connect the emergence of life with favorable conditions such as the presence of water and with some or other contribution from off-world (Kimmerer 2015). Storytelling about origins has always been an important way of knowing and of transmitting a sense of who we are.

The Irreducibly Dilemmatic Character of Space Ethics

Superficially, all three arguments appear to have a common center, i.e., Indigeneity. Upon closer examination they concern Indigeneity *among other things*. They direct our attention to several things in common, but also to differences. Two of them (**The Liberties Argument** and **The Argument from Belonging**) are focused beyond the Earth. The other one (**The Right to Knowledge Argument**) is focused upon terrestrial entitlements. The **Liberties Argument** presupposes human settlement elsewhere, the other two do not. **The Right to Knowledge Argument** appeals to Indigenous knowledge as evidence in support of humans acquiring scientific knowledge through astrobiology, the other two appeal to *ways of belonging* and to the importance of shaping such ways of belonging in the light of Indigenous knowledge about life in extreme environments. Many different sorts of things are going on.

We have yet to see that they pull us in different directions, but we can begin to see this when we bring them to bear upon immediate practical problems such as the pace at which we ought to expand our human activities into the nearby regions of space. Suppose that we ask, "How quickly should we settle elsewhere?" Depending upon which of the three arguments we use, our answer to this simple question will be very different. Each will pull us in different directions. **The Liberties Argument** is about democratization, and this usually pulls us toward a more rapid expansion into space in order to increase levels of inclusion. More rapid processes of expansion will tend to open up more opportunities for excluded groups. Reliance upon a radical redistribution of the existing limited opportunities may be a far less reliable strategy. Any rapid democratization of space is likely to require a comparably rapid expansion of our presence.

By contrast, **The Argument from Belonging** pulls us toward an extension of environmental considerations, and these generally point toward a higher level of caution and a more controlled pace of expansion into space. Fears about destructive activity will lead us toward greater constraint. Rather than going faster, we should go slower to protect what we might otherwise damage. By contrast with the other two, **The Right to Knowledge Argument** is entirely neutral on broader issues of actual human presence, just so long as our actions are consistent with strong forward protection to avoid contaminating sites from which we hope to learn something about where we come from, and how life has emerged. The three arguments pull us toward the views that we should go *faster, slower,* and *it doesn't matter.* They operate in very different ways even when we use them to think about such a simple and practical question. Nor should we expect them to revolve around some single shared approach. Yet each can be illuminating and may help us to arrive at some "all things considered judgement."

More generally, we should not expect appeals to Indigeneity to resolve into a decision procedure, even for issues of this relatively straightforward sort. Nor should we expect such appeals to help us find a new and stable center for space ethics, because space ethics does not have and does not need such a center, a balancing point around which everything else conveniently moves. But we may expect appeals to Indigeneity to deepen our understanding of the dilemmatic nature of ethics and of space ethics in particular (Milligan 2015). Within any practical space ethics, we will always be faced with legitimate and important considerations that pull us in very different directions. Its practicality may even depend upon an acceptance that any imagined center cannot be stable and will not hold.

Acknowledgments

This article is part of a project that has received funding from the European Research Council (ERC) under the European Union's Horizon 2020 research and innovation program (Grant agreement No. 856543).

Note

1. Editor's note: See, in particular, Chapters 3, 9, 14, and 16 of this volume for discussions of issues related to Indigeneity and space.

References

Arendt, Hannah. 1963. "The Conquest of Space and the Stature of Man." In *Between Past and Future: Eight Exercises in Political Thought*, edited by Jerome Kohn, 260–273. London: Penguin, 2006.

Bender, Mark (2019), *The Nuosu Book of Origins: A Creation Epic from Southwest China.* Seattle: University of Washington Press.

Cochrane, Alisdair. 2012. *Animal Rights Without Liberation: Applied Ethics and Human Obligations.* New York: Columbia University Press.

Cockell, Charles. S. 2015. "Freedom in a Box: Paradoxes in the Structure of Extraterrestrial Liberty." In *The Meaning of Liberty Beyond Earth*, edited by Charles Cockell, 47–68. Springer, Cham.

Dick, Steven. J. 2020. *Space, Time, and Aliens: Collected Works on Cosmos and Culture.* New York: Springer.

Elvis, Martin, Alanna Krolikowski, and Tony Milligan. 2022. "Space Resources: Physical Constraints, Policy Choices, and Ethical Considerations." In *Space Resources for Human Exploration*, edited by Volker Hessel, Hirdy Myamoto, and Ian Fisk, 369–388. Weinheim: Wiley-VCH.

Feinberg, Joel. 1974. "The Rights of Animals and Unborn Generations." In *Philosophy and Environmental Crisis*, edited by William Blackstone, 43–68. Athens, GA: University of Georgia Press.

Garner, Robert. 2013. *A Theory of Justice for Animals.* New York: Oxford University Press.

Kimmerer, Robin Wall. 2015. *Braiding Sweetgrass: Indigenous Wisdom, Scientific Knowledge and the Teachings of Plants*, Minneapolis, MN: Milkweed Editions.

Liu, Cixin. 2015. *The Three-Body Problem.* London: Head of Zeus.

Lupisella, Mark. 2020. *Cosmological Theories of Value: Science, Philosophy, and Meaning in Cosmic Evolution.* New York: Springer.

Milligan, Tony. 2015. *Nobody Owns the Moon: The Ethics of Space Exploitation.* Jefferson, NC: McFarland.

Nussbaum, Martha. C. 2001. *The Fragility of Goodness: Luck and Ethics in Greek Tragedy and Philosophy.* Cambridge: Cambridge University Press.

Passmore, John. 1974. *Man's Responsibility for Nature: Ecological Problems and Western Traditions.* New York: Scribner.

Schmitt, Carl. 1954. "Dialogue on New Space." In *Dialogues on Power and Space*, edited by Federico Finchelstein, 51–83. Cambridge, MA: Polity, 2015.

Taylor, Paul W. 1986. *Respect for Nature: A Theory of Environmental Ethics.* Princeton, NJ: Princeton University Press.

Williams, Bernard. 1993. *Shame and Necessity.* Berkeley, CA: University of California Press.

World Bank. 2021. "Indigenous Peoples." https://www.worldbank.org/en/topic/indigenous peoples.

16

On Loving Nonliving Stuff

Daniel Capper

The Value of Abiotic Entities

John Muir (1838–1914), the motive force behind the establishment of the National Park system in the United States, relished a special love for stones despite their lack of life. Without a doubt he also adored living things, too, from his treasured dog friend Stickeen to the sequoia trees that he compared to Jesus. Being fascinated by geology, though, Muir considered stones to be informative history books and excellent science teachers, and on this basis alone argued for the preservation of numerous rocky places. Further, and uniquely, Muir found that "rocks . . . are drenched with spiritual life—with God," so that Muir also considered stones to be teachers of religion (Muir 2013, 54). Of importance to this essay, through his attitudes toward the lifeless world Muir expressed a sense of being in a community with rocky beings, such as when he stated, "The very stones seem talkative, sympathetic, brotherly" (Muir 2003, 326). For Muir, a complete and fulfilled human life demanded a measure of respectful loving of nonliving stuff.

Muir's personal sense of kinship with lifeless entities was exceptional in his time, and today, perhaps, his respectful attitudes toward the abiotic universe still remain out of the mainstream. After all, most of us, whether we are religious or not, are taught by our cultures that life possesses greater value than nonlife. Across societies, humans commonly teach that nonliving things exist merely for our use, not as items of intrinsic value, and since people around us share such views, we do not question them.

As I have discussed in other writings, in the West such attitudes, which value life over nonlife, began with Biblical creation accounts such as Genesis 1:20–31, which granted humans, considered to be superior, hierarchical stewardship or dominion over the nonhuman natural world, including over nonliving things (Capper 2021). Biblical moments that lack the valuation of abiotic entities then fused with various versions of the Greek philosopher Aristotle's delineation of a Great Chain of Being, which explicitly places abiotic entities like

Daniel Capper, *On Loving Nonliving Stuff* In: *Reclaiming Space*. Edited by: James S. J. Schwartz, Linda Billings, and Erika Nesvold, Oxford University Press. © Oxford University Press 2023. DOI: 10.1093/oso/9780197604793.003.0016 water and minerals at the very bottom of a ladder of valuation. Thus, whether religious or not, based on this traditional cultural inheritance, Westerners commonly presume a lack of value in abiotic realities.

Likewise, although differing in manifestation from Euro Americans, Asian citizens typically value life over nonlife as taught through the scriptures of Hinduism, Buddhism, Daoism, and Confucianism (Capper 2021). Moreover, like Tibetan culture, First Nations cultures such as those of North America often contain brilliant expressions of great esteem for some nonliving things, such as Hoʻomana Hawaiʻi reverence for the mountain Mauna Kea or Lakota affection for the Black Hills. Yet, according to the scholar Howard Harrod in *The Animals Came Dancing*, one still finds general valuations of life over non-life in these moral universes (Harrod 2000, 86-87, 111–112). This valuation of life over nonlife in fact extends beyond Earth, since according to an editor of this volume, James S. J. Schwartz, a similar "life bias" in the literature of space exploration both ingrains and expresses valuations of life over nonlife as con-temporary humans travel through our solar system (Schwartz 2020, 124).

Because of the ways that many of us have trained in life-biased worldviews since we were infants, a problem here is that nineteenth century Muir, with his love for nonlife, was more attuned to the ecological realities of the twenty-first century than many of us are. Muir's perspective grasped that since human bodies consist significantly of nonliving elements like water and minerals, we cannot be appropriately grateful for our individual existences without appre-ciating how we live on the shoulders of nonliving things. Further, we cannot properly battle climate change without thoroughly assessing and recognizing the values of nonliving things like the ethically right amount of atmospheric gas elements. We also cannot responsibly sequester atmospheric carbon into rocks until we have appropriately valued those rocks. Hence, we cannot fully appreciate our own bodies or adequately manage climate change without finding a place within us to value things that do not live.

In addition, as humans increasingly leave Earth, even if just via robot, we find that our traditional lack of valuation of nonlife is glaringly obsolete. The only life that we know of at the time of this writing exists on Earth, meaning that we cannot be responsible environmental citizens of our solar system without developing respect for its myriad nonliving realities and entertaining ecologically healthy concern for them in their own rights. If we are to leave our home planet yet avoid being terrible solar system neighbors, we need to reassess our capacity to love nonliving stuff.

Later in this essay, Tibetans, with their terrific regard for the high mountains among them, offer us capable tools for this reassessment. These Himalayan residents teach us how to effectively engage our abilities to revere abiotic ecologies by enthusiastically extending respect to lifeless members through reciprocal actions within an ecological community. This ecological community remains somewhat nebulous, however, unless we first identify and appreciate some of our nonliving solar system group members, which we will do now.

Nonlife in Our Cosmic Neighborhood

When it comes to fondness for extraterrestrial nonlife, take, for instance, the fantastic rings of Saturn. First seen through Galileo's telescope, Saturn's diaphanous rings instantly define the planet for many onlookers because of the rings' majestic nature as well as their incomparability to anything else we know. Remarkably thin yet thousands of kilometers wide, Saturn's eye-catching rings consist of particles of dust, rock, and ice, some of which are as small as a grain of sand while others are as large as a city bus. These particles are shepherded in their orbits around the planet by the presence of numerous moons both within and beyond the rings, with these moons giving each ring its own character while also lending consistency and durability to the ring structure as a whole (Porco and Aulicino 2007, 336). Although Jupiter, Uranus, and Neptune have small rings of their own, the boldly ethereal rings of Saturn particularly stand apart in grandeur and rarity from any other feature of our solar system, and thus deserve our admiration. In the words of philosopher Tony Milligan, the ring structure of Saturn possesses a "special standing, a standing which would make it worth conserving, and not simply for the purposes of scientific enquiry" (Milligan 2015, 21). Because of this special standing, and since the rings of Saturn are nonliving, if we are to care for Saturn properly, we must learn to respect, perhaps even feel affection for, and maybe even love, nonliving entities.

At this point my reader may think that Saturn is difficult to reach and our current hardware is limited in destructiveness toward planetary rings, so who cares about trying to protect Saturn's rings? I offer that we all should care so that our ethics can be clear in formulation and understanding. More vividly, though, there are threatened nonliving places much closer to home that need our attention right now, such as the Malapert Massif. Appearing about 100 km from our Moon's south pole, the collection of peaks called Malapert rises more than 8,000 meters above a sunken crater floor, thus rivalling the Andes Mountains in altitude (Van Susante 2003, 2480). This altitude makes a difference. First, we treasure the Andes Mountains for their own awe-inspiring sakes, and the Malapert Massif in itself should make a similar claim on our esteem. Further, the landscape views from Malapert, stretching 360 degrees over many kilometers, must be absolutely marvelous, making Malapert worth respecting as a place for future Moon science as well as for lunar recreation. In two centuries, humans may enjoy leisure camping on the Moon, and Malapert supplies one of the most dynamic of lunar locations for that activity.

Additionally, altitude engenders near-eternal sunlight on Malapert's slopes. Lacking the season-bringing obliquity, or axial tilt, of Earth, our moon rotates almost vertically relative to our sun. Being a high point that is nearly on top of the Moon's vertical axis, the Malapert region therefore may be sunlit 85 percent of the time, making it a place unlike many others in our galactic neighborhood (Sharpe and Schrunk 2003, 2467). Also, and enchantingly, our Moon's lack of tilt creates another notable Malapert ecological feature: eternally dark craters. Since the sun reaches some deep craters only sideways from the horizon, some crater bottoms have never witnessed sunlight, making them some of the coldest places in our solar system. Hence, just as the often-sunny Malapert slopes offer unbelievable landscape views, these dark craters, being free from a variety of forms of interference, present sky views of our solar system and the center of our galaxy that in their intensity and clarity will incite wonder and amazement even among the most jaded of human beings.

Although sunny high spots and dark craters can be found in various locations at both lunar poles, the especially tall and bright Malapert slopes that are admixed with eternally dark craters make Malapert a special place not just on our moon but also in terms of the rest of our solar system. In its unmatched magnificence and ecological richness the Malapert region remains worthy of our admiration and fond regard. Yet, even as I write, Malapert faces several possible near-term environmental perils. Already, plans and hardware are being developed to place solar power collectors on the sunny highlands of Malapert (Sharpe and Schrunk 2003, 2467). Because of the altitude of the peaks, other people seek to place equipment on Malapert for line-of-sight communication transmissions with the far side of the Moon. Other groups wish to use dark craters for telescope emplacements (Schrunk et al. 2008, 54, 427). Moreover, commercial firms want to employ Malapert as a base for mining water ice from the dark craters and perhaps acquiring the energy source helium-3. For these industrializing and commercializing reasons, international governmental and private ventures like the European Space Agency's Moon Village initiative have already sent or soon will send missions to our Moon's south pole.

In this essay, I lack the space critically to assess, as others have done (Elvis, Krolikowski, and Milligan 2021), these individual proposals for the use of Malapert. Instead, regardless of how the various strategies that I just mentioned are implemented, I focus more foundationally on our prior attitudes. I suggest that before we begin discussions about whether there should be telescopes, communication towers, and the like on Malapert, we must develop the appropriate esteem for that location. Malapert, properly understood, stands apart as a place of ecological value and splendor in its own right, and thereby deserves our caring respect. Because of its terrific scientific and recreational views, not just across the Moon's surface but also of the vibrant lunar sky, perhaps Malapert in fact deserves our ecological affection. Even more, maybe the Malapert Massif should receive our environmental love for its extraordinary characteristics, even if it is a place without life. The wonders of Malapert should encourage us all to love non-living stuff.

Of course, talking about the need to revere the rings of Saturn and the Malapert Massif leaves us with the asteroid in the room: if our cultures are not helpfully guiding us as we journey beyond Earth, how do we learn to love nonliving stuff? Since I have been discussing the high mountains of Malapert, I will answer this question by turning to Tibetans, who are some of the greatest mountain experts on Earth. As we shall see, traditional Tibetan regard and love for mountains spotlights vividly the ways that we all, regardless of our personal beliefs in this moment, can positively revere and interact with treasured abiotic realities through reciprocity within an ecological community. Now let us visit the Himalayas in order to perceive these lessons more vibrantly.

Tibetan Love of Mountains

Tibetans can teach us how to love abiotic things because of the ways that they embody respectful reciprocal relationships with members of their community, which in their perception includes mountains. Rather than solely focus on what is living and what is not, Tibetans, by culture, instead consider the environmental existents around them, including mountains and rivers, to be a part of an interactive social group with them. Keeping practical and ethical order in social groups, as we all know, involves the extension of respect and affection to others in recognition of the reciprocal webs of dependence that bind communities together. By extending such admiration to cherished mountains, Tibetans thereby cultivate robust bonds of love with entities that do not live yet still represent vital members of a thriving community.

This Tibetan cultural love for mountains grows from the religiosity that remains a distinctive and visible element of Himalayan cultures and that arises from the inextricable blending of Buddhism with Indigenous Tibetan spirituality which the Tibetologist Geoffrey Samuel classically portrayed (Samuel 1995). Expressions of this Tibetan fondness for peaks varies from village to village, so here I offer a description of a typical situation that can be diverse as lived. In advance I urge my reader to ponder the cognitive and motivational meanings of these Tibetan attitudes of reciprocity in community without concern for exactly copying Tibetan beliefs and practices.

Regardless of location, Tibetan love for lifeless mountains begins with a deep regard for one's *yul lha*, or local deity. The local deity usually is synonymous, in a practical sense, with a nearby high mountain. In considering gods to be mountains, the typical Tibetan·cultural perception is not that the mountain is personified as a god, but rather that the mountain is the form of manifestation of the god, like the clothing that the god wears. Within this outward divine raiment, the summit of a tall peak provides the residence area for the deity of the mountain and thus is voluntarily forbidden to human presences (Huber 1999, 219), a point to which I will return.

Appearing as a mountain, this deity controls prosperity and misfortune within a region like a valley. One can appreciate that there is a scientific dimension to such beliefs, since high Himalayan peaks literally dominate the weather nearby. More broadly than weather, though, one's life depends upon this local deity, so remaining on good terms with this divine being is paramount. Keeping a beneficial relationship is based upon the notion of respectful reciprocity within a community (Karmay 1998, 426). A couple who is engaged to be married, for instance, will announce their marriage to the local mountain deity, who as a member of the community deserves to know about the union. The couple may then lovingly venerate the local deity so that the god may assist them in having a child. The child later may break her leg, with such a fate perhaps resulting from the mountain deity's punishment of a misdeed. The child's parents nonetheless may continuously provide gifts to the local deity to bring the prosperity needed to heal their daughter today as well as send her to university tomorrow. Obviously, outcomes can be multiplied beyond these. The point is that humans and mountains live in a community together, so that caring for one member means, in a reciprocal way, caring for the other as well (Gagné 2020).

This "ethics of care" with nonliving mountains, as anthropologist Karine Gagné states, institutes a "sense of obligation and responsibility toward the more-than-human" (Gagné 2018, 7). But of course reciprocity works in two directions, since, for example, Himalayan residents may perceive the higher temperatures of climate change as symptomatic of a decline in suitable reciprocal respect of humans for their mountain companions (Gagné 2018, 6). Because, as Gagné says, "landscapes are repositories for morality that constitute and animate ethical dispositions," the realities of climate change in this case demonstrate to some Himalayan residents that humans themselves benefit along with nonhuman environments when humans respectfully recognize their ecological relationships and dependencies with things that lack life (Gagné 2018, 12).

In a practical sense, Tibetans may exhibit their reciprocal regard for the mountains in their communities in numerous manners that start with personal behavior. As I mentioned, one's local deity does not tolerate misbehavior, yet rewards proper acts, so that how one ethically behaves in ordinary life may influence the valence of one's relationship with the mountain deity (Obadia 2008, 120). Extending this notion in terms of community, villages may host important festivals designed to honor their local deity. Another traditional way of magnifying dynamics in the relationship involves making multiday pilgrimages (*nekor*) to holy mountains (Kapstein 2006, 237, 243). During these pilgrimages Tibetans often will camp at the base of a peak while engaging in admiring prayer. Out of reverence, they simultaneously may place a stone on or affix special flags to the mountain deity's *labtse*, or sacred stone cairn. Typically, pilgrims additionally will engage in *korra*, in which they circumambulate the mountain while profoundly reflecting on the holy importance of the site. Of interest, on pilgrimage Tibetans usually avoid the very summits of mountains despite the abilities of many Himalayan mountain climbers to get there. If the summit is the best place on a mountain, Tibetans kindly renounce the summit in order to visibly respect the mountain as special (Karmay 1998, 433). One more related and common expression of Tibetan mountain love involves the burning of *sang*, or incense, perhaps as large juniper branches that are placed in communal fires near, but not at, the summit.

Burning juniper on the open slopes of Malapert's peaks will not be possible because our Moon has no oxygen atmosphere. This is no problem, for I have not described Tibetan beliefs and practices of fondness for nonliving mountains so that they may be strictly emulated. Instead, Tibetan adoration of mountains highlights that we can respect abiotic beings in ways in which we recognize our relationships with those lifeless entities in our mutual community of existents. On this basis, we can extend appreciative and grateful behavior even to ecologies that are not alive. Now it is time to explore this environmental insight in greater extraterrestrial detail.

Affectionate Respect in Community

Despite deriving from specific cultural locations in the Himalayas, this example of Tibetan mountain affection can teach all of us about the importance of respectful relationships of reciprocity within a community that includes both abiotic and biotic entities. In their interactions with mountains, Tibetans do not begin with a discernment between what is living and what is not. Tibetans begin by recognizing community instead. From their point of view, the community does not end with the humans in the group, and their community in fact includes abiotic snow-capped peaks. Being community members, lifeless and living entities affect each other, and these interactions within the community are governed by reciprocity, respectful recognition, and gratitude. Tibetan humans thus show thanks and exhibit goodwill for the benefits and effects that they feel that they receive from the high mountains around them. Mountains, in turn, receive beneficial ministrations from humans within this cultural code. Rather than being treated as meaningless, here nonliving beings like mountains are intentionally treated positively like other persons in the community. In this way Tibetans come to respect, also feel affection for, and quite often love things that are not alive.

Our relationships with nonliving entities in space can follow similar positive tracks regardless of our culture. As anthropologist M. Jane Young has stated, local forms of religion like those explored here "can offer us alternative ways of seeing ourselves in relationship to the natural world and help us to answer the question of what constitutes appropriate behavior--in outer space as well as on Earth" (Young 1987, 269–279). In this case, the lesson is to set aside quick judgments based on a living/not-living binary and instead recognize the need for respectful reciprocity in community. Many of us treat nonhuman pets as parts of our communities; what is needed here is an extension of the idea to include realities in our midst that are not alive.

For instance, the rings of Saturn exist in a solar system community with us, if we will just look at things this way. And those of us who are amazed by the opulence of Saturn's rings have plenty for which to be grateful. In turn, those lovely rings may need to be protected at least from us. Therefore, why not recognize our experience of constructive reciprocity with Saturn's rings by extending caring esteem for them? Even if this inner attitude does not affect Saturn's rings today in a physical sense, it is morally the right thing to do and a good way to train our minds properly to value abiotic realities.

Moreover, future campers and astronomers can have many reasons for gratitude as they enjoy the terrific vistas and overwhelming skies of our Moon's mountains at Malapert. In turn, Malapert, as a member of our solar system community, needs our wise and circumspect treatment today, before the industrialization of the Moon begins without an appropriate environmental plan.[1] We can enact our reciprocal duties toward Malapert on the example of Tibetans, since the rest of us have the same capacities as Tibetans to genuinely respect abiotic things. We all can grow morally by learning from Himalayan residents who conserve nature while preserving positive human relationships with landscapes. Given that a marvel like Malapert is available to amaze us, why not show our thanks by adoring Malapert's abiotic highlands and craters like respectful neighbors should do? We should recognize lunar mountains and their future campers as a part of our community and, for their sakes, caringly honor Malapert's wonderful slopes. So that we can be more complete human beings in our own rights, like John Muir exemplified, as well as solid citizens of our solar system community, let us immediately energize our abilities to love nonliving stuff throughout our captivating cosmos.

Note

1. See Chapter 23 for William R. Kramer's discussion of the importance of environmental planning.

References

Capper, Daniel. 2021. "How Venus Became Cool: Social and Moral Dimensions of Biosignature Science." *Zygon: Journal of Religion and Science* 56, no. 3 (2021): 666–677. https://onlinelibrary.wiley.com/doi/10.1111/zygo.12703.

Elvis, Martin, Alanna Krolikowski, and Tony Milligan. "Concentrated Lunar Resources: Imminent Implications for Governance and Justice." *Philosophical Transactions of the Royal Society A* 379 (2021): 20190563. http://dx.doi.org/10.1098/rsta.2019.0563.

Gagné, Karine. 2018. *Caring for Glaciers: Land, Animals, and Humanity in the Himalayas.* Seattle: University of Washington Press.

Gagné, Karine. 2020. "The Materiality of Ethics: Perspectives on Water and Reciprocity in a Himalayan Anthropocene." *Wiley Interdisciplinary Reviews: Water* 7: e1444. https://wires.onlinelibrary.wiley.com/doi/10.1002/wat2.1444.

Harrod, Howard L. 2000. *The Animals Came Dancing: Native American Sacred Ecology and Animal Kinship.* Tucson: University of Arizona Press.

Huber, Toni. 1999. *The Cult of Pure Crystal Mountain: Popular Pilgrimage and Visionary Landscape in Southeast Tibet.* New York: Oxford University Press.

Kapstein, Matthew T. 2006. *The Tibetans.* Malden: Blackwell.

Karmay, Samten G. 1998. *The Arrow and the Spindle, vol. 1.* Kathmandu: Mandala Books, 1998.

Milligan, Tony. 2015. *Nobody Owns the Moon: The Ethics of Space Exploitation.* Jefferson: McFarland and Company.

Muir, John. 2003. *My First Summer in the Sierras.* New York: Random House.

Muir, John. 2013. *John Muir: Spiritual Writings*, ed. Tim Flinders. Maryknoll, NY: Orbis Books.

Obadia, Lionel. 2008. "The Conflicting Relationships of Sherpa to Nature: Indigenous or Western Ecology." *Journal for the Study of Religion, Nature, and Culture* 2, no. 1: 116–134. https://doi.org/10.1558/jsrnc.v2i1.116.

Porco, Carolyn C. and Barbara Aulicino. 2007. "Cassini: The First One Thousand Days." *American Scientist* 95, no. 4: 334–341. https://www.jstor.org/stable/27858995.

Samuel, Geoffrey. 1995. *Civilized Shamans: Buddhism in Tibetan Societies*. Washington, D.C.: Smithsonian.

Schrunk, David G., Burton L. Sharpe, Bonnie L. Cooper, and Madhu Thangavelu. 2008. *The Moon: Resources, Future Development, and Settlement*. Chichester: Praxis Publishing.

Schwartz, James S. J. 2020. *The Value of Science in Space Exploration*. New York: Oxford University Press.

Sharpe, Burton L. and David G. Schrunk. 2003. "Malapert Mountain: Gateway to the Moon." *Advances in Space Research* 31, 11: 2467–2472. https://doi.org/10.1016/S0273-1177(03)00535-0.

Van Susante, P. J. 2003. "Study towards Construction and Operations of Large Lunar Telescopes." *Advances in Space Research* 31, no. 11: 2479–2484. https://doi.org/10.1016/S0273-1177(03)00563-5.

Young, M. Jane. 1987. "'Pity the Indians of Outer Space': Native American Views of the Space Program." *Western Folklore* 46, no. 4: 269–279. https://www.jstor.org/stable/1499889.

17

Reclaiming Space

On Hope in a Jar, a Bear in the Sky, and the Running Red Queen

Kathryn Denning

Prelude: The Red Queen's Advice, and Reclaiming Space

> *"Now, here, you see, it takes all the running you can do, to keep in the same place.*
>
> *If you want to get somewhere else, you must run at least twice as fast as that!"*
>
> —**Spoken by the Red Queen to Alice, on a life-sized chess board in Wonderland**[1]

I first heard of "The Red Queen Hypothesis" in the context of evolutionary biology (Van Valen 1973), rather than directly from the hallucinatory book that inspired the expression: *Alice in Wonderland/Through the Looking Glass,* by the weirdly inventive and controversial Lewis Carroll. It's stuck with me for years, this idea of constantly running as fast as possible just to stay in the evolutionary game, in a particularly futile version of an arms race.

When I began to think about "reclaiming space," this idea danced in my head. The Red Queen's pronouncement captures something important about the time we're in, and about the challenges before us. Technological change is accelerating faster than the corresponding governance frameworks. Earth's climate is changing faster than we are adapting. And space development is moving much faster than most people know.

Contributors to this volume consider the subject of "reclaiming space" in many different ways, all of them valuable. Personally, in the near term, I do not support an aggressive timeline for human expansion beyond Earth; rather, I am strongly in favor of robotic exploration and robust scientific research.

Kathryn Denning, *Reclaiming Space* In: *Reclaiming Space.* Edited by: James S. J. Schwartz, Linda Billings, and Erika Nesvold, Oxford University Press. © Oxford University Press 2023. DOI: 10.1093/oso/9780197604793.003.0017

Therefore, for me, "reclaiming space" is a set of different, overlapping tasks, on timescales ranging from a moment in time to centuries. It means pushing for multilateralism and peace wherever possible—and working toward global space governance that is more proactive than reactive, takes multiple stakeholders into account, and manages solar system resources and places in equitable and cautious ways. It means emphasizing rational planning instead of wishful thinking, including many expertises, learning from history, and encouraging citizens everywhere to become more informed about space issues that affect our shared future, so they can truly participate in dialogue and planning. It means listening to hopes and fears about space, being realistic, and honoring old stories while finding new ones to tell. It means acknowledging our collective grief and apprehension about Earth's deteriorating climate and sky, bearing witness to the losses, and finding the best ways to channel that energy. And it means doing all this as fast as we can, and for a long time, because "reclaiming space" will be a multigenerational marathon.

One View in ~8 Billion

Anthropologists often start work by describing ourselves and from where, and how, we see the world. We know we don't have perfect objectivity, and that our knowledge is influenced by our social, cultural, and historical contexts. By sharing a little about who we are and how we see, we hope to encourage our readers to engage with different ways of seeing the world.

So: I was born nine months after men first walked on the Moon, and composed my first-year university essays on a typewriter, which was normal then. I've now been involved in outer-space-related work for nearly two decades, which still surprises me, since that was definitely not my plan in 1999 when I finished my PhD on archaeological theory and public science. I reckon it is ultimately Carl Sagan's fault. Currently, I am torn between amusement that some apparently identify me as a "thought leader" on some space matters—in a recent invitation's words, not mine—and feeling a weighty responsibility to try to fill that role well. I try to support diverse and younger voices in space discussions, an endeavor aided by receiving tenure over a decade ago, with its blessings of job security and intellectual independence. Finally, I am a female, white, first-generation Canadian of Welsh heritage, working in Canada at a large, highly multicultural university.

My views on space are profoundly shaped by my political and historical situation. I live as a settler in a country grappling with the truth of its oppression of Indigenous peoples and the difficult processes of Reconciliation, and as a descendant of people an ocean away who sometimes experienced cultural and economic subjugation in their own homeland, and as a member of an academic community which includes many newcomers to Canada. So I think often about how history can put anyone on either the lucky or unlucky side of power and place, and how colonial, extractive, wealth-building enterprises can damage and destroy worlds in ways that reverberate for centuries.

This resource-rich "new world" of North America has much for which I am grateful. But I often wish: that the trees in my Ontario home's landscape had not all been cut 150–200 years ago for the purposes of displacing people, growing crops for export, and building British ships; that the abundant oil reserves in the ground had gone unnoticed; and that tall grey smokestacks did not today tower over the First Nations community next door. I wish that the 400-year-old Catholic mission that I helped excavate thirty years ago as a student had not culminated in the unmarked graves of Indigenous children at twentieth-century residential schools and so much devastating grief today. I am acutely aware that those who did all this—who forced people of the First Nations from their land, obtained timber rights, planted wheat, loaded ships, drilled the earth, built petroleum processing facilities, and preached from an ancient book—often believed, fervently and confidently, that this was progress and the best path for the future. But so often they were wrong, generally they were short-sighted, and frequently their actions had unintended negative consequences which are painfully clear today.

Given what I know of human beings after several decades of study, I cannot suppose that we are now inherently wiser or better than our predecessors, or that our foresight skills are superior. Rather, I believe that we must work hard and continually to plan our shared futures better, while learning from the past and attempting damage control in the present. And that is why I think we must *also* look to outer space even while we plant trees, convert to renewable energy, reduce plastics, press for environmental justice, absorb Truth and Reconciliation Commission reports,[2] listen carefully to those who know how, and whom, government systems fail, and learn from those whose traditions suggest better ways forward.

On that subject of damage control and mitigation and loss, archaeologists and anthropologists like me know it well. Many of us try to reduce or at least document damage and loss to human cultures, health, heritage, languages, and homelands, caused by colonization and its sequelae. Sometimes we can advocate for human rights and equity or help with cultural preservation and revival. Sometimes we can only be witnesses to what Wade Davis called the "waterfall of destruction unprecedented in the history of our species." As he observed, "In our lifetime half of the voices of humanity are being silenced" (2009, 166). Which voices, then, will be included in a space-age human "we"?

And then there are the other areas of anthropology, including our closest relatives, the nonhuman primates. Most primate species will likely be extinct in the wild soon (Estrada et al. 2017). But many individuals continue to live in labs for experimental purposes—including for space-related research—and need our continual advocacy. And as for the matter of archaeology: our ancient human ancestors are dead already, but that doesn't mean they're beyond harm, or that they're socially inert. Teaching archaeology today is also a constant challenge to equanimity.

All those factors and realizations have inevitably shaped me and how I see humanity's prospects in space. Many academics (especially social scientists, humanists, and biologists) share some of these perspectives. You, dear reader, may think similarly or differently, but that's the point—our views about space may resonate or diverge, but we both have something to contribute anyway. We need to think *together* about space as citizens of the solar system, like we do about our challenges of coexistence, stewardship, and future-building on Earth.

For me, simple optimism about space isn't an option, and neither is complacency. Humans have incredible potential for technological innovation, but as that capacity grows, so does our need for social systems and knowledge to steer and limit our technologies in socially responsible, fair, and sustainable ways (Farmer et al. 2020). Space development holds the potential for unimaginable warfare and destruction, and for disparities in wealth and power which would break today's graphs. It also holds the possibility for global cooperation, responsible resource development, scientific discovery, and benefits to Earth's biosphere. It's all in one package. One box. One jar.

And the lid is off now. But it wasn't always that way.

". . . if they are written in the stars they can be read and remembered forever"—First Woman

There is an astonishingly beautiful traditional Diné concept that the stars were placed in patterns for a reason by the first beings on Earth, who were also the progenitors of humans. In this story, the stars and stardust were made of quartz, chipped into small pieces by the flint knives of First Woman and First Man and their compatriots. When they had the pile of stars ready,

First Woman said: "I will use these to write the laws that are to govern mankind for all time. These laws cannot be written on the water as that is always changing its form, nor can they be written in the sand as the wind would soon erase them, but if they are written in the stars they can be read and remembered forever." (retold in Newcomb 1990, 83)

If you looked up and knew that the stars were placed in the sky long ago to tell you something, would it change the way you see the sky—or how you live your life? In the Diné tradition, the stories associated with the constellations "provide moral guidance, reminding the Earth Surface People to adhere to the values essential to the establishment and maintenance of harmony in their lives and in the universe" (Griffin-Pierce 1992, 142).

This is one profound tradition of a multitude.[3] For many people, the stars are vitally important cultural mnemonics, as well as sacred. There are incredible variations, but also amazing commonalities like The Seven Sisters story about the Pleiades, which may indicate very ancient shared mythologies, perhaps even 100,000 years old (Norris 2020; Norris and Norris 2021). Likewise, in many cultural traditions worldwide, the Milky Way is a path for the souls of the dead on their journey to the Otherworld or land of the ancestors (Lebeuf 1996; Hall 1997). One Ojibwe expression for death means "no more light, the light moves on"; the spirit is referred to as starlight (Lee et al. 2014, 28). The Milky Way is also associated with birth and fertility, i.e., literally the milk of mother goddesses (Lebeuf 1996). And the Moon has myriad significances, which too many of us have long forgotten.

No wonder that some Indigenous people and traditional knowledge-keepers report a deep sense of loss due to the visual interference in the night sky from proliferating satellites. Layered as this is upon myriad other losses, past and present, sky-related and Earth-bound (Mortillaro 2019; Taylor 2019), the concern is great enough that even during a global pandemic, Indigenous representatives joined astronomers and allies to respond urgently to SpaceX's Starlink satellite constellations (Venkatesan and Lowenthal 2021; also, Venkatesan et al. 2020; Ferreira 2021; Lawrence 2021).

Indigenous people are not alone in grieving a deteriorating environment and sky. They often have more and longer experience with such losses—and also more knowledge of survivance—but more and more people from settler-colonial backgrounds now feel it too. The sky some of us took for granted as eternal, isn't. And the future is coming so fast.

Climate Change, Ecological Grief, and Space

In Sept 2021, UN Secretary-General António Guterres said to the General Assembly and the world:

> I am here to sound the alarm: the world must wake up. We are on the edge of an abyss—and moving in the wrong direction. Our world has never been more threatened. Or more divided. We face the greatest cascade of crises in our lifetimes. (Guterres 2021).

That sums up the situation effectively, as does the Doomsday Clock's position at 100 seconds to midnight (Mecklin et al. 2022). There is much cause for distress.

As I alluded to earlier, many researchers and advocates with social and environmental expertises suffer an emotional toll because of the overwhelming nature of what they study, and the immensity of the problems they are trying to fix. I've seen this in many social scientists and biologists. But recently, I've also started to see it in conversations about outer space, with experts, students, journalists, filmmakers, and members of the public. The concept of "ecological grief" seems to approximately match, i.e., "mourning of the loss of ecosystems, landscapes, species and ways of life" (Comtesse et al. 2021, 1). One form is known as "solastalgia," or "the distress caused by the transformation and degradation of one's home environment" (Galway et al. 2019, 1). We may need a new word that applies to the inner solar system and our attachments to it.

Seventy years ago, Aldo Leopold famously wrote that "One of the penalties of an ecological education is to live alone in a world of wounds"—that is, a mind trained in certain ways sees, and grieves over, the problems early (Ellis and Cunsolo 2018). Today, ecologists certainly have more company in their distress. In fact, this is now deeply serious. Young people worldwide are highly affected: many feel that governments are failing the younger generation through inaction on climate (Thompson 2021), a sentiment which is entirely commensurate with the intergenerational inequity forecast for the decades ahead (Thiery et al. 2021).

We need social research on how this may connect to space. But I can share my own observation, that in my recent space-related conversations, many of my interlocutors of all ages often express a loss of some kind. Sometimes it's the fading hope that astrobiologists could thoroughly explore Mars robotically before private human missions land, sometimes it's the memory of darker skies, sometimes an acute concern about orbital debris or the militarization of space, or potential appropriation of lunar territory, or the disturbing realization that space isn't as fairly governed as it should be.[4] I regularly encounter others' weariness, sadness, anger, frustration, and disbelief at the pace and extent of the changes and the feeling of powerlessness, and the strain of trying to keep up. Perhaps the Red Queen was right.

The ecological grief literature shows some ways forward and is worth exploring, since the strains are likely to worsen. For now, I will suggest simply that we should be careful about where this grief and fear and exhaustion lead us, and mindful about the stories we hear and tell about how things will get better.

Life, Death, Earth, Space, and the Return of Captain Kirk

When William Shatner returned from his brief voyage past the Karman Line on October 13, 2021, he clambered out the capsule door that Jeff Bezos opened, and emoted tearfully about Earth being the realm of life and space being the realm of death. Shortly after, Shatner spoke to the cameras about the climate crisis, and of being "overwhelmed by sadness and empathy for this beautiful thing we call Earth . . . Burying your head another instant about global warming and the destruction of the planet is suicide" (2021a). On one of many TV shows in two days, Shatner emphasized "the fragility of this planet, the coming catastrophic events, and we all have to clean this act up NOW" (2021b).

Although I haven't been to space, I agree with Shatner on those points, and that shortly before the 2021 United Nations Climate Change Conference (COP26) was an excellent time to emphasize them. Equally obviously, comments such as these from astronauts/cosmonauts are nothing new: Yuri Gagarin famously said in 1961, "Orbiting Earth in the spaceship, I saw how beautiful our planet is. People, let us preserve and increase this beauty, not destroy it!" (in Launius 2019, 195). However, the phenomenon of the *nonastronaut* celebrity spacefarer *is* quite novel, and bears watching in a social media environment saturated with celebrities and algorithms that amplify them. Given the magnitude of the climate crisis, remarks from future space tourists could frequently relate to climate explicitly.

And there's an important twist: because these high-fliers are strongly self-selected believers in the transformative potential of space, and are also guests (paying or otherwise) of a private space corporation, they may magnify claims like those made by Bezos and compatriots, i.e., that private space initiatives will solve problems because we can move all heavy industry off planet, stop mining on Earth, etc. At least, that sequence played out after Shatner's voyage. In response to Prince William's much-publicized remark that "We need some of the world's greatest brains and minds fixed on trying to repair this planet, not trying to find the next place to go and live" (in Bowdon 2021), Shatner responded with the Bezos position that

> this is a baby step into the idea of getting industry up there, so that all those polluting industries, especially, for example, the industries that make electricity... off of Earth . . . We've got all the technology, the rockets, to send the things up there . . . You can build a base 250, 280 miles above the Earth and send that power down here, and they catch it, and they then use it. (Gawley 2021)

Of course, one reaction to environmental stress and grief is to look upwards for salvation. But this vividly shows how today's deep, authentic concern about climate change and environmental destruction can be directly co-opted into boosterism for private space industry and its long-term goals. We should be attentive to those conflations, for multiple reasons. First, even if some fliers really do have profound personal epiphanies, those don't necessarily help other people or our planet (Deudney 2020). Second, space tourism flights and much larger eventual goals are connected in space corporations' business plans, but we shouldn't kid ourselves that those business plans are fundamentally philanthropic in nature, that the "baby steps" are *only* in service of goals we'd support—or that the space barons know what's best for the rest of us. Meanwhile, launch emissions are inadequately regulated and have a real environmental cost, a subject of growing concern among scientists (Ross and Vedda 2018; Marais 2021; Mann 2010; Gammon 2021; Dallas et al. 2020). Third, some scenarios Bezos proposes are highly contestable: some analysts assert that these would "add to the environmental pressure on Earth without adding to the quality of human life on Earth" (Timperley 2021; see also Pultzarova 2021). Fourth, space resources are often suggested to be a way to *increase* our various consumptions, not make them more sustainable: the logic of capitalism and economic growth generally dictates that new niches will be exploited *in addition to*, not *instead of*, old ones.

The disproportionate media attention given to celebrities and billionaires is another reason for concern. Bezos' July 2021 brief suborbital trip garnered nearly as much attention on American broadcast morning television (e.g., *Good Morning America*) as the climate crisis did *during the entire year of 2020* (Cooper, in Timperley 2021). Similarly, a climate scientist's scheduled on-air interview during a killer heatwave was bumped by Richard Branson's first brief space flight: as reporter Molly Taft (2021) said, "networks by and large chose to do PR for a billionaire rather than cover the most pressing story of our time."

In fairness, both Musk and Bezos have recently announced significant funding for environmental initiatives, and perhaps some good will come of that. More generally, environmental concern and space activities are certainly not inherently at odds. Space-based tools are invaluable to climate science, for example, and can serve sustainable development well. Similarly, some technologies that hold great promise for space could also benefit Earth. But space activities do vary *enormously* in their cost/benefit ratios, environmental impacts, justifications, timelines, and distribution of risks and rewards . . . and actual goals.

So we're left with the question: how can we do better than to channel environmental concern into unquestioning support for the space agendas of the hyperwealthy? What might better public awareness or more citizen involvement achieve? There are many ways forward, but all require endurance. . . just like on Earth. So, what will keep us going?

So, What of Hope?

Even hope isn't simple anymore. But perhaps it never really was.

For example, the tale of Pandora's box has not become clearer over time. One wonders: How did a five-foot-tall clay jar become a dainty locked box? Why was Hope bottled up with nasty things like labor, sickness, vice, madness, and death? Is "Elpis" (the Greek) best translated as "hope," or should it be "expectation"? Why did Zeus compel Pandora to slam the lid shut and leave Hope in the jar? When Hope/Elpis was the one thing left trapped in the jar, available to humans on demand, was that meant to be a blessing, or a particularly twisted curse?[5]

Hmmm. Meanwhile, back in the here and now, Elon Musk was right about something:

> those who attack space
> maybe don't realize that
> space represents hope
> for so many people

—Elon Musk on Twitter, 11:05 pm, July 12, 2021

Yes indeed. But that's another jar with weird contents, since a startling amount of that space-hope is factually muddled. Many people encounter more space fiction or propaganda than facts. So it's not surprising that much space-hope is under-informed about important things like: the lethal tendencies of Mars or the need for much more astrobiology research there; well-founded doubts about humans' ability to reproduce beyond Earth; the improbability of crewed interstellar flight; the fact that generally, "habitable" exoplanets aren't habitable for *us*; the immense challenges for governance of anything and everything in space; the contested legality of asteroid mining; the nigh-inevitable unequal distribution of profits from space resources; the risks of space militarization; and, to cap it all off, what Daniel Deudney (2020) describes as the "catastrophic and existential threats" to humanity on Earth that could be *caused* (not avoided) by expansion into the solar system.[6] And there's the question of *hope for what*, and *hope for whom*? Some space-hope is ultimately and explicitly about an exit plan for a few, not for all (Valentine 2012).

Of course, space-hope matters. But we can ask exactly *when* and *why* and *how* hope about space matters, and *which* hopes about space should matter the most. Hope does its best work when strongly tethered to facts. Musk, Bezos, and their compatriots stand to benefit considerably from under-informed popular hope about space as a panacea for Earth's troubles. And in summer 2021, Musk's category of "those who attack space" probably included a lot of people who are quite appreciative of the possibilities of space but nonetheless opposed in principle to Earth's two richest men—unelected and undertaxed entrepreneurs with unprecedented wealth—making decisions for our solar system (e.g., Mack 2021; Reich 2021; Giridharadas 2018; McCormick 2021).

Greta Thunberg got it right about hope too, but differently, in her scathing speech at the Youth4Climate Summit in fall 2021:

> We can no longer let the people in power decide what is politically possible or not. We can no longer let the people in power decide what hope is . . . Hope is not passive. Hope is not blah, blah, blah. Hope is telling the truth. Hope is taking action. Hope always comes from the people. And we, we the people, want a safe future. We want real climate action and we want climate justice. (2021)

And so Thunberg echoes philosopher Hannah Arendt, who wrestled with the concept of hope in some of the twentieth century's darkest days, and concluded that instead it is in *natality*—"the ability to break with the current situation and begin something new" (Hill 2021)—that possibilities for action and for a better future lie. Whatever one calls it, the key is to do something if one can.

For me, the hope for space is not that it will provide us with an off-Earth backup for humanity, or resources without end, or journeys to the stars. Instead I hope that we can find our way to multilateral solutions for shared problems, to fairness, and to effective compromises regarding the commons of space. I find inspiration in groups like SpaceEnabled, the Committee on the Peaceful Uses of Outer Space (COPUOS), the United Nations Office for Outer Space Affairs (UNOOSA) including Space4Women and SpaceGeneration, the JustSpace Alliance, and the Secure World Foundation.

Anyone becoming interested in issues involving space policy or ethics may find space law bafflingly complex. I certainly do sometimes. But mercifully, the organizations I just mentioned are there to help people learn, and many space lawyers regularly write accessible pieces which non-specialists can understand. In a world where mega-corporations and the hyperwealthy often wield inordinate influence in shaping rules, or effectively out-maneuver regulation, or just break the rules and pay the fines (e.g., Arnold et al. 2020; Whitehouse and Stinnett 2017), it is easy to wonder whether laws and guidelines really matter . . . *but they do*. And those laws and guidelines are continually being remade, which constitutes an opportunity for input, and international and multistakeholder alliances are essential.

Ram Jakhu, Kuan-Wei Chen, and Bayar Goswami (2020) of McGill University's Institute of Air & Space Law provide specific suggestions for the laws and treaties that matter most for preserving space for peaceful purposes; they acknowledge that the necessary international efforts will take time, but they are clear that there is no other option. They add these words of encouragement:

> In various parts of the world, political establishments are tending toward the inward-looking and nationalistic overtones of the political right. Perhaps, the exploration and use of outer space under a governance regime that is rules-based and built on common interests and shared concerns may appear difficult to realize. However, populistic ideas are expected to swing toward favoring a rules-based global governance system. The drivers for this change will be the global youth, who are fully equipped with advance technologies and access to extensive networks of information technology, and who are acutely aware of and strive to develop a sense of community based on a shared sense of humanity. It is critical to remain optimistic that though there are pretentious voices and rhetoric parroting narrow-minded national security interests and the need to assert dominance outer space, the voices of restraint, calm, and rationality will prevail. (Jakhu et al. 2020, 39)

They're right. And yet, looking to youth to solve global problems is also "all wrong", as a then-sixteen-year-old Thunberg (2019) pointed out about climate in an unforgettable speech to the UN Climate Action Summit: Thunberg and her friends should have been able to stay peacefully in school instead of demonstrating in the streets to keep their elders focused on climate. But it is what it is. As for space, even if we had an effective multilateral space governance system that was fully equal to today's problems, there will be new challenges ahead, to be solved by those who are young today or not yet born.

Fortunately, there is a venerable history of activists of all ages devoted to the seemingly impossible. Some win, some lose, and some fights don't end. But there, inspiration may be found. Rebecca Solnit's wonderful, uplifting book *Hope in the Dark* describes many grassroots movements that worked, and concludes:

> How do we get back to the struggle over the future? I think you have to hope, and hope in this sense is not a prize or a gift, but something you earn through study, through resisting the ease of despair, and through digging tunnels, cutting windows, opening doors, or finding the people who do these things. (2016, 142)

Hope can also be found in words that change people's minds. Young poet Amanda Gorman said it all:

> Hope—
> we must bestow it
> like a wick in the poet
> so it can grow, lit,
> bringing with it
> stories to rewrite
> —from "In This Place (An American Lyric)"
> (Gorman 2017)

Coda: Great Bear, interrupted

Poet Robert Bringhurst (2009) combined ancient stories of the Bear constellation into one extraordinary epic in English, Cree, Greek, and Latin: *Ursa Major*. I just opened it again because recently I saw the Great Bear shining in a dark sky: She prowled around the North Star high above my tent in the forest, and was mirrored by the still water next to my canoe. I was so grateful to the traditional caretakers of the territory, and to those who had the foresight decades ago to preserve this northern sanctuary where land, sky, and water meet. That night, a wolf howled, an owl called, the Milky Way glowed, and time collapsed. My "now" stretched back millennia, and my "here" spanned the ocean.

Then artificial satellites slowly pierced the Bear's body, the spell broke, and I felt lost.

I am in the first generation of humans born into a world with these new lights in the sky, and I have not yet heard a story that I can live with, that tells me what they will really mean for us. In the Judeo-Christian tradition—a recent arrival in this hemisphere—a new fixed star means a savior is born, while new falling stars mean the end is nigh. But what of myriad new stars endlessly circling and proliferating?

Bringhurst reminds us that in ancient traditions,

Humans can eat and sleep with the gods,
and bear their children. Still, they can be just a breath away
from being rocks and trees and wolves and deer
and bears and stars and darkness. Just a breath away
from deathlessness, and just a breath away
from all that darkness in between the stars.

Bringhurst, *Ursa Major*, 49)

Today, humans and our technology move steadily outward into a realm that was never ours—a realm once reserved for gods, celestial changelings, the not-yet-born, and the dead: spirits outside of time. And so a clock is ticking now. Now, when I look at the Moon, I know my generation of humans may be the last to see her without permanent bases and residents. When I look at Mars, I wonder how long it will be before the first boots bring all that is human to her rocky plains.

Somewhere in all that darkness—between the stars and planets and asteroids, the satellites and orbital debris, the space law and the technical reports, the conferences and assemblies, the ceremonies and the telescopes, the headlines and the tweets—there is still the problem of what it all *means* now. But there is, it seems, no going back. And so we must find new stories to tell each other, along with the old, under our shared sky, while we run as fast as we can.

Acknowledgments

Errors are my own. I am thankful for conversations about space in multiple countries over the past two decades, with people whose words, actions, wisdom, and energy changed the way I see and live. Colleagues, students, writers, journalists, scientists, engineers, lawyers, filmmakers, artists, loved ones and friends have all influenced me deeply, as have people whose names I never learned, who stayed to talk after a public event. I am particularly grateful to editors Jim Schwartz, Linda Billings, and Erika Nesvold, for hope and for helping me run faster.

Notes

1. Chapter 1 of *Through the Looking Glass*, by Lewis Carroll (1872). In the end, it was all a dream, and the Red Queen was actually Alice's kitten . . . oh, never mind.
2. Truth and Reconciliation Commission Reports, Canada: https://nctr.ca/records/reports/.
3. For more, see Maryboy and Begay (2010), Penprase (2017), Neilson (2019), and the wonderful resource built by Annette Lee and colleagues, Native Skywatchers (Lee et al. 2007): www.nativeskywatchers.com.
4. See Cheney et al. (2020) regarding planetary protection for Mars, International Dark Sky Association (darksky.org), Marin (2011), and Declaration in Defense of the Night Sky and the Right to Starlight at www3.astronomicalheritage.net/images/astronomicalheritage.net/documents/Starlight_Declaration-En.pdf, Elvis et al. (2016), and Deudney (2020).
5. The ever-cheerful Nietzsche noted: "Zeus did not wish man, however much he might be tormented by the other evils, to fling away his life, but to go on letting himself be tormented again and again. Therefore he gives man hope—in reality it is the worst of all evils, because it prolongs the torments of man" (2006[1878], 53).
6. For some accurate information and careful thinking on some of those subjects, see Scharf (2020), Millis et al. (2018), Cockell ed. (2014), Jakhu et al. (2020), and Deudney (2020).

References

Arnold, Dennis G., Oscar Jerome Steward, and Tammy Beck. 2020. "Financial Penalties Imposed on Large Pharmaceutical Firms for Illegal Activities." *JAMA* 324, no. 19: 1995–1997. DOI: 10.1001/jama.2020.18740.

Bowdon, George. 2021. "Prince William: Saving Earth Should Come Before Space Tourism." *BBC*, Oct 14, 2021. www.bbc.com/news/uk-58903078.

Bringhurst, Robert. 2009. *Ursa Major: A Polyphonic Masque for Speakers and Dancers.* Kentville, NS: Gaspereau Press.

Carroll, Lewis. 1872. *Through the Looking Glass.* Project Gutenberg 1991 ebook edition. Available at www.gutenberg.org/files/12/12-h/12-h.htm#link2HCH0002.

Cheney, Thomas, et al. 2020. "Planetary Protection in the New Space Era: Science and Governance." *Front. Astron. Space Sci.* 7. November 13, 2020. https://doi.org/10.3389/fspas.2020.589817.

Cockell, Charles S., ed. 2014. *The Meaning of Liberty Beyond Earth.* Springer.

Comtesse, Hannah, et al. 2021. "Ecological Grief as a Response to Environmental Change: A Mental Health Risk or Functional Response?" *Int J Environ Res Public Health* 18, no. 2: 734. https://www.ncbi.nlm.nih.gov/pmc/articles/PMC7830022/.

Dallas, J.A. et al. 2020. "The environmental impact of emissions from space launches: A comprehensive review." *Journal of Cleaner Production* 255: 120209. https://doi.org/10.1016/j.jclepro.2020.120209.

Davis, Wade. 2009. *The Wayfinders: Why Ancient Wisdom Matters in the Modern World*. House of Anansi Press.

Deudney, Daniel. 2020. *Dark Skies: Space Expansionism, Planetary Geopolitics, and the Ends of Humanity*. Oxford University Press.

Ellis, Neville and Ashlee Cunsolo. 2018. "Hope and mourning in the Anthropocene: Understanding ecological grief." *The Conversation*, April 4, 2018. https://theconversation.com/hope-and-mourning-in-the-anthropocene-understanding-ecological-grief-88630.

Elvis, Martin, Tony Milligan, and Alanna Krolikowski. 2016. "The peaks of eternal light: A near-term property issue on the moon." *Space Policy* 38: 30–38.

Estrada, Alejandro et al. 2017. "Impending Extinction Crisis of the World's Primates: Why Primates Matter." *Science Advances* 3, no. 1: e1600946. DOI: 10.1126/sciadv.1600946.

Farmer, Doyne, et al. 2020. "Collaborators in Creation." *Aeon*, Feb 11, 2020. https://aeon.co/essays/how-social-and-physical-technologies-collaborate-to-create.

Ferreira, Becky. 2021. "SpaceX's Satellite Megaconstellations Are Astrocolonialism, Indigenous Advocates Say." *Vice/Motherboard*, May 10, 2021. www.vice.com/en/article/k78mnz/spacexs-satellite-megaconstellations-are-astrocolonialism-indigenous-advocates-say.

Galway, Lindsay P., et al. 2019. "Mapping the Solastalgia Literature: A Scoping Review Study." *Int J Environ Res Public Health* 16, no. 15: 2662.

Gammon, Katharine. 2021. "How the Billionaire Space Race Could Be One Giant Leap for Pollution." *The Guardian*, July 19, 2021. www.theguardian.com/science/2021/jul/19/billionaires-space-tourism-environment-emissions.

Gawley, Paige. 2021. "William Shatner Reacts to Prince William's Disapproval of Space Race." *ET Online*, Oct 14, 2021. www.etonline.com/william-shatner-reacts-to-prince-williams-disapproval-of-space-race-exclusive-173739

Giridharadas, Anand, 2018. *Winners Take All: The Elite Charade of Changing the World*. New York: Vintage Books.

Gorman, Amanda. 2017. "In This Place (An American Lyric)" at Splitthisrock.org: www.splitthisrock.org/poetry-database/poem/in-this-place-an-american-lyric.

Griffin-Pierce, Trudy. 1992. *Earth Is My Mother, Sky Is My Father: Space, Time, and Astronomy in Navajo Sandpainting*. Albuquerque: University of New Mexico Press.

Guterres, António. 2021. Statement to the UN General Assembly, Sept 21, 2021. Available: https://www.un.org/press/en/2021/sgsm20918.doc.htm.

Hall, Robert L. 1997. *An Archaeology of the Soul: Native American Indian Belief and Ritual*. Urbana, IL: University of Illinois Press.

Hill, Samantha. 2021. "When Hope Is a Hindrance." *Aeon*, October 4, 2021. https://aeon.co/essays/for-arendt-hope-in-dark-times-is-no-match-for-action.

Jakhu, Ram S., David Kuan-Wei Chen, and Bayar Goswami. 2020. "Threats to Peaceful Purposes of Outer Space: Politics and Law." *Astropolitics* 18, no. 1: 22–50. https://doi.org/10.1080/14777622.2020.1729061.

Launius, Roger. 2019. *Reaching for the Moon: A Short History of the Space Race*. New Haven, CT: Yale University Press.

Lawrence, Andy. 2021 *Losing the Sky*. Photon Productions.

Lebeuf, Arnold. 1996. "The Milky Way: A Path of the Souls." In *Astronomical Traditions in Past Cultures*, edited by V. Koleva and D. Kolev, 148–161. Institute of Astronomy, Bulgarian Academy of Sciences.

Lee, Annette. 2007-2022. *Native Skywatchers: Indigenous Astronomy Revitalization*. Available at: www.nativeskywatchers.com.

Lee, Annette, William Wilson, Jeffrey Tibbets, and Carl Gawboy. 2014. *Ojibwe Giizhig Anang Masinaa'igan/Ojibwe Sky Star Map Constellation Guide: An Introduction to Ojibwe Star Knowledge*. North Rocks, CA: Lightning-Source Ingram Spar.

Mack, Eric. 2021. "The problem with Jeff Bezos and other billionaires going to space." *CNET*, July 21, 2021. www.cnet.com/news/the-problem-with-jeff-bezos-and-other-billionai res-going-to-space/

Mann, Adam. 2010. "Space tourism to accelerate climate change." *Nature News*, October 22, 2010. https://doi.org/10.1038/news.2010.558.

Marais, Eloise 2021. "Space Tourism: Rockets Emit 100 Times More CO_2 Per Passenger than Flights—Imagine a Whole Industry." *The Conversation*, July 19, 2021. https://theconversat ion.com/space-tourism-rockets-emit-100-times-more-co-per-passenger-than-flights-imag ine-a-whole-industry-164601.

Marin, Cipriano. 2011. "Starlight: A Common Heritage." Proceedings of the International Astronomical Union, 5(S260), 449–456. doi:10.1017/S1743921311002663.

Maryboy, Nancy and David Begay. 2010. *Sharing the Skies: Navajo Astronomy*. Tucson: Rio Nuevo Publishers.

McCormick, Ted. 2021. "The Billionaire Space Race Reflects a Colonial Mindset that Fails to Imagine a Different World." *The Conversation*, Aug 15, 2021. https://theconversation.com/ the-billionaire-space-race-reflects-a-colonial-mindset-that-fails-to-imagine-a-different-world-165235.

Mecklin, John, ed. 2022. "At Doom's Doorstep: It Is 100 Seconds to Midnight." 2022 Doomsday Clock Statement, Jan 20, 2022. Science and Security Board, Bulletin of the Atomic Scientists. https://thebulletin.org/doomsday-clock/current-time/

Millis, Marc, et al. 2018. "Breakthrough Propulsion Study: Assessing Interstellar Flight Challenges and Prospects." NASA report. June 1, 2018. https://ntrs.nasa.gov/citations/2018 0006480.

Mortillaro, Nicole. 2019. "'We Come from the Stars': How Indigenous Peoples are Taking Back Astronomy." *CBC News*, March 30, 2019. www.cbc.ca/news/science/indigenous-astronomy-1.5077070.

Neilson, Hilding. 2019. "Indigenizing Astronomy Reading List." https://medium.com/@hil dingneilson/indigenizing-astronomy-reading-list-66cdec04a8af.

Newcomb, Frank J. 1990. *Navaho Folk Tales*. Albuquerque: University of New Mexico Press.

Nietzsche, Friedrich. 2006 [1878]. *Human, All-Too-Human*. Parts One and Two. Translated by H. Zimmern and P. Cohn. New York: Dover Publications.

Norris, Ray. 2020. "The World's Oldest Story? Astronomers Say Global Myths about 'Seven Sisters' Stars May Reach Back 100,000 Years." *The Conversation*, Dec 21, 2020. http://thec onversation.com/the-worlds-oldest-story-astronomers-say-global-myths-about-seven-sist ers-stars-may-reach-back-100-000-years-151568.

Norris, Ray, and Barnaby Norris. 2021. "Why Are There Seven Sisters?" In *Advancing Cultural Astronomy*, edited by E. Boutsikas et al., 223–235. Springer.

Penprase, Bryan. 2017. *The Power of Stars*, second edition. Springer International Publishing.

Pultzarova, Tereza. 2021. "The Rise of Space Tourism Could Affect Earth's Climate in Unforeseen Ways, Scientists Worry." *Space.com*, July 26, 2021. www.space.com/environmen tal-impact-space-tourism-flights.

Reich, Robert. 2021. "The Fake Heroism of Space Billionaires." Blog post. July 19, 2021. https:// robertreich.org/post/657190772027817984.

Ross, Martin and James Vedda. 2018 "The Policy and Science of Rocket Emissions." Center for Space Policy and Strategy. https://aerospace.org/sites/default/files/2018-05/RocketEmission s_0.pdf.

Scharf, Caleb. 2020. "Death on Mars." *Scientific American*, Jan 20, 2020. https://blogs.scientifica merican.com/life-unbounded/death-on-mars1/.

Shatner, William. 2021a. "'Star Trek' Legend William Shatner Sets Record in Space with Blue Origin Mission." Feature by Chris Cuomo, CNN, Oct 15, 2021. Available: www.youtube. com/watch?v=qCFJktZnjjE

Shatner, William. 2021b. "William Shatner Reacts To Seeing Earth From Space: 'It's So Fragile.'" Interview on Today Show, Oct 14, 2021. Available: www.youtube.com/watch?v=bx_C dBcRexc

Solnit, Rebecca. 2016. *Hope in The Dark*, third edition. Chicago: Haymarket Books.

Taft, Molly. 2021. "Why TV Is So Bad at Covering Climate Change." *Gizmodo*, July 14, 2021. https://gizmodo.com/why-tv-is-so-bad-at-covering-climate-change-1847283248

Taylor, C. 2019. "Relearning The Star Stories Of Indigenous Peoples." *Science Friday*. www. sciencefriday.com/articles/indigenous-peoples-astronomy/.

Thiery, Wim, et al. 2021. "Intergenerational Inequities in Exposure to Climate Extremes." *Science* 374(6564): 158–160. DOI: 10.1126/science.abi7339.

Thompson, Tosin. 2021. "Young People's Climate Anxiety Revealed in Landmark Survey." *Nature News* 597(7878): 605. Sept 22, 2021. www.nature.com/articles/d41586-021-02582-8

Thunberg, Greta. 2021. "Greta Thunberg's Full Keynote Speech at Youth4Climate Pre-COP26." Sep 29, 2021. Available at: www.youtube.com/watch?v=n2TJMpiG5XQ

Thunberg, Greta. 2019. "Transcript: Greta Thunberg's Speech At The U.N. Climate Action Summit." Sept 23, 2019. NPR staff. www.npr.org/2019/09/23/763452863/transcript-greta-thunbergs-speech-at-the-u-n-climate-action-summit.

Timperley, Jocelyn. 2021. "Billionaire Space Race: What Does It Mean for Climate Change and the Environment?" *BBC Science Focus*, Aug 12, 2021. www.sciencefocus.com/news/billiona ire-space-race-what-does-it-mean-for-climate-change-and-the-environment/.

Valentine, David. 2012. "Exit Strategy: Profit, Cosmology, and the Future of Humans in Space." *Anthropological Quarterly* 4(85): 1045–1067.

Van Valen, Leigh. 1973. "A New Evolutionary Law." *Evol. Theory* 1:1–30.

Venkatesan, Aparna, et al. 2020. "The Impact of Satellite Constellations on Space as an Ancestral Global Commons." *Nat Astron* 4: 1043–1048. https://doi.org/10.1038/s41550-020-01238-3.

Venkatasan, Aparna and James Lowenthal. 2021. "Community Engagement Working Group." Presentation at Press Conference: Conclusions from Satellite Constellations (SATCON2) Workshop, July 16, 2021. https://noirlab.edu/satcon2/ Video available: www.youtube.com/ watch?v=VFd0Kob8QLY. NoirLab and American Astronomical Society 2021.

Whitehouse, Sheldon and Melanie Stinnett. 2017. *Captured: The Corporate Infiltration of American Democracy*. New York: The New Press.

PART 4

BEING ACCOUNTABLE IN
THE PRESENT

18

Contact Zones and Outer Space Environments

A Feminist Archaeological Analysis of Space Habitats

Alice Gorman

Introduction

The struggles of women to attain equal participation in an enterprise argued to be a fundamental human urge—the desire to explore space—are well known (e.g., B. Lovell 2021; Nolen 2004; Penley 1997; Shayler and Moule 2005; Weitekamp 2004). Their exclusion has been justified across a number of fronts. These include the weakness of the female body, temperamental unsuitability, the unpredictable effects of menstruation, impacts on future fertility, distracting male crew, and women's supposed lack of interest (e.g., Healey 2018). Even in 2021, women comprise approximately 20% of space industry globally (UN Affairs 2021), and only 12.5% of all space travelers.[1] The 1967 Outer Space Treaty's proclamation of space as "the province of all mankind [*sic*]" is admirable but hollow when half the world's population of eight billion are effectively absent (Gorman 2021a; Steer 2020).

The "colonization" of other worlds has been a persistent dream of the Space Age. Just like Isabel Burton, the wife of nineteenth-century adventurer Sir Richard Burton, women were supposed to "pay, pack and follow" (M. Lovell 1998) once the frontier had been opened by heroic male explorers (e.g., Sage 2009, 160). During the Apollo era of the 1960s, astronauts and administrators in the US space program would make sometimes explicit, sometimes oblique, references to women's future role as propagators of the "human race" on other worlds, as if they were no more than incubating machinery to support masculine aspirations of conquest. The presence of women in space is still often reduced to a discussion of reproductive functions.

Alice Gorman, *Contact Zones and Outer Space Environments* In: *Reclaiming Space*. Edited by: James S. J. Schwartz, Linda Billings, and Erika Nesvold, Oxford University Press. © Oxford University Press 2023.
DOI: 10.1093/oso/9780197604793.003.0018

Running parallel to this strand of outer space discourse is a utopian vision which, I suggest, has its roots in the influential cosmist philosophy of Russian space theorist Konstantin Tsiolkovsky and the (misplaced) postwar optimism of the 1950s and 1960s. Although largely unstated in most space literature, there is an expectation that equality will reign in future space habitats in orbit and on other planetary bodies. For example, there is no need to design a space habitat to reduce opportunities for sexual assault, as such gendered violence is assumed eradicated. In the meantime, there is a massive data gap about women's physical and social experiences of space.

To date, all space vehicles and habitats have been designed around the default male body. Recent events, such as the lack of medium-sized spacesuits when two women were planned to do an EVA (Extra Vehicular Activity or spacewalk) in 2019 and the delivery of the first female-friendly toilet after twenty years of continuous occupation of the International Space Station (ISS) in 2021 have highlighted just how ill-prepared the space agencies and aerospace corporations are for female bodies.

To achieve gender equality in space, it's necessary to interrogate how technology reinforces or subverts existing power imbalances. This means examining how the materiality of space cultures, predominantly Russian and American, enacts certain forms of gender, using a feminist archaeological approach (e.g., Gero and Conkey 1991; Sorensen 2013). In this chapter, through fictional, historical, and contemporary examples of space material culture, I explore some of the intersections of design, gender, and social behavior.

Early Conceptions of Women Living in Space

Space is coded masculine now, but this wasn't always the case. Prior to the advent of the Space Age, scientists and fiction writers delved into both the science and the social requirements of space travel. Two examples serve to illustrate the difference.

US writer Edward Everett Hale's 1869 *Brick Moon* (Hale 1899) is acknowledged to be the earliest fictional depiction of a space station. A close reading of the novella from the perspective of the social expectations of a space society is revealing, as women are involved in every aspect of the mission. The Brick Moon—constructed of bricks because of their heat resistance—is funded by subscription (or what we would now call crowdfunding). Many of the subscribers are women. The construction of the Brick Moon is accomplished by several families, including that of the narrator. His young daughters, Alice and Bertha, having studied civil engineering, take part in surveying the launch site and building the brick space station.

The spacecraft is designed to accommodate families in its thirteen chambers, with such success that several of them move into completed sections while continuing the construction. When the Brick Moon is accidentally launched, it takes this tiny population with it. Civil society is played out in a microcosm of thirty-seven people; there is a wedding, and two babies are born during the satellite's sojourn in orbit. The regulation of sexual and reproductive behavior is provided by religion (the dissenting Sandemanian church).

The Brick Moon, however, does not amplify the emancipatory tendencies of its architects. The evolving space society replicates the gender and labor practices of Earth, as we find out when the women arrange for private communication with Earth away from the ears of men: "The brick-women explained at once to our girls that they had sent their men round to the other side to cut ice, and that they were manning the telescope, and running the signals for themselves, and that they could have a nice talk without any bother about the law-books or the magnetic pole" (Hale 1899).[2] The subject of this discussion was the disposition of daily life. Mrs. Belle Brannan observes to the Earth women, "Of women's work, as in all worlds, there are twenty-four in one of your days, but for my part I like it." She is referring to the rapid succession of day and night as the Brick Moon orbits Earth approximately every hour; but it's also a metaphor for the greater labor of women, while the menfolk are occupied with science and governance. This division of labor, she seems to imply, is invariant across the universe.

Perhaps not as promising for the future of space as we might hope; but what strikes me as interesting about Hale's story is that women do not have to be "included" in the Brick Moon. They are there from the beginning, in the funding, construction, and daily life of the station. It was never intended that they should be excluded; therefore, there are no "accommodations" or tensions arising from their presence.

Writing from the late 1890s up to the 1930s, Konstantin Tsiolkovsky rarely discussed women explicitly in his accounts of life in space habitats, but he predicted that the advantages of life without gravity would induce large populations to abandon Earth and take up their abode in orbit. He posited that microgravity would promote an egalitarian society (Gorman 2018; Gorman 2020). In *Outside the Earth*, however, he speculates about how the lack of gravity in a space habitat will affect female mobility: "The ladies have tied tape round the lower part of their skirts, because they had little use for their legs, and also because the situation was embarrassing. Some were wearing male clothing – emancipation of a kind!" (Tsiolkovsky 1920, 76). Trousers would have enabled female occupants to take full advantage of the freedom of movement, which Tsiolkovsky equates with a degree of social emancipation. "Unisex" clothing is the result of this adaptation to microgravity. Some accounts of Tsiolkovsky's life suggest that he regularly discussed his ideas with his wife Varvara Sokolova over the dinner table (Gorman 2021b). It is tempting to think that this very practical observation on how microgravity conditions affect women may have originated with her.

How did women get written out of plans for life in space? One explanation I propose is that after the 1930s, speculation about satellite and space habitat technologies was overtaken by the realities of rockets as weapons of war, a domain from which women in Europe and the US were largely excluded. The bodies associated with space vehicles became military bodies, rather than religious or familial bodies; space as a peaceful enterprise undertaken by a whole society gave way to a masculinized conquest or defense of new territory as the Second World War metamorphosed into decades of Cold War. In early space industry women were "computers" performing astrodynamic and ballistic calculations, or "seamstresses" creating the space suits and sun shields—expert and important roles, but more in line with the Victorian "separate spheres" ideology (Cordea 2013; Shoemaker 2014), than part of a liberating utopia.

Accommodating Bodies

As NASA prepared for its first intake of female astronauts in the 1970s, financial analyst Glenda Callanen prepared a RAND report on the opportunities for women. Her observations, while overwhelmingly positive, also reveal a divergence in perception. She noted: "It is assumed that modern technology is capable of solving the relatively minor problems of adapting space vehicles to the specific needs of women in the areas of personal hygiene and waste management" (Callanen 1975, 5).

Similarly, psychologist Albert Harrison noted in a 1978 report on psychosocial factors in space that there was "a growing recognition that technical systems are as easily engineered to meet women's needs as men's needs" (Harrison 1978, 15). This ease is undermined by Callanen's subsequent observation that "It is generally held that some costs will be increased when women are admitted to the space program. Some modifications of hardware will be required of testing and training facilities, as well as the spacecraft itself" (Callanen 1975, 5).

This attitude is exemplified in the experience of astronaut Rhea Seddon, who said of space suits: "They could never get one to fit me, as I was too small, and they questioned whether they should invest millions of dollars into redesigning the suit to fit all sorts of women's size and shape differences" (Quoted in Shayler and Moule 2005, 191). This response, of course, ignores the fact that male bodies also come in all sorts of sizes and shapes, as well as the idea present almost from the beginning of space travel that smaller female bodies would be better suited to the constricted habitable volumes of spacecraft. Accommodating astronauts over six feet tall does not seem to have attracted the same level of resistance.

A similar attitude was evident in the Soviet Union as the first female astronaut cadre was training in the early 1960s. Following Valentina Tereshkova's successful spaceflight in 1963, there was a proposal for an all-female mission. Historians David Shayler and Ian Moule (2005, 66) state that "Zvezda (the suit manufacturer) was opposed to an all female flight, and refused to fabricate a special EVA suit."

But was the cost of developing female-adapted space suits higher than the cost of *not* developing them? Joe Kosmo, a NASA engineer who worked on spacesuit designs from the Mercury program through to the Space Shuttle, believed that if women had succeeded in gaining access to the astronaut program in the 1960s, an in-suit waste management system would have been introduced at that stage rather than waiting for the Space Shuttle generation of suits (Shayler and Moule 2005, 90). Reframing the narrative in this way highlights faulty design principles which are incapable of adapting to different bodies, and the failure to take into account the true cost of living in space to begin with. The problem, one might say, was never the presence of women; it was rather their absence.

Proxemics and Hygiene

Callanen (1975) flagged the distinct personal hygiene requirements of female bodies and the implications for waste management systems (which can be read as menstrual blood and the imagined higher mucosity of female urine [Foster 2011, 147]). In the early days of the US human spaceflight program, concepts of hygiene were entangled with anxiety about the proximity of male and female bodies introducing a taint of sexuality into the pure enterprise of space exploration. The Russian response to this was the ongoing exclusion of women from space. The fifth Russian woman in space, Yulia Peresild, commented in an interview that "Here in Russia there's this belief that space is not a proper place for women" (Lodderhose 2021; see also B. Lovell 2021, 67).

These concerns are reflected in a recollection from NASA Deputy Administrator Robert Seamans: "On occasion I was asked in congressional testimony for my views on women astronauts, particularly for the Gemini programme. I explained that a flight in Gemini was equivalent to a flight in the front two seats of a Volkswagen for periods up to fourteen days. During that time the cockpit could get pretty grungy" (quoted in Shayler and Moule 2005, 93). Here, lack of space entails sweaty, smelly bodies, which, it appears, would be offensive to female astronauts or difficult for them to cope with. Why this would be is clearly held to be self-evident. Drawing on a plethora of sexist tropes, reasons could include higher standards of female cleanliness, more delicate female olfactory systems, and the assumption that women (often only of a certain "race" or class) do not have bodily functions themselves—ironic given the concern about how leaking female bodies could overwhelm the machines used to cleanse the spacecraft.

Stanley White, who designed life support systems for NASA, opined that "accommodations for both sexes onboard the two-person Gemini [capsule] would have been near impossible due to lack of space" (quoted in Shayler and Moule 2005, 93). In fact, he did not mean lack of space, as, if anything, female bodies are on average smaller than male: he meant lack of distance. This was a statement about proxemics, or cultural perceptions of space between bodies (Hall 1966). Even separated by spacesuits, there were risks in placing men and women adjacent in the same space. Again, there is much left unstated here: was the risk to the safety of the woman, for whom there would be no escape route from the threat of sexual violence from a man incapable of self-control, or to the male astronaut, whose self-control could be compromised by so much woman-flesh only two spacesuit widths away? And what *was* the proper distance apart? It is noteworthy that the solution to this problem was never to have a female-only crew.

The more spacious US Skylab space station, first launched in 1973 and deorbited in 1979, was a different story. "There is no question in my mind that Skylab could have been configured for a mixed sex crew," Seamans said (Shayler and Moule 2005, 97). Skylab was meant to be a "home in space," to develop the technologies and capabilities to live rather than just survive (Hitt et al. 2008).

The Space Shuttle also offered more space. As former director of NASA Johnson Space Center Carolyn Huntoon said, "It was going to have more space in it for the crews. It was going to have some of the conveniences of home that previous space capsules had not had" (2002). Women brought domestication with them to space (Sage 2009, 160), as opposed to the dirty locker room vibe of all-masculine spaces. Indeed, when the first women were admitted to the astronaut program in 1978, NASA had to build a women's locker room in the gym (Healey 2018, 601).

Anxiety about the spatial and social separation of men and women on spacecraft is partially because the sexualization of the female body is so normalized in Western patriarchal cultures. Larger spacecraft allayed this anxiety, which we should note in these examples is not being raised by astronauts themselves, but by designers and administrators. The regulation of social and sexual relations is a function of the interior space within a spacecraft. On the Brick Moon, Hale achieves this (according to his standards) by religious practice and the institution of marriage. Underlying much of this discussion is an attempt to erase heteronormative sexuality, coupled with an implied acknowledgement of sexual violence.

Sexual Violence in Space

While the possibility of sexual activity in space inspires huge public interest and endless speculation, there is remarkably little serious scholarship on this topic, as Simon Dubé et al. (2021) note in their call for a sexology of space. From the perspective of feminist critique and the vantage point of the present, the investigations that do exist often read as bizarrely naïve. Sexuality is subsumed under studies of psychological or social factors which contribute to mission success, particularly for long duration missions. I will eschew the temptation to analyze the discourse here, simply noting that "jealousy" as a threat to crew harmony has often been their focus (e.g., Harrison 1978, 16).

The general taboos around discussing sexuality in space, while essential to protecting crew privacy, also act to mask sexual assault or harassment; there is virtually no literature on this topic from which to assess its prevalence in off-Earth settings, or to gauge the efficacy of solutions.

Here, I am interested in material or technological responses to this threat. There is one published account of sexual harassment and assault which took place on the Sphinx-99 analogue mission in Moscow in 1999 (Lapierre 2007; Oberhaus 2015). This series of incidents highlighted aspects of the material environment as enabling or inhibiting certain behaviors. One solution to help female crew feel safe, after a male cosmonaut had attempted to force a Canadian crew member to kiss him, was to place locks on the intervening doors (Oberhaus 2015). This segregated the space by gender/ethnicity and would have had impacts on the movement of crew between internal spaces.

On the Sphinx-99 mission, sexually abusive discussions of female crew by the male crew took place off camera (Lapierre 2007), so there was no record of it although they were overheard. Camera surveillance can be seen as a safeguard, but this could also increase the danger in places where cameras don't reach. How sound carries within the corridors of a spacecraft is also a factor (Walsh and Gorman 2021). In terrestrial settings, increased lighting is often a design solution to making women feel safe (Boomsma and Steg 2014). However, surveillance is also a disincentive to consensual relationships or activity. Privacy is both a protection and a danger; and achieving a balance between these is surely a crucial consideration for space habitat design.

However, there is little data about how crew use spaces inside space stations. One of the few studies is by archaeologist Justin Walsh et al. (2021), using images from NASA's publicly available Flickr account as a proxy for more systematic locational data on crew movement. An analysis showed that women are photographed inside the European Space Agency's cupola module on the ISS at twice the frequency they should be given an equal statistical distribution. The cupola is a seven-sided window providing a spectacular view of the blue-and-white Earth rolling beneath. There are many limitations to these data, but they do provide an early indication of a pattern which clearly requires more data, and theory, to explain.

Knowing how different genders prefer to use spaces within a habitat is a first step to making them safer for everyone. This encompasses understanding how spaces are adapted by crew to maximize privacy and safety and using this information in preventative design. Ultimately, however, the real issue is not how women keep themselves safe from the predations of men (acknowledging that intragender harassment and assault also occurs), but how space culture can divest itself of the patriarchal values which have plagued its evolution. This includes positioning women as inferior or lacking and replicating traditional gender roles around work and leisure.

Domesticity and the Labor of Science

Women's role in space as domestic workers has been the subject of many "jokes." A satirical poster labelled "Moon Maid," from a 1973 protest against NASA holding a beauty pageant, shows a woman standing on the Moon with a mop and bucket (Maher 2017), reflecting the Sisyphean task of cleaning moon dust and hence the never-ending cycle of domestic labor. In 1982, an apron was presented to Svetlana Savitskaya, the second Soviet woman in space, by her Salyut 7 crewmates. The apron was a material culture symbol of domestic servitude and a reminder that women were not really part of the crew.

Cleaning and "housework" did not become issues until people started living in space. In his imagined microgravity orbiting habitats, Tsiolkovsky noted that "All sorts of odds and ends are drifting about the rocket, dust doesn't settle – how are we to wipe it away?" and that the state of the communal lavatories was "just awful," although he was silent on the subject of who was responsible for maintaining cleanliness (1920, 235). On the Brick Moon, Hale assumed a traditional gendered division of labor; and from Belle Brannan's comments, there was plenty of it to be had.

The separation of living and working spaces is a key factor in the structuring and valuing of labor. In Western industrial nations before the era of COVID-19, what counted as work took place outside the home; and labor undertaken inside the home, generally by women, was not counted as work. This labor was left out of indices of economic wealth and development (Ferguson 2020). Space habitats can make visible the labor that on Earth is invisible, but this is another area where there seems to be virtually no analysis of how social expectations affect the use of space and equipment.

An aspect of the spatial separation of male and female work is that the house is a locus of leisure for men, who leave work behind when they return home (Wearing and Wearing 1988). By contrast it is the locus of work for women, from dawn until sleep ("women's work is never done"). There are both spatial and chronological elements to this, as the Brick Moon's Belle Brannan emphasized in her comparison of the different terrestrial and orbital relationships with day and night (Hale 1899).

In most Russian and US space stations from Salyut 1 to the ISS, the "separate spheres" have been rolled into one as crew live, work, and take their leisure in the same spaces; indeed, they are living inside a laboratory and are part of the experimental apparatus themselves. Hence, the spatial disposition of space station interiors is closer to women's experience of the home than it is to the traditional male experience.

On the ISS, Saturday mornings are set aside for cleaning tasks, using vacuums to clean filters and vents, and disinfectant wet wipes for surfaces and handholds. Cleaning is given a scientific complexion, as one purpose is to control the proliferation of molds and bacteria which could be harmful for crew health. While this is scheduled, it is also outside regular work schedules, and it seems there is a fair bit of latitude in how thoroughly it is done, which would not be acceptable in external maintenance or in the execution of experiments. Anecdotal evidence suggests that some crew do the bare minimum of cleaning (Anderson 2015; [identity redacted], pers. comm.).

Old attitudes die hard. As astronaut Clayton C. Anderson said in 2015 when responding to a Quora question about how the ISS was cleaned, "If I answer this question, please don't tell my wife! If she finds out I actually DO cleaning, it could spell 'doom' for me here in Houston!" Again, this is framed as a "joke"; but reveals deeply ingrained beliefs about the status of different kinds of work and their gendered associations.

In a study of ISS life, anthropologist Jack Stuster asked ten NASA crew members to keep journals during their expeditions. The crew included women, although how many is not clear as the responses are not identified by gender. One answer, though, is revealing, as it's hard to imagine a male crew member being in this position: "[The Russians] save the really low-skill work for me—changing filters, cleaning fan grills, etc. I don't mind—they schedule plenty of time, so I'm not rushed, and it's somewhat relaxing" (Stuster 2010, 11). One of the recommendations from this study was to "Distribute tedious and housekeeping tasks as evenly as possible among the crew," suggesting that this was far from the case during the study period (Stuster 2010, 55).

Some space habitat concepts separate work and living areas, such as Herman Noordung's 1929 concept for a space station, where work takes place in microgravity modules and the spinning of the main wheel provides a more convenient environment for sleeping, dining, and washing. The proposed NASA Lunar Gateway space station also separates residence from work. The function of the Habitation and Logistics Outpost (HALO) module of the Lunar Gateway is primarily accommodation for crew preparing for and re-turning from lunar surface expeditions (Thales Alenia 2021). Given the gen-dered differences of how "home" is experienced, how this relates to labor, and the challenges of microgravity hygiene, it seems there may be something to be learnt from applying a gendered analysis to these interior spaces.

Reinscribing the Feminine

As opposed to becoming "one of the boys," one way in which women have asserted their agency in these spaces is by performing traditionally feminine crafts, which have a complex relationship to work and leisure (Parker 1984; Stalp 2015).

In 2012, NASA astronaut Don Pettit took unpaired knitting needles man-ufactured from different polymers in his personal crew preference kit to the ISS. The intention was to demonstrate the behavior of water drops when the needles were charged, to make an educational video (Futurism 2014).

Nonetheless, the presence of knitting needles in space caused delighted reactions from knitters and crafters on Earth (e.g., Hobson 2012).

In 2013, NASA astronaut Karen Nyberg sewed an 18 cm square quilt block, testing the challenges of cutting fabric patterns in microgravity. Some of these were overcome by adapting materials commonly in use on the ISS, such as Velcro and ziplock bags, to secure her equipment and materials. She was not able to bring specialist sewing scissors and had to use those already on board, which were less than ideal (Pearlman 2014). Her quilt block turned into a collaborative project between Earth and space with the invitation for others to contribute. The resulting quilt, comprising over 2,400 blocks from all over the world, was displayed at the 2014 International Quilt Festival in Houston (White 2015).

Using the needles and thread she'd brought for the quilting, Nyberg also made a stuffed dinosaur toy, as a gift for her son, from discarded materials already on board the station. In an interview she said: "It is made out of velcro-like fabric that lines the Russian food containers [that are] found here on the International Space Station . . . It is lightly stuffed with scraps from a used t-shirt" (Pearlman 2013).

These are materials which, in the normal course of daily life, would have been destined for trash and destroyed. The thriftiness of recycling materials in this fashion is also a traditional feminine skill (König 2013), and likely one that will be vital in more remote space habitats of the future. Scholarship in recent decades has critiqued the distinctions between, and relative status of, craft, science, and art (e.g., Flannery 2001; Hein 2007) and repositioned expertise formerly derided for its association with women's work, such as the sewing skills utilized to manufacture spacesuits.

Nyberg's crafting activities were very visible; she talked about her intentions prior to launch and made videos of her quilting and dinosaur while on orbit. However, the results of her labor were returned to Earth and did not stay on board as a physical reminder of her "subversive stitches" (Parker 1984). What other evidence there may be of creativity, adaptation and symbolism deriving from similarly female-coded practices is unknown, as these questions have not yet formed part of the study of human factors in space travel.

A Contact Theory of Gender in Outer Space

The interaction of men and women in confined and constructed environments off-Earth could be framed as a form of cultural contact, a concept drawn from the field of historical archaeology. "Contact" refers to the processes of cultural change when a powerful invading group—usually Europeans in the context of settler societies—subjugate an Indigenous population (this term is also used sometimes for alien/human interactions in SETI). The models used to describe the resulting material culture responses include acculturation, accommodation, domination and resistance, and creolization (e.g., Birmingham 1992; Deagan 1990; Miller et al. 1995).

Acculturation, in the archaeological context, broadly refers to how objects or artefacts change under the influence of exposure to a hegemonic culture (Rubertone 2000). At the colonial mission site of Wybalenna in Australia, archaeologist Judy Birmingham (1992) interpreted this as "accommodation," arguing that the adoption of dominant cultural traits was a selective and mosaic process rather than "a steady and irreversible acquisition" (Lydon and Ash 2010, 14). Domination and resistance models of material culture focus on the ways in which mundane objects and practices are mobilized into a symbolic system of expressions of power. The concept of creolization "allows the voice of the periphery to be heard, and it acknowledges changes in the dominant as well as the dominated" (Källén 2001, 63), hence making a space to understand creativity and the emergence of new cultural forms.

These approaches can be applied to many situations where there is a power differential. They also emphasize the subtle ways that the less powerful exercise agency to produce novel and hybrid forms of material culture. Following this, we might ask how women have adapted to "a world designed for men" (Criado-Perez 2020) —and also how the presence of women has resulted in change, both physical and ideological.

In historical archaeology the zone of contact is generally geographic, such as a frontier or entrepôt, but also personal, when it comes to the relationships between members of the dominant and subjugated groups. Material culture is used to negotiate identity and power, with the accumulation of personal choices manifesting in statistical patterns that can be discerned archaeologically. Archaeologist Stephen W. Silliman (2001) has emphasized that culture contact is not a moment in time, but an ongoing process. Different aspects of this process can manifest across different scales of space and time.

Inside spacecraft or space habitats, the zone of contact may be spatial, physical, and technological. Surfaces and technological interfaces (like waste management systems and space suits) are cyborgized zones of contact which are designed for some and not for others—against the backdrop of microgravity to which all must acculturate. The translation to space of 1G Earth bodies, objects, and practices has the effect of "making the familiar unfamiliar" (Buchli and Lucas 2001, 9). Common routines and everyday objects become the subject of intense planning and design. The inclusion of women brings formerly buried or ignored aspects of spaceflight to the fore: the realities of human bodies, sexual violence, and the labor involved in maintaining a habitat.

When women were finally allowed into space vehicles, they faced a process of "accommodation." This worked in two directions: the alterations made to existing technologies and processes to "accommodate" the presence of women, and the accommodations made by women in order to "fit in" to a system not designed for them, such as using poorly designed toilets for twenty years on the ISS.

As the examples discussed in this chapter show, there are numerous cultural assumptions about the proper spatial disposition of genders that have impacted on access to space. Larger spacecraft have given more access to women, while, it could be argued, creating more "housework." The distance between men and women is subtly expected to reduce sexual violence, without its underlying causes in male entitlement being acknowledged. The use of locks between spaces in the Sphinx-99 analogue habitat shows how material environments and objects are used to negotiate expressions of domination and resistance. Nyberg's very visible use of craft demonstrates a resistance to dominant narratives of masculine science and an exercise of agency that inheres in the objects she made, using the actual fabric of life in space to create new forms.

Looking at space vehicles and habitats as zones of contact between genders (and potentially other cultural categories and identities), offers perspectives for analyzing material culture structures and choices in space. While the application of archaeological contact theory in this chapter is preliminary and exploratory, it also opens avenues of reconceptualizing spacecraft design. Fundamentally, however, we should also question the categories of gender which are replicated or subverted in space environments and ask how they may be rewritten without the constraints of deep Western, Christian, and patriarchal traditions.

NASA's commitment to a gender binary has historically emphasized differences rather than similarities. Questions about this were present from the admission of women to the NASA astronaut program. Harrison (1978, 23) talked of the benefits of the "androgynous" crew member who could combine the "autonomous, independent, somewhat dominating and aggressive, and emotionally inhibited" gender expectations of men with the "warm and nurturant" characteristics of women who "openly display their feelings." But, as we know, it's not that simple. Sage (2009, 161) points to the multiple masculinities enacted in space contexts and Armstrong (2020), looking at science museum space galleries, notes the instability and continuous evolution of constructs of gender. Using categories of analysis, such as male and female, which one is also attempting to dismantle, is a fraught undertaking.

The movement into space has been accompanied by a belief that gendered bodies are subsumed under the banner of "humanity." This assumption only serves to highlight the fractures exposed by living in the space environment. Despite the optimism of the international space community that they will be beacons of gender equality, the history of gender relations in space so far is less than stellar. As multiple new space stations are under development, as well as plans for lunar and Martian surface habitats, the dearth of data and of feminist analysis is glaring. How can space habitat design support rather than undermine equality? This is a question that we're not yet in a position to answer.

Notes

1. These figures exclude suborbital flights. Numbers of nonbinary people or people of marginalized genders in space industry are not available, although surveys of space scientists show that between 0.5 and 13 percent of people identify as nonbinary (Strauss et al. 2020).
2. Quotes taken from Project Gutenberg edition: https://www.gutenberg.org/cache/epub/1633/pg1633-images.html.

References

Anderson, Clayton C. 2015. "The Secret Behind How the ISS Gets Cleaned." *Quora*, May 27, 2015. https://www.quora.com/International-Space-Station/How-is-the-interior-cleanliness-maintained-aboard-the-International-Space-Station.

Armstrong, Eleanor. 2020. "Exploring Space(s): Queer Feminist Approaches to Understanding Pedagogy in Science Museum Galleries." Unpublished PhD thesis, University College London.

Birmingham, Judy. 1992. *Wybalenna: The Archaeology of Cultural Accommodation in Nineteenth-Century Tasmania*. Sydney: Australian Society for Historical Archaeology.

Boomsma, Christine and Linda Steg. 2014. "Feeling Safe in the Dark: Examining the Effect of Entrapment, Lighting Levels, and Gender on Feelings of Safety and Lighting Policy Acceptability." *Environment and Behavior* 46, no. 2: 193–212.

Buchli, Victor and Gavin Lucas. 2001. "The Absent Present: Archaeologies of the Contemporary Past." In *Archaeologies of the Contemporary Past*, edited by Victor Buchli and Gavin Lucas, 3–18. London: Routledge

Callanen, Glenda. 1975. "Future Space Exploration: An Equal Opportunity Employer?" Rand Paper Series P-5492.

Cordea, Diana. 2013. "Two Approaches on the Philosophy of Separate Spheres in Mid-Victorian England: John Ruskin and John Stuart Mill." *Procedia-Social and Behavioral Sciences* 71: 115–122.

Criado-Perez, Caroline. 2020. *Invisible Women: Exposing Data Bias in a World Designed for Men*. London: Random House UK.

Deagan, Kathleen. A. 1990. "Accommodation and Resistance: The Process and Impact of Spanish Colonization in the South-East." In *Archaeological and Historical Perspectives on the Spanish Borderlands East. Columbian Consequences*, Vol. 2, edited by D. H. Thomas, 297–314. Washington, DC: Smithsonian Institution Press.

Dubé, S., M. Santaguida, D. Anctil, L. Giaccari, and J. Lapierre. 2021. "The Case for Space Sexology." *The Journal of Sex Research*. https://doi.org/10.1080/00224499.2021.2012639.

Ferguson, Sarah. 2020. *Women and Work: Feminism, Labor, and Social Reproduction*. London: Pluto Press.

Flannery, M. C. 2001. "Quilting: A Feminist Metaphor for Scientific Inquiry." *Qualitative Inquiry* 7, no. 5: 628–645.

Foster, Amy E. 2011. *Integrating Women into the Astronaut Corps: Politics and Logistics at NASA, 1972–2004*. Baltimore: Johns Hopkins University Press.

Futurism. 2014. "Knitting Needles, a Sheet of Paper and a Water-Filled Syringe Make an Interesting Science Experiment." November 19, 2014. https://futurism.com/water-droplets-orbiting-a-knitting-needle-in-space.

Gero, Joan and Margaret Conkey, eds. 1991. *Engendering Archaeology: Women and Prehistory*. Hoboken, NJ: Wiley-Blackwell.

Gorman, Alice C. 2018. "Looking Up a Century Ago, a Vision of the Future of Space Exploration." *The Conversation*, January 19, 2018. https://theconversation.com/looking-up-a-century-ago-a-vision-of-the-future-of-space-exploration-89859/.

Gorman, Alice C. 2020. "Geometry and the Uncanny in the International Space Station." In *Interior Spaces. A Visual Exploration of the International Space Station*, edited by Roland Miller and Paolo Nespoli, 53–57. Bologna: Damiani.

Gorman, Alice C. 2021a. "Moonwalking: When Other Worlds Belong to Women." *Griffith Review* 74: 152–162.

Gorman, Alice C. 2021b. "Between the House and the Stars: The Life of Varvara Sokolova Who Married Konstantin Tsiolkovsky." *Space Age Archaeology*, January 17, 2021. https://zoharesque.blogspot.com/2020/05/varvara-sokolova.html.

Hale, Edward Everett. 1899. *The Brick Moon and Other Stories*. Boston: Little, Brown, and Company.

Hall, Edward T. 1966. *The Hidden Dimension*. Garden City, NY: Doubleday.

Harrison, Albert. 1978. "Preliminary Report on Social Psychological Factors in Long Duration Space Flights: Review and Directions for Future Research." NASA—CR-15238.

Healey, Devlin. 2018. "'There Are No Bras in Space': How Spaceflight Adapted to Women and How Women Adapt to Spaceflight." *The Georgetown Journal of Gender and the Law* 19: 593–617.

Hein, H. S. 2007. "Redressing the Museum in Feminist Theory." *Museum Management and Curatorship* 22, no. 1: 29–42.

Hitt, David, Owen Garriott, and Joe Kerwin. 2008. *Homesteading Space: The Skylab Story*. Lincoln, NE: University of Nebraska Press.

Hobson, Rachel. 2012. "Water Droplets Dancing around a Knitting Needle in Space." *Make*, February 7, 2012. https://makezine.com/2012/02/07/water_droplets_dancing_around/.

Huntoon, Carol. 2002. Interview with C. Butler, Johnston Space Center Oral History Project, June 5, 2002. http://www.jsc.nasa.gov/history/oral_histories/herstory.htm.

Källén, A. 2001. "Creolised Swedish Archaeology." *Current Swedish Archaeology* 9: 59–76.

König, A. 2013. "A Stitch in Time: Changing Cultural Constructions of Craft and Mending." *Culture Unbound* 5, no. 4: 569–585.

Lapierre, J. 2007. "Operational, Human Factors and Health Research and Issues for Mars 500 Based on Challenges of the Russian SFINCSS Study in 1999: SFINCSS-99-00 Final Report, Simulation of Flight of International Crew on Space Station Confinement Study." Marc Heppener, European Space and Technology Research Center.

Lodderhose, Diana 2021 "'The Challenge': Russia's Klim Shipenko and Yuliya Peresild Talk Shooting First Film in Space—'It's a Four-Dimensional World Up There.'" *Deadline*, November 4, 2021. https://deadline.com/2021/11/the-challenge-russia-klim-shipenko-yulia-peresild-first-film-space-four-dimensional-reality-1234867822/.

Lovell, Bronwyn D. 2021. "Sex and the Stars: The Enduring Structure of Gender Discrimination in the Space Industry." *Journal of Feminist Scholarship* 18: 61–77.

Lovell, Mary S. 1998. *A Rage to Live: A Biography of Richard and Isabel Burton*. New York: W.W. Norton.

Lydon, Jane and Jeremy Ash. 2010. "The Archaeology of Missions in Australasia: Introduction." *International Journal of Historical Archaeology* 14, no. 1: 1–14.

Maher, Neil M. 2017. *Apollo in the Age of Aquarius*. Cambridge, MA: Harvard University Press.

Miller, Daniel, Michael Rowlands, and Christopher Tilley, eds. 1995. *Domination and Resistance*. London: Routledge.

Nolen, Stephanie. 2004. *Promised the Moon: The Untold Story of the First Women in the Space Race*. New York: Thunder's North Press.

Noordung, Hermann. 1929. *The Problem of Space Travel: The Rocket Motor*, edited by E. Stuhlinger and J. D. Hunley. Washington, DC: NASA SP-4206.

Oberhaus, Daniel. 2015. "Sexism in Space." *Motherboard*, April 2, 2015. https://www.vice.com/en/article/539498/sexism-in-space.

Parker, Roszika. 1984. *The Subversive Stitch: Embroidery and the Making of the Feminine*. London: The Women's Press.

Pearlman, Robert. 2013. "'Made in Space!' Astronaut Sews Dinosaur Toy from Space Station Scraps." *CollectSpace*, September 27, 2013. http://www.collectspace.com/news/news-092713c.html.

Pearlman, Robert. 2014. "Astronaut's Sewn-in-Space Star Shines at Quilt Festival." *Space.com*, November 6, 2014. https://www.space.com/27671-astronaut-space-quilt-festival.html.

Penley, Constance. 1997. *NASA/Trek: Popular Science and Sex in America*. London: Verso.

Rubertone, Patricia E. 2000. "The Historical Archaeology of Native Americans." *Annual Review of Anthropology* 29, no. 1: 425–446.

Sage, Daniel. 2009. "Giant Leaps and Forgotten Steps: NASA and the Performance of Gender." In *Space Travel and Culture: From Apollo to Space Tourism* edited by Martin Parker and David J. Bell, 146–163. Blackwell Publishing.

Shayler, David and Ian Moule. 2005. *Women in Space: Following Valentina*. Chichester: Praxis/Springer.

Shoemaker, Robert Brink. 2014. *Gender in English Society 1650–1850: The Emergence of Separate Spheres?* London: Routledge.

Silliman, Stephen W. 2001. "Agency, Practical Politics, and the Archaeology of Culture Contact." *Journal of Social Archaeology* 1, no. 2: 190–209.

Sørensen, Marie Louise Stig. 2013. *Gender Archaeology*. Hoboken, NJ: John Wiley & Sons.

Stalp, M.C. 201.5" Girls Just Want to Have Fun (Too): Complicating the Study of Femininity and Women's Leisure." *Sociology Compass* 9, no. 4: 261–271.

Steer, Cassandra. 2020. "'The Province of all Humankind' - A feminist analysis of space law." In *Military and Commercial Uses of Outer Space*, edited by Stacey Henderson and Melissa de Zwart, 169–188. New York: Springer.

Strauss, Beck E., Schuyler R. Borges, Thea Faridani, Jennifer A. Grier, Avery Kiihne, Erin R. Maier, Charlotte Olsen, Theo O'Neill, Edgard G. Rivera-Valentin, Evan L. Sneed, Dany Waller, and Vic Zamloot. 2020. *Nonbinary Systems: Looking Towards the Future of Gender Equity in Planetary Science. A State of the Profession White Paper for Planetary Science and Astrobiology Decadal Survey 2023–2032.* https://arxiv.org/abs/2009.08247.

Stuster, Jack. 2010. *Behavioral Issues Associated with Long Duration Space Expeditions: Review and Analysis of Astronaut Journals.* Anacapa Sciences Inc, Santa Barbara, California.

Thales. 2021. "HALO: First Component of the Gateway Cislunar Space Station Takes Shape." February 15, 2021. https://www.thalesgroup.com/en/worldwide/space/news/halo-first-component-gateway-cislunar-space-station-takes-shape/.

Tsiolkovsky, Konstantin. 1920. "Outside the Earth." In *The Call of the Cosmos*, edited by V. Dutt, 161–332. Moscow: Foreign Languages Publishing House.

UN Affairs. 2021. "Only around 1 in 5 Space Industry Workers Are Women." October 4, 2021. https://news.un.org/en/story/2021/10/1102082/.

Walsh, Justin, Rao H. Ali, Alice C. Gorman, and Amir K. Kashefi. 2021. "A First Approximation of Population Distributions on the International Space Station." SocArXiv. November 18, 3032. https://doi.org/10.31235/osf.io/ra4c3.

Walsh, Justin and Alice Gorman. 2021. "A Methodology for Research in Space Archaeology: The International Space Station Archaeological Project." *Antiquity* 61: 1331–1343.

Wearing, B. and S. Wearing. 1988. "'All in a Day's Leisure': Gender and the Concept of Leisure." *Leisure Studies* 7, no. 2: 111–123.

Weitekamp, M. A. 2004. *Right Stuff, Wrong Sex: America's First Women.* Baltimore: Johns Hopkins University Press.

White, Maura. 2015. "The Space Quilt Challenge." *Bullock Museum.* https://www.thestoryofte xas.com/discover/texas-story-project/space-quilt-houston/.

19

Occupy Space

Will Disabled People Fly?

Sheri Wells-Jensen

Introduction

There is no one alive who would doubt that Colonel Yuri Alekseyevich Gagarin, hero of the Soviet Union, Order of Lenin, Pilot Cosmonaut of the USSR, was a man's man, a dude's dude, a *piervyi iz pieryvih*. According to the worldwide press, which followed his every move during the early 1960s, nothing ever bothered him, and nothing dislodged that signature charismatic smile. Cheerfully, he hopped into centrifuges, hiked through bitter Arctic terrain, submitted himself to oxygen deprivation, and spent ten long days locked alone in an anechoic chamber. This last, by the way, is a time-tested way to go about breaking the spirits of enemy spies, demonstrating that not only was he physically tough and very brave, but he also possessed copious amounts of internal equanimity. Apparently, just all in a day's work for our—or everybody's—hero.

This was the man who, on April 12, 1961, strapped himself into the Vostok 1 spacecraft and blasted off for a 108-minute ride, becoming the first human to reach space. He completed one orbit around the Earth, dutifully documenting the experience in his notebook (of the paper variety) until he let go of his pencil and the traitorous thing floated out of reach, lodging itself behind some equipment. Buckled securely into his seat as he was, there was nothing he could do about that, so that particular accomplishment (the first space diary) was cut short, and that might actually have been the only documented instance in which he failed to surpass all expectations. He was the one who first went boldly where literally no man (only a couple apes and some extremely unfortunate dogs) had gone before, and he did it with style!

After his solo spin in space, Gagarin plunged back into the atmosphere, belting out the traditional Russian folk song "The Motherland hears, the Motherland knows" as his little craft nose-dived toward the Russian

Sheri Wells-Jensen, *Occupy Space* In: *Reclaiming Space*. Edited by: James S. J. Schwartz, Linda Billings, and Erika Nesvold, Oxford University Press. © Oxford University Press 2023. DOI: 10.1093/oso/9780197604793.003.0019 countryside. He bailed out (on schedule) around 10,000 feet, to land (probably on his feet) in a potato field, just in time for lunch, reportedly explaining to the startled farmer and her granddaughter on the scene, "Don't be afraid. I am a Soviet Citizen, like you, who has descended from space, . . . And I have to find a telephone to call Moscow."

Self-effacing and smiling, he went on to international acclaim, receiving uncountable honors at uncountable ceremonies on the face of the globe he had so recently orbited (for more about the life of Yuri Gagarin, consult Jenks 2019). He was followed just two years later by the first woman in space, Major General Valentina Tereshkova, but for the next twenty years, humankind mostly sent Yuri after Yuri, *pieryvih* after *pieryvih*: steely eyed missile men whose charisma and goodwill were exceeded only by their superhuman accomplishments. Handsome, likable, outrageously fit incarnations of the fictional Buck Rogers, James T. Kirk, and John Carter of Mars.

As time went on, though, this splendid monotony began to lose its cultural appeal. Earthlings took a breath of hope in the 1980s as changes began to happen. In 1980, cosmonaut Arnaldo Tamayo Mendez, who was born in Guantanamo, Cuba, became the first person of African heritage in space. In 1982, Svetlana Savitskaya became the second woman in space, followed in 1983 by astronaut and physicist Sally Ride, the first American woman in space. Subsequently, Savitskaya was the first woman to perform a spacewalk, in 1984.

New swaths of humanity recognized themselves in these pioneers, and when Dr. Mae Jemison, the first African American woman in space, stepped aboard the space shuttle *Endeavor* in 1992, the relief and celebration were palpable: there might still be a ways to go to achieve complete equality, but barriers to space did seem to be vanishing, one by one, burnt to unrecognizable cinders with each vessel launched.

Space began to seem, more and more, like a place real people could actually go, and designers began to get the idea that they could no longer model everything destined for space on the admirable physique of the ninety-fifth percentile of military males. It was clear that we at least had to begin creating smaller spacesuits sized to fit women, and new possibilities, along with new designs, began to emerge in the early twenty-first century as private companies entered the scene.

But not everyone had equal cause to celebrate. While many people did begin to look hopefully toward the skies, the majority of physical flight requirements still hovered stubbornly around the Buck Rogers level. People with disabilities—in fact, people with even small medical imperfections—remained grounded.

To date, NASA's application process specifies that candidates must pass the "NASA Long Duration Astronaut Physical," which includes vision correctable to 20/20 in each eye, blood pressure less than 140/90 in a sitting position, and standing height between 62 and 75 inches. In addition, applicants must complete military water survival training, pass a flying syllabus, and become SCUBA qualified; it's worth noting that all of these have their own physical qualifications built in. That is, all skills necessary to become a pilot in the twenty-first century are assumed in NASA's requirements. Covering all bases, the nondiscrimination policy assures applicants that NASA will not eliminate anybody based on gender, religion, (the rest of the usual things) and *nondisqualifying* physical or mental disabilities (emphasis added) (Stierwalt 2020, NASA 2011, NASA 2021).

Given that one in five people on Earth—arguably including some of our best scientists and most creative artists—is disabled, it's reasonable to ask what motivated—and still motivates—these wholesale exclusions. After all, we want the very best people in space, so why are we eliminating so many without even looking at their résumés? It seems self-defeating. What is motivating this nonsense? Why haven't disabled people taken their place in orbit with everybody else? There are a couple of answers to this question.

Before proceeding, it's worth adding a terminological note. The careful reader may have noticed that I have referred to the subjects of this chapter as both "disabled people" and "people with disabilities." The former, known as "identity-first language," is generally preferred by disability scholars and activists within the disability sphere, while the latter, "person-first language," is often required by professional style guides and journal editors. As a blind person myself, I prefer identity-first language while remaining strategically agnostic about what other disabled people wish to call themselves, so you will read both in this chapter. If the details are of interest, my philosophy on this was articulated nicely in Vaughan (2009).

Historically, the Answer Is NO!

Until quite recently, this wasn't even a question. No one would imagine sending a disabled person into space; that'd be as ridiculous as, well, sending a disabled person into space! The pervasive cultural narrative about disability tells us in no uncertain terms that disabled people are weak, dependent, and passive, useful mostly as a source of "inspiration" or to add a dash of pathos or humor to some narrative or other. If they are heroic, it is only that their circumstances are so unrelentingly dismal that they are more or less heroes just by getting up in the morning and eating breakfast.

Tiny Tim and Mr. Magoo keep company with the pencil-selling blind beggar and swarms of embittered, disfigured comic-book villains. Along with these, generating chilling echoes of the tens of thousands of actual disabled people killed in concentration camps in World War II, there are the disabled protagonists who fight "valiantly" for the right to kill themselves, either because they could obviously never be happy (consult the tropes above) or because their friends and family would be better off without them. And then there are the Rain Men and the Forrest Gumps of movie fame, and while these are charming, inspirational even, they are most assuredly not people you would trust with anything important. You might admire them all you want, but you would never hire them to rewire your house or babysit your child.

If art were not nature's mirror, we could dismiss all this without concern, but the lived experience of disabled people reflects these realities. Disabled people encounter these attitudes on a daily basis, and there are serious professional, financial, and interpersonal repercussions. Expectations of disabled people are low, and it is not generally assumed that we are, or should be, capable professionals. For example, I am an associate professor of linguistics with over twenty years of teaching experience and a thriving research program, but when I call a cab for a ride to my university, often the driver will ask me, in the tone of voice one might use for a very small child—or a very small dog—"is there a special program on campus for you today?"

Surely, this fetid cultural guano contributes to the lack of disabled astronauts, but it is certainly not the sole cause, or perhaps even the most important. After all, while there are in fact successful, brilliant disabled scientists, artists, and military leaders, not a single disabled person has managed to slip into any professional astronaut corps anywhere in the world. Is there something special about being an astronaut?

Medically, the Answer Is DEFINITELY NO!

While the first factor above is thinly veiled bigotry, the second masquerades as reasonableness. Space is, after all, very definitely dangerous (NASA 2021). In fact, it's not too much to say that space—with its vicious temperature swings from near absolute zero to millions of degrees, its searing radiation, its lack of air, and its debilitating microgravity—is ready and willing to kill any living creature foolish enough to poke its nose above the atmosphere. We are simply not bred for space. Our bodies evolved in a kinder, gentler place, where the temperature is (mostly) reasonable, and the atmosphere and magnetic fields keep vast amounts of radiation from reaching our delicate cells. The comfortable one g (32 feet per second squared) holds that atmosphere snugly in place and pulls helpfully on our muscles and bones, keeping them strong and keeping our feet securely on the ground. And the sunlight and the wind and the sky and the company of our beloveds keep us from the perils of isolation and monotony. But in space these protections are forfeit. Dangers range from the small and unpleasant (nausea, backaches, and rashes) through the truly troubling (bone loss, muscle loss, and diminished vision and cognitive function) to the possibly fatal (profound depression, cancer, and explosive decompression).

We can try (and have been trying) to mitigate these effects, protecting our more recent Yuris through the use of vigorous daily exercise, increased shielding, and treatments designed to counteract the more devastating aspects of exposure to microgravity, but we do have to admit, at some point, that we cannot ever be entirely successful. Our Yuris are going to be hurt, and some will become disabled, as we venture further and further from home (Wells-Jensen 2018). Given all these dangers, it might seem prudent to send only the healthiest people available, as these splendid specimens might be best suited to resist the inevitable.

On the other hand, if these effects are inevitable, an alternate approach would be to prepare both habitat and inhabitants for what may come, disregard the excessive physical requirements, and choose the best scientists, doctors, and pilots we can find and trust them to do their jobs. And if those people happen to be disabled, maybe that's an advantage. Who better to manage muscle weakness, fatigue, or waning visual acuity in space than the very people who manage these inconveniences on a daily basis on Earth? Disabled people may not be superheroes, but we do strategize daily about how to get around unexpected difficulties and efficiently organize backup plans when things go awry. This situational awareness is exactly the sort of problem-solving mindset spacefarers need.

If we acknowledge that disability may occur in space, then all vessels, environments, equipment, control systems, policies, and daily routines should be designed so that disabled crews will be able to live and work safely without returning to Earth. These crews could then carry out their mission despite changes brought about by the disabling effects of space.

The process of creating environments that disabled and able-bodied people can both use is aptly called "universal design" (Steinfeld and Maesel 2012), and the result is an environment that is safer and more convenient for everyone. For example, on Earth, curb cuts (originally designed for wheelchair users) and video captions (designed for d/Deaf users) have become common-place. We expect to be able to roll our shopping carts and strollers easily from the sidewalk down to the street and back up again, and people take for granted that their media will have helpful running captions. Want to watch with the sound off and still know what's going on? No problem.

Similarly, spacecraft designed with disabled crews in mind will be safer and more user-friendly for everyone, there will be no a priori reason to ex-clude disabled people from being recruited as astronauts in the first place, and we could easily rewrite the regulations that prevent them from applying and gaining their rightful place among the stars.

If these were the only two factors at play, disabled people would be pop-ping up in professional astronaut corps all over the planet. But they are not. Perhaps, then, there is still another reason.

Logically, the Answer Is: Let's Learn How!

The third barrier to full inclusion in space is surprisingly mundane, and it's one that science is able to address. We could agree to free ourselves from stereo-types and open the astronaut corps to the most highly qualified applicants regardless of disability. And we could acknowledge that it's wise to employ the principles of universal design in the fabrication of all space vessels, control systems and procedures. But if we don't know *how* to do these things—that is, if we don't know what modifications would be ideal—we can do nothing, even if it is our sincere intention to get things done.

We have to find out what the features of a universally designed space en-vironment are, and what changes (if any) we need to make in standard procedures to bring the benefits of universal design into space. We obviously cannot rely on our planet-bound regulations and experience. For example, there is no reason to ensure that doorways aboard space vessels are wide enough to permit wheelchairs; an Earth-made wheelchair would be fantas-tically ridiculous, useless, and probably dangerous in space. Similarly, there is no sense in outfitting computer systems with braille devices used on Earth when we have no idea whether the mechanisms that raise and lower the pins will work outside Earth's gravity field.

We need systematic, repeatable observations of disabled people in micro-gravity, until we can determine what problems exist and design solutions to these problems. A useful mechanism for some of these observations would be zero-gravity parabolic flights, during which selected participants could experience short periods of microgravity without the danger or expense of traveling to space. During these flights, a disabled crew could work interactively to address questions specific to each kind of disability.

Here are some things we might want to know. This is, of course, not a wholly satisfactory list. It assumes that every individual has only one disability, and it omits some categories of disability entirely. But even if we knew only these things, we would be well on our way toward opening space for all.

Blind crew:

- How do fully blind people stay oriented in microgravity when there is no physical sense of what is "down"?
- How do they maintain orientation if they begin to rotate?
- Since objects that are released in microgravity do not fall but remain (silently, elusively) floating nearby, how will a blind person locate and retrieve them?

D/deaf crew:[1]

- How can alert systems aboard space vessels be modified for the use of Deaf crew members?
- How would American Sign Language (ASL) be modified in microgravity?
- Will using ASL cause the signer to involuntarily move about in microgravity?
- What happens to the signing space and to comprehension when one conversational partner is upside down with reference to another?

Crew with mobility disabilities:

- How might an astronaut with no voluntary control over her legs move about the cabin? How might she maintain her position when working?
- What happens to various kinds of prosthetic limbs in zero-gravity? What modifications to the standard flight suit might be useful to astronauts with prosthetics?

Next Steps and Conclusions

In fall 2021, AstroAccess, an initiative of SciAccess, began this work. A crew of twelve disabled people, comprising members with vision, hearing, and mobility disabilities, embarked on a parabolic flight with the Zero Gravity (Zero-G) Corporation. Although results of their work are not available at this writing, the crew carried out multiple demonstrations and observations in microgravity and hypergravity.

Although this is a wonderful move in the right direction, it's clear that the use of a parabolic flight as an analog for longer space missions has three immediate drawbacks.

First, unlike the zero-gravity portion of a suborbital experience, these flights are extraordinarily loud, making communication difficult. Although it is quite possible for Zero-G coaches to communicate with flyers, casual conversation is very difficult, and the usual sound cues used by blind people to navigate are obliterated.

Second, although the resultant microgravity is real, each parabola lasts for only fifteen to twenty-five seconds, depending on weather and other conditions during the flight. The first few seconds of each parabola are taken up with getting off the cabin floor and into a suitable working position, while the last few seconds are consumed in the (sometimes frantic) attempt to return to the cabin floor safely before the return of gravity. This scenario leaves very little time to conduct any kind of investigation.

Finally, the existing prohibition against disabled people in space has meant that they have not had the benefit of the extensive mentorship and experience afforded other would-be astronauts. Only one of the disabled crew aboard this important inaugural flight had any previous zero-gravity experience. Disabled researchers have some serious catching up to do; the first few flights may well be fully occupied with the process of learning how to move and work in microgravity before useful data can be gathered about accommodations.[2]

Just as the movement away from a largely homogeneous human occupation of space felt inevitable in the early 1980s, the current trend in both public and private space companies is edging toward inclusion of disabled people. In early 2021, for example, the European Space Agency announced its "parastronaut" program, which will recruit candidates with specific physical disabilities. These include people with lower limb amputations, people with legs of differing lengths, or people of short stature. This is a prudently cautious first exploration, and because accommodations needed by this group are minor, it has every chance of success. If all goes as planned, ESA may well send the first physically disabled person into Earth orbit or beyond.

The question of whether disabled people will travel in space was answered long ago. Space itself has answered it. If crews live above the atmosphere for long enough, disability will become a part of their experience, and we may find eventually that the resulting mix of abled and disabled crew represents a new way for humans to live and work together. In fact, as space becomes more accessible, allowing all kinds of humans to live and work there, we may find

Earth becoming more accessible, as lessons learned in constructing accessible habitats in orbit are used to improve accessibility for humans who will never go to space.

Notes

1. It should be remembered that, although deafness appears in lists of disabilities, many Deaf people do not consider themselves "disabled."
2. Later, similar studies of activities in Lunar and Martian gravity could be done (as the gravitational force experienced by the participants is controlled by the angle at which the aircraft dives). These would have the same limitations as zero-gravity parabolas, but they could still be of great interest to anyone designing Lunar or Martian colonies or free-floating space structures which are spun to create "artificial" gravity.

References

Jenks, Andrew L. 2019. *The Cosmonaut Who Couldn't Stop Smiling: The Life and Legend of Yuri Gagarin*. Dekalb, IL: Northern Illinois University Press.

NASA. 2011. "Astronaut Selection and Training." *NASAfacts*. https://www.nasa.gov/centers/johnson/pdf/606877main_FS-2011-11-057-JSC-astro_trng.pdf.

NASA. 2021. "The Human Body in Space." *NASA*, February 2, 2021. https://www.nasa.gov/hrp/bodyinspace

Steinfeld, Edward and Jordana Maisel. 2012. *Universal Design: Creating Inclusive Environments*. Hoboken, NJ: John Wiley & Sons.

Stierwalt, Sabrina. 2020. "Do You Have What It Takes to Be an Astronaut?" *Scientific American*, June 21, 2020. https://www.scientificamerican.com/article/do-you-have-what-it-takes-to-be-an-astronaut/

Vaughan, C. Edwin. 2009. "People-First Language: An Unholy Crusade." *Braille Monitor* 52, no. 3.

Wells-Jensen, Sheri. 2018. "The Case for Disabled Astronauts." *Scientific American*, May 30, 2018. https://blogs.scientificamerican.com/observations/the-case-for-disabled-astronauts/

20

Protecting Labor Rights in Space

Erika Nesvold

The recent growth of the commercial space industry has shifted the public narrative about space away from the Cold War–era battle between governments, reframing it as a capitalist race for profit. Space enthusiasts idolize and/or demonize famous space entrepreneurs like Elon Musk, Jeff Bezos, and Richard Branson. Sometimes labeled "visionaries," these businessmen are not unique in their interest in increasing humanity's presence in space, but they do command unusually large amounts of financial resources, enabling them to pursue these goals while increasing their wealth.

Recent years have also seen a growing public fascination with the other side of our nascent space economy: the customer. NASA notably shifted into this role after the Space Shuttle program ended in 2011, leaving them in search of an alternative method for transporting astronauts to the International Space Station. But other space-loving members of the ultra-wealthy class have also stepped into the limelight to become some of the first space tourists, taking the world's most elite and exclusive adventure vacations. Beyond individual space tourists, more publicly minded space advocates like to advertise the benefits of the space economy for significantly less-wealthy customers, such as rural internet users for whom satellite fleets may provide their only internet access. And of course, much of the public enthusiasm for the private space industry depends on encouraging middle-class space enthusiasts to picture *themselves* as the eventual customers of space companies: to imagine taking a suborbital flight instead of a long intercontinental airplane ride for a business trip, or perhaps one day selling their terrestrial belongings for a ticket to exotic retirement on Mars. Elon Musk is already attempting to entice customers for his future colony ships to Mars, describing passengers enjoying the voyage on luxury spaceliners before patronizing the first "pizza joint" on the Martian surface (Davenport 2016).

This framing of space as a new arena for capitalist competition and a fertile ground of untapped customers too often leaves out the third vital component of the space economy: the workers. As writer Nicole Dieker points out,

Erika Nesvold, *Protecting Labor Rights in Space* In: *Reclaiming Space*. Edited by: James S. J. Schwartz, Linda Billings, and Erika Nesvold, Oxford University Press. © Oxford University Press 2023. DOI: 10.1093/oso/9780197604793.003.0020 someone will need to wash dishes in that Martian pizza joint, and someone will need to clean the rooms of the interplanetary spaceliner while the passengers are enjoying dinner with a view of the stars (Dieker 2016). Besides tourism, an even larger industry in the predicted space economy is resource extraction. Asteroid mining will not be performed by miners in spacesuits with pickaxes, but someone will need to monitor and repair the equipment. Space colonization advocate Robert Zubrin thinks space workers will have the advantage in negotiations in space due to an anticipated labor shortage (Zubrin 2011). But how will workers actually be treated as the space industry grows? What can we learn from labor rights throughout history, and the struggles of space industry workers in particular, about the future of work in space?

The first manufactured object to enter space—to cross the 100 km "Kármán line"—was a German A-4 rocket (later called the V-2) during a vertical test launch in 1944. This successful launch soon led to large-scale manufacturing, requiring large-scale labor to construct more rockets. This labor force could therefore be considered to include some of the first space workers, along with the rocket development team who designed and built the test rockets. But while the lead designer of the V-2, Wernher von Braun, dedicated his career toward advancing technology with the goal of human spaceflight, the V-2 rockets were not designed for science, exploration, or even colonization: they were weapons built for a global war, used to indiscriminately bomb British civilians. And the workers who built these space weapons were prisoners in Nazi concentration camps who were beaten, hanged, starved, and worked to death in torturous underground manufacturing facilities (Neufield 2007).

Three-quarters of a century later, the most well-known space workers today are not rocket manufacturers (aside from celebrity CEOs like Musk and Bezos), but the astronauts who live and work in space, conducting scientific research, engineering tests, and space station construction on behalf of their governments. Humanity's astronauts are celebrated and elite: they represent the "best of the best," the ultra-qualified "right stuff." While their work is physically demanding and potentially fatal, each astronaut not only volunteered to go into space, but competed aggressively for the honor. Children dream of becoming astronauts one day, and many adults live vicariously through these ambassadors to space.

But most of the people contributing labor to the space industry today, both publicly and privately funded, will never cross the Kármán line. The engineers, designers, programmers, manufacturers, and others who support humanity's presence in space are not celebrated at the level of astronauts, because they don't share the risk or require the same qualifications as astronauts. As humanity extends its civilization beyond this planet, this class of laborers will begin to live and work in space, stretching the definition of "astronaut," or perhaps requiring a new category representing the ordinary workers who happen to live off-Earth. Where will they fall on the long spectrum from the imprisoned V-2 manufacturers in the 1940s to the elite astronauts of today, in terms of how they are treated and what rights they are granted by their employers?

As with any new venture, space enthusiasts often reach for ways to describe their plans for space using analogies with the past, to help their audience construct a model of what life will be like in their vision of the future. But we should also interrogate these analogies for clues about how the less powerful members of the space economy will be treated.

Space as the Transcontinental Railroad

For example, advocates of large-scale space infrastructure—habitable space stations, fuel depots, orbital construction platforms, reusable rockets, etc.—often draw parallels between these plans and the transcontinental railroad that connected the east and west coasts of the United States beginning in the nineteenth century (Launius 2014). Elon Musk, discussing his dream of an Interplanetary Transport System at the International Astronautical Congress in 2016, made this comparison explicitly: "The goal of SpaceX is really to build the transport system. It's like building the Union Pacific Railroad" (Robertson 2016). The Union Pacific Railroad company, along with the Central Pacific Railroad, was the first to connect San Francisco to the eastern network of US rail lines, bringing customers, materials, and laborers to areas of the US that had previously been difficult and expensive to reach and invigorating the local economies along its path. In the process, it also made the owners of the railroad companies fantastically wealthy. Similarly, entrepreneurs in the space infrastructure business hope to open the skies for new economic opportunities while making their own profit. As Musk has also predicted, "Once that transport system is built, then there's a tremendous opportunity for anyone to go to Mars to build something. That's really where a tremendous amount of entrepreneurship will flourish" (Patel 2016).

Indeed, the construction of the first transcontinental railroads in North America was an infrastructure investment that not only provided safe and cheap access to the western half of the continent for US settlers from the east but incentivized a rapid increase in settlement and created profitable new trade routes (White 2012). But this project also required a massive investment of labor, first to build the railroads and later to operate them. During construction, workers tore a path for the railroad through some of the harshest regions of the continent, struggling through the deadly heat of the Colorado Desert and the biting cold of the Nevada mountains. Workers risked being swept away by avalanches or buried in rubble while blasting tunnels with unstable explosives, or simply falling from steep cliffs. Once the railroads were built and in operation, new difficulties and dangers arose for railroad employees. The link-and-pin coupler between railcars required a worker to stand between two moving cars to attach the coupler. This frequently led to lost fingers, hands, or even death if the worker did not move quickly enough. Brakemen, working in the deadliest occupation in the country in the late nineteenth century, scrambled along the tops of moving railcars to operate the brake wheels. While the risk of falling between cars was always a possibility, the danger increased while working at night, during storms, or when the railroad companies attempted to decrease staffing to save money, increasing the number of cars that each brakeman would need to operate (Licht 1983).

The physical hazards of space will exceed those experienced by railroad builders and operators, but the greater danger for space workers may come from labor inequalities and power imbalances that parallel those faced by railroad workers in the nineteenth century. For example, much of the construction work of the transcontinental railroad was performed by Chinese immigrant laborers, but they also experienced significant Sinophobia and pay disparities compared to their white counterparts, leading to a strike in June of 1867. The strike was ultimately broken when the railroad company cut off the food supply to the workers, although they did eventually raise wages (Chang, Fishkin, and Obenzinger 2019, 14–15). In other words, the railroad companies took advantage of the existing marginalization of their primary workers to decrease wages, then used the workers' physical isolation and dependence on fragile supply lines against them when they protested. It's unclear what kind of marginalization and prejudices will exist when humanity is regularly living and working in space in the future, but such power imbalances, combined with the harsh and dangerous environment, will make workers vulnerable to similar labor abuses.

Space as the California Gold Rush

The rapidly growing private spaceflight industry has also been compared favorably to the California Gold Rush of the mid-nineteenth century. The discovery of gold in 1848 spurred a mass migration of primarily young men to the western coast of the United States, all hoping to find and extract enough gold to make it worth the trip. For those lucky miners who landed on a rich distribution of ore, they could earn as much money in a matter of weeks as a well-off farmer back east could make in a year (Rohrbough 1997). The idea of a "gold rush to space," in pursuit of valuable resources like rare-Earth minerals, platinum, and water, rather than gold, is an appealing story for entrepreneurs considering investment in the nascent space mining industry.

The mythology of the California Gold Rush appealed particularly to the American values of individuality and the value of hard work: the idea that if you were bold enough to travel to distant and unseen lands, and you were willing to work hard once you got there, you could rapidly make a better life for yourself and your family. This "rags-to-riches" story, however, depended on one's ability to afford the trip west in the first place. As historian Daniel Cornford notes, "probably only a small portion of ordinary workers and farmers caught in an emergent industrial revolution could avail themselves of the opportunity of joining the Gold Rush. Few of those who came were among the abject poor" (Conford 1999, 83). Elon Musk has predicted the space settlement will become feasible once the cost of a ticket to Mars falls to about $200,000, or "roughly equivalent to a median house price in the United States" (2017, 47). Even this low price, relative to the current cost of transporting material to Mars, will be an insurmountable obstacle to today's space-hopeful abject poor.

The Gold Rush transformed the state of California as the incoming flood of miners greatly increased the population of American settlers while displacing and, not infrequently, killing the Indigenous population. Support workers soon followed to earn a living cooking the miners' food, washing their laundry, and entertaining them on cold, dark nights. Aside from the opportunities for wealth, Gold Rush advocates at the time also celebrated what they saw as the civilizing effect of the increased white settler population in California, and they recruited miners from the eastern United States using "the ethnocentric belief that they were also serving the higher purpose of fulfilling the nation's Manifest Destiny." Space settlement advocates hope that a gold rush to space will similarly develop the infrastructure and technology needed for long-term human habitation in space, but parallels can also be drawn between the American myth of manifest destiny that motivated Gold Rush hopefuls and the common space industry rhetoric of "humanity's manifest destiny in the stars" (Armus 2020).

Several space advocates, including entrepreneur Peter Diamandis, have predicted that "Earth's first trillionaire will be made in space" (Wei 2015). But the first millionaire of the California Gold Rush was not a miner who pulled $1 million worth of ore from the ground; it was Samuel Brannan, an entrepreneur who happened to own a store near the location of the first discovery of gold by American settlers in 1848. Brannan quickly bought up mining equipment and supplies, sold them to miners at inflated prices, then used the profits to open more stores and eventually expanded his business into hotels and land speculation. Besides price gouging, social isolation, widespread illness, risk of injury, and the literally back-breaking nature of the work, Gold Rush miners also struggled to survive a steady drop in income as the years passed. The gold that was easy to access and extract by single miners or small teams was quickly depleted, and large investments of capital for projects like redirecting rivers was required to continue producing profit. Within a decade after the discovery of gold, a majority of gold miners were wage laborers rather than independent prospectors (Conford 1999, 93).

A space mining gold rush may very well bring in large profits for a lucky few investors and entrepreneurs, just as the California Gold Rush bestowed fortunes on some of its miners. But copying the Californian template into space could also result in even larger numbers of exploited workers: lured to the mining fields of the Moon, Mars, or the asteroid belt with the promise of vast riches in exchange for boldness and hard work, expected to fund their own journey, then exploited upon arrival by local merchants, only to see their incomes fall over time as accessible resources decrease and the labor pool grows.

Space as International Waters

Debate continues over the nature of property rights in space: The Outer Space Treaty of 1967 forbids nations from appropriating territory in space, but a capitalist system of resource extraction requires some form of private property ownership. In recent years, legislation like the US Space Act of 2015 has begun moving us toward a future space mining industry where no one can claim ownership to territory beyond Earth's surface, but companies extracting resources can claim ownership of, and therefore sell, those resources. One frequently used terrestrial analog for this model is the fishing industry—specifically, fishing outside national borders (Wall 2015). As Luxembourg's economic minister, Etienne Schneider, explained in 2016, "The situation is equivalent to the rights of a trawler in international waters. Fishermen own the fish they catch but they do not own the ocean" (Amos 2016).

While not quite the get-rich-quick analogy of the Gold Rush, the comparison between space mining and fishing in international waters appeals because it suggests that with the right transportation and equipment, anyone can make a living working in space. Unlike a gold rush, where prospective miners may have to stake a claim, fishing boats care only about access and use rights, rather than having to compete for exclusive property ownership. Casting space mining as fishing in the open seas also evokes an image of space resources as sustainable and essentially infinite, an obviously poor analogy upon examination, given that fish populations will replenish themselves, given the chance, while minable space resources like waters and precious metals will not.

But the fishing industry carries its own legacy of labor exploitation, rooted in certain physical characteristics of the ocean that will also make working in space dangerous. For example, the sheer size of the ocean makes monitoring the fishing industry for labor abuses challenging, and enforcement of labor laws is complicated by the international nature of the industry. The ocean's vastness and the isolated nature of fishing work also means that it can be literally impossible for workers to escape abusive situations. Today's Thai fishing industry is a prime example of the potential for labor rights violations at sea: employers frequently withhold their workers' identity papers or passports to keep them from quitting or leaving the boat. At sea, workers are dependent upon the skipper or vessel owner for food, water, shelter, and eventually transportation back to their home port. Migrant workers are especially vulnerable to such abuses, as language barriers, discriminatory laws, and fear of deportation prevent them from organizing and standing up to their employers (Human Rights Watch 2018).

Workers transported into space by their employers will be similarly dependent on their employers for return transportation, not to mention uniquely dependent on them for food, water, and life support. Conflicting labor laws in different spacefaring nations could make enforcement of labor protections difficult, and monitoring the working conditions of space laborers will be an even greater challenge. Without deliberate efforts to protect workers in space from abuse, these future workers may find themselves trapped in even worse conditions than today's migrant fishers in Thailand.

Protecting Future Space Workers

The danger for future space workers is not that these comparisons with terrestrial industries are poor analogies, but that they are excellent ones: accurate predictors of how laborers in similar physical environments and economic systems will be treated in space. Many of the physical characteristics of space will leave workers particularly vulnerable to abuse: the difficulty in traveling to space will make monitoring working conditions and enforcing labor laws challenging and expensive, while the difficulty of returning safely to Earth will make escaping abusive situations difficult if not impossible. Space is uninhabitable without extensive resources and centralized life support technology, so employers who control the life support system where their employees live and work will have significant control over the workers' activities and potential organizing. Finally, space holds the promise of vast, untapped profits for private spaceflight and space mining companies. This potential for wealth increases both the motivation for and the ability of spaceflight companies to evade regulations, ignore safety concerns, and exploit workers in pursuit of a fortune for their shareholders.

Given these risks, and cautionary tales from past and future terrestrial industries, what can we do now to help protect future space workers? First, we can continue to fight against labor exploitation and abuses on Earth today. This will help us develop better legal, monitoring, and enforcement systems that we can later adapt for space. But it will also ensure that the cultural legacy we leave for our descendants, the ones who will be building the future space economy, prioritizes and values workers' rights.

More specifically, we need to be aware of and address labor rights issues in today's space industry. Two of the most prominent private space companies today, Elon Musk's SpaceX and Jeff Bezos' Blue Origin, have been criticized for their labor practices. SpaceX has faced accusations of violating California labor laws, in a class action suit settled in 2017 (Patel 2017) and discriminating against non-US citizens in its hiring practice, in a US Department of Justice investigation still ongoing in 2021 (Sheetz 2021). Blue Origin angered employees by pressuring them to travel for a test launch in April 2020 during the first wave of the coronavirus pandemic (Grush 2020), and a group of current and former employees published an essay in 2021 accusing the company of fostering a workplace environment that is permissive of sexism and dismissive of safety concerns (Roulette 2021).

Both SpaceX and Tesla are also funded partly by their founder's other companies, Tesla and Amazon, which have each also been criticized for labor rights issues. Elon Musk's electric car manufacturer Tesla was fined for more than fifty workplace safety violations between 2014 and 2018, more than the ten biggest US car manufacturers combined (Ohnsman 2019). In 2019, the US National Labor Relations Board ruled that Tesla had violated labor laws by interrogating, threatening, and firing workers to dissuade them from unionizing (Eidelson 2021). Meanwhile, Jeff Bezos' online marketplace, Amazon, has also been cited by the Nation Labor Relations Board for interfering with a unionization vote among its employees (Glaser and

Kaplan 2021), and criticized for unsafe working conditions by warehouse employees, who experience injury rates up to three times the national average (Sainato 2020).

To ensure a better future for workers in space in the future, we must improve the protections for workers in the space industry today and deliberately prepare to protect workers' rights beyond the surface of the Earth. If the popular comparisons with historical and contemporary terrestrial industries turn out to be anywhere close to accurate, we might otherwise find ourselves repeating the labor abuses of our past and present as we expand our civilization into space.

References

Amos, Jonathan. 2016. "Luxembourg to Support Space Mining." *BBC News*, February 3, 2016. https://www.bbc.com/news/science-environment-35482427.

Armus, Teo. 2020. "Trump's 'Manifest Destiny' in Space Revives Old Phrase to Provocative Effect." *The Washington Post*, February 5, 2020. https://www.washingtonpost.com/nation/2020/02/05/trumps-manifest-destiny-space-revives-old-phrase-provocative-effect/.

Chang, Gordan H., Shelley Fisher Fishkin, and Hilton Obenzinger. 2019. Introduction. In *The Chinese and the Iron Road: Building the Transcontinental Railroad*, edited by Gordan H. Chang and Shelley Fisher Fishkin, 1–26. Stanford, CA: Stanford University Press.

Conford, Daniel. 1999. "'We All Live More Like Brutes than Humans': Labor and Capital in the Gold Rush." In *A Golden State: Mining and Economic Development in Gold Rush California*, edited by James J. Rawls and Richard J. Orsi, 78–99. Berkeley, CA: University of California Press.

Davenport, Christian. 2016. "Elon Musk on Mariachi Bands, Zero-G Games, and Why His Mars Plan Is Like 'Battlestar Galactica.'" *The Washington Post*, September 28, 2016. https://www.washingtonpost.com/news/the-switch/wp/2016/09/28/elon-musk-on-mariachi-bands-how-he-plans-to-make-space-travel-fun-and-why-his-mars-plan-is-like-battlestar-galactica/.

Dieker, Nicole. 2016. "Mars: The Rich Planet." *Medium*, September 28, 2016. https://medium.com/the-billfold/mars-the-rich-planet-e6be98360f2b/.

Eidelson, Josh. 2021. "Tesla Is Ordered to Rehire Worker, Make Musk Delete Tweet." *Bloomberg*, March 25, 2021. https://www.bloomberg.com/news/articles/2021-03-25/tesla-illegally-fired-worker-and-must-kill-musk-tweet-nlrb-says.

Glaser, April and Ezra Kaplan. 2021. "Amazon Violated Labor Law in Alabama Union Election, Labor Official Finds." *NBC News*, August 2, 2021. https://www.nbcnews.com/tech/tech-news/amazon-violated-labor-law-alabama-union-election-labor-official-finds-rcna1582.

Grush, Loren. 2020. "Jeff Bezos' Space Company Is Pressuring Employees to Launch a Tourist Rocket During the Pandemic." *The Verge*, April 2, 2020. https://www.theverge.com/2020/4/2/21198272/blue-origin-coronavirus-leaked-audio-test-launch-workers-jeff-bezos/.

Human Rights Watch. 2018. *Hidden Chains: Rights Abuses and Forced Labor in Thailand's Fishing Industry*. https://www.hrw.org/report/2018/01/23/hidden-chains-rights-abuses-and-forced-labor-thailands-fishing-industry#7694/.

Launius, Roger. 2014. "The Railroads and the Space Program Revisited: Historical Analogs and the Stimulation of Commercial Space Operations." *The International Journal of Space Policy & Politics* 12: 167–179. https://doi.org/10.1080/14777622.2014.964129.

Licht, Walter. 1983. *Working for the Railroad: The Organization of Work in the Nineteenth Century.* Princeton, NJ: Princeton University Press.

Musk, Elon. 2017. "Making Humans a Multi-Planetary Species." *New Space* 5, no. 2: 46–61. https://doi.org/–.–/space.2017.29009.emu.

Neufield, Michael J. 2007. *Von Braun: Dreamer of Space, Engineer of War.* New York: Random House.

Ohnsman, Alan. 2019. "Inside Tesla's Model 3 Factory, Where Safety Violations Keep Rising." *Forbes*, March 1, 2019. https://www.forbes.com/sites/alanohnsman/2019/03/01/tesla-saf ety-violations-dwarf-big-us-auto-plants-in-aftermath-of-musks-model-3-push/.

Patel, Neel V. 2016. "SpaceX Wants to Become the Union Pacific Railroad to Mars." *Inverse*, September 27, 2016. https://www.inverse.com/article/21490-elon-musk-mars-spacex-union-pacific-railroad-mars/.

Patel, Neel V. 2017. "SpaceX Must Pay $4 Million for Thousands of Underpaid Employees." *Inverse*, May 11, 2017. https://www.inverse.com/article/31478-spacex-settles-underpaid-workers-lawsuit-for-4-million.

Robertson, Adi. 2016. "SpaceX Wants to Be the Railroad of the Future." *The Verge*, September 27, 2016. https://www.theverge.com/2016/9/27/13080970/spacex-elon-musk-mars-expedit ion-railroad-of-the-future/.

Rohrbough, Malcolm J. 1997. *Days of Gold: The California Gold Rush and the American Nation.* Berkeley, CA: University of California Press.

Roulette, Joey. 2021. "'Rife with Sexism': Employees of Jeff Bezos' Blue Origin Describe 'Toxic' Workplace Culture." *The Verge*, October 1, 2019. https://www.theverge.com/2021/9/30/ 22702335/jeff-bezos-employees-essay-blue-origin-toxic-workplace.

Sainato, Michael. 2020. "'I'm Not a Robot': Amazon Workers Condemn Unsafe, Grueling Conditions at Warehouse." *The Guardian*, February 5, 2020. https://www.theguardian.com/ technology/2020/feb/05/amazon-workers-protest-unsafe-grueling-conditions-warehouse.

Sheetz, Michael. 2021. "Justice Department Investigating Elon Musk's SpaceX following Complaint of Hiring Discrimination." *CNBC*, January 28, 2021. https://www.cnbc.com/ 2021/01/28/doj-investigating-spacex-after-hiring-discrimination-complaint-.html.

Wall, Mike. 2015. "New Space Mining Legislation Is 'History in the Making.'" *Space.com*, November 20, 2015 https://www.space.com/31177-space-mining-commercial-spaceflight-congress.html.

Wei, Willi. 2015. "Peter Diamandis: The First Trillionaire Is Going to be Made in Space." *Business Insider*, March 2, 2015. https://www.businessinsider.com/peter-diamandis-space-trillionaire-entrepreneur-2015-2.

White, Richard. 2012. *Railroaded: The Transcontinentals the Making of Modern America.* New York: W. W. Norton & Company.

Zubrin, Robert. 2011. *The Case for Mars: The Plan to Settle the Red Planet and Why We Must.* New York: Simon & Shuster.

21

Reclaiming Lunar Resources

Paving the Way for An International Property Rights Regime for Outer Space

Ruvimbo Samanga

The Moon Is Made of Cheese

There's a global race to the center of our universe and our first pitstop is cheese. That's correct, the Moon is made up of cheese, and the lunar outpost is as dynamic as cheese comes. See, cheese is incredibly versatile, differing in value, texture, aroma; it's almost an enigma of sorts. No wonder there's clamor to get even one sample of lunar cheese.

But you know another curious fact about this delicacy? One slight alteration to the cheese-making process and you get a totally different outcome each time; the style is surely what you make of it. For the most part, cheese-making is much to the discretion of the connoisseur, and what a great number of them we have across the globe. A large, curious, voracious cheese-eating globe.

Perfect, until you consider that the Moon being made of cheese is a metaphor for human gullibility. It would be gullible to think that the Moon is some infinite resource. It would be incredulous to believe that each "connoisseur" is at liberty to determine their own processes. And it would be absurd to consider that everyone will have equal claims to this collective heritage.

That said, the Moon is not in fact made of cheese, but rather a delicate combination of mainly oxygen, silicon, magnesium, calcium, and iron (Valencia 2019, 1). According to lunar petrologist Sarah Valencia (2019, 2), the Moon began as a molten conglomeration of elements during what is known as its Lunar Magma Ocean (LMO) stage. Bear with me through the geological exposition which follows.

As the LMO began to solidify, common rock-forming minerals—the most common being plagioclase—rose to the top of the lunar crust, while denser minerals (olivine and pyroxene) sank to the bottom. These are the so-called "compatible" elements, distinguished from the "incompatible" elements

Ruvimbo Samanga, *Reclaiming Lunar Resources* In: *Reclaiming Space*. Edited by: James S. J. Schwartz, Linda Billings, and Erika Nesvold, Oxford University Press. © Oxford University Press 2023. DOI: 10.1093/oso/9780197604793.003.0021 known as KREEP (Potassium [K], rare earth elements [REE], and phosphorus [P]), which formed the sedimentary layer of the LMO.

Compatible elements are enriched in their solid state, while incompatible elements are those that are inclined to a liquid state. This has an impact on where these elements are more commonly found, i.e., compatible elements along the Moon's crust (surface elements as it were), and incompatible elements deeper down toward the mantle.

This analysis is vital when one begins to consider that industrial development has long hinged on the availability of material resources, and accordingly, space industry development toward an eventual in-space economy would be no different. In the words of planetary scientist Ian Crawford (2015, 137), "as the Earth's closest celestial neighbour, the Moon seems likely to play a major role." That said, understanding the geological make-up and location of these elements points toward their potential for viable extraction and use.

The Rush to Space

What is fueling the current rush to space is essentially a competition to secure some form of property ownership over space resources. Property rights in terrestrial regimes come in tangible and intangible forms, with tangible property relating to those assets that can be perceived by the five senses, and intangible assets forming the converse.

Despite not being explicitly defined in space treaties, examples of tangible property in this context would be satellite components, satellite systems, or launch vehicles, while intangible property would include the patents, copyrights, insurance, contracts, partnership interests, and investments associated with these, otherwise known as intellectual property.

Having these rights over property directly correlates to the potential to make some form of monetary gain through commercial transfer, by ascribing a value to the asset or concomitant right in question. While international commercial laws regarding terrestrial sales are well regulated (particularly under the International Convention on the Sale of Goods),[1] the transferring of property rights, and whether this is yet permissible in the context of space resources (aside from developing state practice), is still unclear.

Consider that the right to transfer a satellite's jurisdiction to a non-launching State is yet to be articulated in international treaty laws, and the claim to commercial rights over space resources is even more contested (Othman 2017, 68). This *lacunae* in the law, while the main focus of this discussion, has not hindered progress in lunar development, as the following discourse will demonstrate.

In an interesting development, NASA has agreed to compensate four space companies that will collect and transfer ownership of space resources from the Moon. In a press release, NASA (2020) reported a $25,001 contract would be split between Lunar Outpost ($1), ispace Japan ($5000), ispace Europe ($5000), and Masten Space Systems ($15,000). This contract is awarded in recognition of the role which space resources will play in supporting NASA's Artemis lunar development program and future space endeavors (NASA 2020).

In facilitating such a contractual award, NASA is setting two core precedents: (1) that the appropriation of nonland-resources on the Moon is permissible and (2) that the commercial transaction of space resources is permissible (Schingler 2021). This would indicate that the international space law treaty system grants sufficient powers to states to enact national laws, especially concerning space resource utilization.

Determining the limits of such exercise becomes the focal point of ensuring sustainable practices in outer space, and reducing the risk of developing customary international laws which infringe upon the rights of all stakeholders, developing countries, and underserved communities especially. To this end, the author proposes that the time is ripe to reclaim lunar resources on an equitable basis for all, by establishing a robust property law system.

To Mine or Not to Mine?

There are three main reasons for pursuing lunar resource extraction, namely, supporting (1) lunar development in a process otherwise known as in situ resource utilization (ISRU); (2) pace development in cislunar orbit; and (3) resource development to directly support the global economy (Crawford 2015, 139).

The available resources in question? Solar-wind-implanted volatiles, helium-3 (whose abundance is frequently exaggerated), water (in the form of water ice or water-producing molecules), oxygen, metals (including iron and the iron-loving siderophile elements, titanium, and aluminum), silicon, rare-earth elements and potentially even uranium and thorium (Crawford 2015, 142–153). In light of all of this perceived value: To mine or not to mine? That is the question.

Despite express provisions on space mining itself, international space law does not operate in a legal vacuum, quite the contrary. The so-called

Magna Carta or constitution of the outer space legal framework is the 1967 United Nations Treaty on Principles Governing the Activities of States in the Exploration and Use of Outer Space, including the Moon and Other Celestial Bodies.[2]

It is generally understood that this international framework is guided by three core principles, namely, freedom of exploration and use of space; peaceful uses of space; and state responsibility (Masson-Zwaan 2017, 3). It necessarily follows that any activities, even those concerning resource utilization, will have to align with these fundamental provisions.

In addition to customary laws arising through state practice, and various principles adopted to guide states, these provisions form the *corpus* of international space law. A much broader outlook will reveal that general international law also has a bearing on space activities, since space law belongs to the public international branch, as it pertains to the governance of state activities (Outer Space Law, n.d.). Accordingly, gleaning principles from international commercial laws will give insight on how to manage the gap in property rights concerning space resources.

While doing so, one must note that international space law treaties were drafted at a time when many contemporary space activities could not have been envisioned, which necessitates new legal project management. Nevertheless, as with all treaty systems, it can be agreed that the Outer Space Treaty balances both rights and obligations (Masson-Zwaan 2017, 4). Therefore, the remainder of this article will focus on reasonable recommendations for the extension of a balancing of rights and obligations in a multistakeholder system of space resource utilization. But first, an exposition on existing laws.

The very first article of the Outer Space Treaty gives us a glimpse into the extent to which space mining may be permissible in terms of international space laws. It begins by asserting that space activities must have societal benefit at heart in its "province of all mankind" declaration, emphasizing that scientific, economic, and developing country status may not be an impediment to enjoying the fruits of space endeavors. The second paragraph reads: "Outer space, including the Moon and other celestial bodies, shall be free for exploration and use by all States without discrimination of any kind, on a basis of equality and in accordance with international law, and there shall be free access to all areas of celestial bodies."

Three important points can be derived from this provision: (1) states are entitled freedom of access to explore and use outer space, and implicitly its resources; (2) such freedom of access must be unfettered, equitable and determined by internationally accepted standards; and (3) no area of celestial terrain shall be deemed off limits for use and exploration. In rounding off this provision, a clarion call for international cooperation is made, prompting states to facilitate greater interoperability in achieving this intricate goal.

By virtue of Article I of the Outer Space Treaty one can conclude that space commercialization, and more specifically space mining, does indeed fall within acceptable use and exploration of outer space, as "use" in this context is generally agreed by scholars to include economic and non-economic exploitation (Hobe 2009a, 35).

The second point, however, is not so straightforward, as it would seem this would be the first indication that there is a gap in terms of the current treaty system on the existence of an internationally accepted framework for space resource utilization.[3] In the absence of distinct rules on ownership, determining the limits of non-discriminatory and equitable uses of space becomes a haphazard affair.

The third point is even more complicated, considering the emergence of safety zones on the Moon associated with areas of cultural and environmental significance. What claims to property would enforcers of safety zones have, and to what extent then would another stakeholder have freedom to access such zones, given that no area of a celestial body shall be barred from exploration and use.

Considering increased private sector participation in space activities evident in recent years (Hobe 2009, 4), this exercise can only be expected to become more onerous, as national regulations, particularly those pertaining to doing business in space, require revamping, environmental law continues to be extended to new scenarios as we consider human exploration of extraterrestrial terrain, and even international law has a bearing on the way States select laws of application in dispute resolution (Scandinavian Institute of Maritime Law, n.d.). The first step to alleviating this emerging challenge would be to clarify a few misconceptions.

Who Really Owns Space Resources?

So now that resource extraction capabilities are beginning to dawn on us, who owns the Moon? By international space law standards, it would be no one in particular and everyone at large. After all, the Outer Space Treaty designates outer space resources as the "province of all mankind." Article II of the Outer Space Treaty points toward collective ownership: "Outer space, including the Moon and other celestial bodies, is not subject to national appropriation by claim of sovereignty, by means of use or occupation, or by any other means."

Masson-Zwaan (2017, 46) argues that the lengthy elaboration of the treaty is done to signify that non-appropriation not only extends to physical bodies, but also to void space. So not only does no one own the Moon, but its cis-lunar surrounds as well. During the 1968 drafting process, however, a definitional lack occurred in determining what constitutes a celestial body, and whether this is limited to large bodies such as planets or would extend to include smaller units such as asteroids.

Nevertheless, in 1969, the Apollo 11 mission touched down on the Moon approximately 20 kilometers south-southwest of the Sabine D crater, in the southwestern part of *Mare Tranquillitatis* (the Sea of Tranquility). The mission's objectives? To walk on, observe, measure, collect and examine lunar surface phenomena (*Apollo 11: "One Giant Leap"*, n.d.).

According to the Apollo Lunar Surface Journal, astronaut Neil Armstrong collected the bulk sample of lunar material, which consisted of 15 kilograms of rock and soil, collected in a total of twenty-two or twenty-three scoops. (NASA 1977, 14). In addition to forty-eight rocks, fine lunar regolith was returned, composed of basalts (volcanic rock), breccias (mineral sediments), and soil (consisting of a number of components including igneous rock and mineral fragments, breccia, glass spheres, microanorthositic/plagioclase fragments, and meteoritic material; NASA 1977, 29–35). This was the first mission to carry geological samples of the Moon to Earth (Lunar and Planetary Institute, n.d.).

From 1970 to 1976, Russia followed suit and launched four successful robotic lunar sample return missions: Luna 16, 20, 23, and 24 (NASA, n.d.). Nearly five decades later, China became the next nation to obtain a lunar sample, after its Chang'e-5 spacecraft returned 1,731 grams of Moon material to Earth (Associated Press 2020). Thus, despite sufficient regulation thereto, claims over nonland resources have become custom in governmental space endeavors. This has hardly remained a state-centric ambition, however. In 2016, a private company, Moon Express, was for the first time granted US government permission to land on and extract resources from the Moon (Wall 2016).

Space exploration is no longer a solely government affair. Several private companies have expressed interests in space mining, including Shackleton Energy Company, Golden Spike (which closed its doors in 2013), Astrobotic Technology, Deep Space Industries, and Planetary Resources (the latter two companies were purchased by other corporations and are no longer in the business of space mining). So while the international governance framework lacks an appropriate property regime, especially with regards to material derived from outer space, state practice would indicate that sufficient custom is developing to allow for private and commercial transfer of space resources for economic benefit. A balancing of rights and interests is required to ensure the societal leg of multistakeholder interests in outer space resources are enforced accordingly.

An important aspect of ensuring that these activities remain consistent with societal interests is upholding international treaty laws. While nongovernmental actors are not directly liable to the obligations under treaty law (Gerhard 2009, 111), the last sentence of Article VI of the Outer Space Treaty explicitly states that parties to the Treaty are to assure that "national activities are carried out in conformity with the provisions set forth" (Gerhard 2009, 116). It is then the responsibility of states to authorize activities permissible for nongovernmental actors, in the context of the broad obligations expressed in the Treaty systems, both space law and general international law, the latter of which would include international property laws.

What this implicitly creates is a responsibility on member states to enact enabling national policies which not only speak to the desired outcomes of national space programs (Othman 2017, 56), but also delineate legal sanctions for actions which go against the normative international law system. That said, if a nation's intention is to foster the use of space resources for economic or even scientific purposes, an accompanying property regime must be established in conformity with internationally accepted standards to manage the scope of claims over such resources.

In keeping with this responsibility, nation states have begun to enact national legislation pertaining to space resource utilization. On June 15, 2021, Japan became the fourth country to introduce a national legal, regulatory and policy framework for the exploration and exploitation of space resources ("Japan Fourth Country in the World" 2021).[4] While these acts are yet to delve into the intricacies of defining space property, establishing norms for the transfer of such property, or determining how disputed claims can be resolved, they at the very least reinforce the notion that stakeholders are at liberty to use and explore space resources for both economic and noneconomic purposes.

As states and private entities continue to sample portions of the lunar surface, questions will arise as to what constitutes ownership over such property, and related issues concerning transfer, dispute resolution and security of tenure. This chapter, however, is focused on space resources and not territorial claims over land or lunar zones. The latter is sufficiently exhausted by virtue of the nonappropriation principle expressed in Article II of the Outer Space Treaty, wherein it is agreed that states may not lay claims of sovereignty over any celestial body, which lays to rest the age-old query, "Can one sell the

Moon?" No, one cannot. And while Russia's claims to Venus in 2020 may have been a show of technological prowess, as opposed to a declaration of outright ownership, it brought to light important considerations on the limits of sovereignty over celestial bodies.

Reclaiming Lunar Resources

Crawford (2015, 159) agrees that the development of lunar resources will require international legal interventions to (1) spur large-scale investments in prospecting and extraction activities; (2) manage geopolitical interests; and (3) sustain equitable resource benefits in the interests of all stakeholders. Investments, state interests, and benefits sharing are all extensions of property rights. Therefore, in a nutshell, lunar resource development will neither achieve full potential nor derive societal benefit without an equitable and comprehensive property rights regime.

This is where analysis of the meaning of ownership becomes particularly intricate. Existing frameworks do not provide sufficient definition as to what constitutes ownership in the context of space assets in general, and space resources in particular. However, what is already clear is that deriving monetary benefit from the sale of space resources on the basis of intellectual/property rights would be difficult to read into and sustain in the wording of the current treaty system, and thus requires a separate regime.

Such a novel framework must, at its core, distinguish between ownership and possession, where ownership is the ability to exclude nonowners from the property (Katz 2008). On the basis of the nonappropriation principle, it is clear that exclusive claims and transfers of ownership would not be permissible in a space resources property regime.

On the other hand, possession, which is a Roman law principle found in a number of civil and common law jurisdictions, denotes physical domination over a thing, with a responsibility to safeguard against loss or damage, while enjoying a concomitant right to its benefits while it is in possession (NVS 2018). This definition would be more aligned with a property regime geared toward a global commons, as it does not violate the nonappropriation principle by granting sovereign title of a resource. Instead, it recognizes and protects the right to make economic use of property and thus aligns with the freedom to use space resources. The rights of the possessor are lesser than those of the owner, who has sovereign title to exclude use of the property.

Thus, if outer space belongs to all, it would be consistent with ownership falling in the hands of the collective, which can then exclude or authorize the right to possess certain resources. An appropriate international property regime for space resources is one which will then lay the basis of legal claims for space resources on the right to possession (that is, use and enjoyment of benefits), while vesting the right to own outer space resources in the hands of all humankind.[5]

Conclusion

Professor of public and international law Stephan Hobe (2009b, 2) notes that during the initial drafting of the Outer Space Treaty, interests of developed versus developing states differed, leading to calls for a new international economic order on behalf of the latter, which ultimately had an influence on the final document. At this turning point in space development, similar sentiments begin to arise, as we consider the geopolitical impact of terrestrial resource development, and what that may look like off-world. We must begin to consider the lessons we hope to carry forward, and the mistakes better left for the cold archives of history.

We must also consider the role those different regions will assume in international cooperative efforts toward lunar industry development. Africa, with all its expertise in managing multistakeholder interests in the terrestrial mining industry, has a large role to play in fostering research toward an international framework on space mining, based on Africa's rich extractive industry experience. Such a framework can be derived from the Africa Mining Vision,[6] which supports the principle of subsidiarity. This principle, the author believes, is at the heart of the outer space treaty system, a principle which extends rights to nation states, to manage their own affairs under a predetermined and accepted international framework.

According to legal scholar R.K. Vischer (2001, 106), subsidiarity is a devolutionary principle that reallocates social function from higher to lower government bodies, and from government to nongovernmental bodies. This principle stems from constitutional law, and is adopted to regulate authority within a political or legal order (Follesdal 2011, 4). As such, subsidiarity would find application in space governance, which has a particularly unique political order, requiring sound democratic underpinnings.

In the words of legal scholar Frans von der Dunk (2020), outer space "belongs to no state, and is not subject to appropriation, though its resources are." He also opines that the governance of such a global commons cannot be left to rest on a single nation alone. And it is these thoughts which bring the author to conclude that the lack of a central administrative mechanism for enforcement of space governance protocols will continue to impede the development of international legal instruments, such as those proposed for property rights, satellite mega-constellations, space debris mitigation, and space traffic management, to name a few.

The principle of subsidiarity supports the building of a constitutional framework for outer space by providing certain and uniform rules by which all stakeholders can abide and adapt according to needs, which is why it is often referred to as a "structuring principle."[7] By so doing, outer space governance can embrace the centrality of sovereign states, meaning that public international law plays a supporting role to domestic law. Furthermore, a state may still consent to creating legal obligations, and is afforded a "Margin of Appreciation" to enforce obligations and exercise national remedies (Follesdal 2011, 7). These pillars remain consistent with state responsibility as envisioned in Article 6 of the Outer Space Treaty, while acknowledging the overarching need for an effective administrative body, which oversees the progress of the decentralized units. Such polycentric governance would prove appropriate for the outer space context.

Polycentricity is defined by three core features, namely, (1) multiple governance centers or hierarchies; (2) an overarching, single system of rules; and (3) spontaneity in the development of new legal orders and institutions (Kuhn 2021). These features make the polycentric governance model a useful intervention for advancing international frameworks for outer space.

The need for international frameworks cannot be understated. Hobe (2009b, 5) opines that space law drafting is currently in the grips of a crisis in the medium to long term. It will not continue to suffice that nonbinding instruments and bilateral agreements replace international normative and binding rules for outer space activities.

In conclusion, the recognition of an international property law on space resources will have a positive bearing on (1) regulation of the global commons (by establishing a uniform normative framework); (2) coordination of transboundary property rights (such as in the case of transfer of lunar resources across terrestrial and extraterrestrial jurisdictions); (3) adoption of global policies to protect against specific harm (i.e., export and import controls on materials which may pose a threat of harmful contamination, or prohibition on the transfer of possession of artifacts/resources of cultural/environmental significance); and finally, (4) protection of human rights (in safeguarding the property interests of marginalized, vulnerable, Indigenous, underserved, or developing country communities; Sprankling 2012, 472).[8]

Ultimately, the instrumental value of establishing property rights in space resources will create doctrine that is certain, administrable, teachable, and adaptable in the context of a future which is as expansive as the universe itself.

Notes

1. This is a multilateral treaty that establishes a uniform framework for international commerce.
2. This Treaty is further supported by the Agreement on the Rescue of Astronauts, the Return of Astronauts and the Return of Objects Launched into Outer Space, 1968 (Rescue Agreement); Convention on International Liability for Damage Caused by Space Objects, 1972 (Liability Convention); the Convention on Registration of Objects Launched into Outer Space, 1976 (Registration Convention); and the Agreement Governing the Activities of States on the Moon and Other Celestial Bodies, 1984 (Moon Agreement).
3. For a proposal on a possible international economic framework for space resource utilization see author's (Ruvimbo Samanga) publication on "ISRU: Lessons Learnt from Mining Governance in Africa" (2021) Available at https://www.openlunar.org/library/isru-lessons-learnt-from-mining-governance-in-africa/.
4. The National Diet of Japan passed the Law Concerning the Promotion of Business Activities Related to the Exploration and Development of Space Resources. On May 21, 2015, the US House of Representatives passed a revision of the then ASTEROIDS Act, which is now referred to as the Space Resource Exploration and Utilization Act. This would become the first national bill on space resource utilization. Luxembourg followed shortly with its own legal intervention, enacting the Space Resources Law on August 1, 2017, in accordance with which any stakeholder may conduct a space resource mission for commercial purposes upon written mission authorization from the Ministers in charge of economic and space affairs. In December 2019, the UAE enacted its Federal Law No. 12 of 2019, on Regulation of the Space Sector, which, amongst other things, made provision for the utilization of space resources, namely in Article 4, which regulates the collection or trade of meteorites which fall into the UAE.
5. As the regime develops, additional considerations can be made concerning the transfer of possession, such as the setting of a fixed, nominal cost for transfer to curb over-exploitation of space resources, as was precedented by the NASA contractual awards.
6. Available at https://au.int/en/ti/amv/about/.
7. See Vischer (2001).
8. The law of property can also serve to clarify other areas of space development, such as the allocation of geostationary satellite orbits, which can be treated akin to long-term leases.

References

Apollo 11: "One Giant Leap." n.d. Viewed 7 January 2022, https://sciowa.org/upl/downloads/library/apollo-11-fact-sheet.pdf.

Associated Press. 2020. "China Collects Moon Samples to Study on Earth." *The Washington Post*, December 2, 2020. https://www.washingtonpost.com/lifestyle/kidspost/china-colle cts-moon-samples-to-study-on-earth/2020/12/02/1a22977e-2ad2-11eb-92b7-6ef17b3fe3b 4_story.html.

Crawford, I. A. 2015. "Lunar Resources: A Review." *Progress in Physical Geography* 39, no. 2: 137–167. https://journals.sagepub.com/doi/full/10.1177/0309133314567585.

Follesdal, A. 2011. "Global Governance as Public Authority: Structures, Contestation, and Normative Change." Working Paper, December 2011. Available at https://jeanmonnetprog ram.org/wp-content/uploads/2014/12/JMWP12Follesdal.pdf.

Gerhard, M. 2009. "Article VI Outer Space Treaty." In *Cologne Commentary on Space Law*, edited by S. Hobe, B. Schmidt-Tedd and K. U. Schrogl, 103–123. Köln: Carl Heymanns Verlag.

Hobe, S. 2009a. "Article 1." In *Cologne Commentary on Space Law*, edited by S. Hobe, B. Schmidt-Tedd and K. U. Schrogl, 25–43. Köln: Carl Heymanns Verlag.

Hobe, S. 2009b. "Historical background." In *Cologne Commentary on Space Law*, edited by S. Hobe, B. Schmidt-Tedd and K. U. Schrogl, 1–24. Köln: Carl Heymanns Verlag.

"Japan Fourth Country in the World to Pass Space Resources Law." 2021. *Spacewatch*, June 16, 2021. https://spacewatch.global/2021/06/japan-fourth-country-in-the-world-to-pass-space-resources-law/.

Katz, L. 2008. "Exclusion and Exclusivity in Property Law." *University of Toronto Law Journal* 58, no. 3: 275–315. https://www.utpjournals.press/doi/abs/10.3138/utlj.58.3.275/.

Kuhn, L. 2021 "Introduction to Polycentricity." Blog post. *Open Lunar Foundation*, May 19, 2021. https://www.openlunar.org/library/introduction-to-polycentricity/.

Lunar Planetary Institute. n.d. "Apollo 11 Lunar Samples." Accessed January 21, 2022. https://www.lpi.usra.edu/lunar/missions/apollo/apollo_11/samples/.

Masson-Zwaan, T. 2017. "The International Framework for Space Activities." In *Handbook for New Actors in Space*, edited by C. D. Johnson, 2–50. Washington, DC: Secure World Foundation.

National Aeronautics and Space Administration (NASA). n.d. "Lunar Reconnaissance Orbiter." Accessed January 19, 2022. https://www.nasa.gov/mission_pages/LRO/multimedia/lroima ges/lroc-20100316-luna.html.

National Aeronautics and Space Administration (NASA). 1977. "Apollo-11 Lunar Sample Information Catalogue (Revised)." Accessed January 7, 2022. https://www.hq.nasa.gov/off ice/pao/History/alsj/a11/A11SampleCat.pdf.

National Aeronautics and Space Administration (NASA). 2020. "NASA Selects Companies to Collect Lunar Resources for Artemis Demonstrations." Accessed January 23, 2022. https://www.nasa.gov/press-release/nasa-selects-companies-to-collect-lunar-resources-for-arte mis-demonstrations/.

NVS. 2018. "Occupation, Possession, Transfer and the Passing of Risk in a Property Transaction." Blog post. July 4, 2018. https://nvsinc.co.za/occupation-possession-transfer-and-the-passing-of-risk-in-a-property-transaction/.

Othman, M. 2017. "National Space Policy and Administration." In *Handbook for New Actors in Space*, edited by C. D. Johnson, 54–87. Washington, DC: Secure World Foundation.

Scandinavian Institute of Maritime Law. n.d. "Outer Space Law." Accessed January 8, 2022. https://www.jus.uio.no/nifs/english/research/areas/outer-space-law/.

Schingler, J. K. 2021 "Lunar Resource Precedents and the Definition of 'Collected.'" *Moon Wonk*. Blog post, February 2. https://getproofed.com/writing-tips/how-to-cite-a-blog-post-in-harvard-referencing/.

Sprankling, J. G. 2012. "The Emergence of International Property Law." *North Carolina Law Review* no. 90: 461–509.

Valencia, S. 2019. "Top 5 Elements on the Surface of the Moon." PDF Presentation. International Year of the Periodic Table of Chemical Elements 2019. Accessed January 5, 2022. https://www.lpi.usra.edu/education/IYPT/Moon.pdf.

Vischer, R. K. 2001. "Subsidiarity as a Principle of Governance: Beyond Devolution." *Indiana Law Review* 35, no. 103: 103–142. https://mckinneylaw.iu.edu/ilr/pdf/vol35p103.pdf.

Von der Dunk, F. 2020. "Structuring the Governance of Space Activities Worldwide." *Georgia Journal of International and Comparative Law* 48, no. 3: 645–659.

Wall, M. 2016. "Private Company Cleared for Moon Landing in 2017." *Scientific American*, August 3, 2016. https://www.scientificamerican.com/article/private-company-cleared-for-moon-landing-in-2017/.

22

Starlink or Stargazing

Will Commerce Outshine Science?

Tanja Masson-Zwaan

The Problem and the Outer Space Treaty

Everyone has heard about pollution of Earth's land, air, seas, and oceans by plastics, chemicals, and other sorts of waste. Action to address such pollution is undertaken at various levels by governments, politicians, industry, and civil society. Most are also familiar with "light pollution" caused by city lights in built-up areas. A lesser-known phenomenon is that of pollution of the night skies from outer space. Yet, there is a growing concern that the observation of stars and planets by professional and amateur astronomers, both in terms of ground-based optical astronomy and radio astronomy, will be severely hampered by the huge number of artificial satellites in outer space. Optical astronomy suffers from the light pollution caused by the reflection of the Sun on these satellites, whereas radio astronomy suffers from the radio interference they cause.

Several private space companies plan to launch extremely large numbers of small satellites into low Earth orbit (LEO) to provide communications services all over the globe. These projects are commonly referred to as mega-constellations, or large constellations.[1] The four main constellations known in detail as of mid-2021 would comprise more than 50,000 satellites. They are:

- SpaceX, with its Starlink constellation (30,000 satellites),
- Amazon, with its Kuiper constellation (7,774 satellites),
- OneWeb constellation (6,372 satellites), and
- China with its Guangwang constellation (12,992 satellites) (McDowell, n.d.).

By mid-2021, SpaceX had already launched some 1,700 Starlink satellites. That is around 50 percent of all active satellites that are currently in orbit (Mohanta 2021). And that is not all. In November 2021, several companies

Tanja Masson-Zwaan, *Starlink or Stargazing* In: *Reclaiming Space*. Edited by: James S. J. Schwartz, Linda Billings, and Erika Nesvold, Oxford University Press. © Oxford University Press 2023. DOI: 10.1093/oso/9780197604793.003.0022 filed requests with the US Federal Communications Commission (FCC) for new or expanded broadband networks, asking for approval of nearly 38,000 satellites. The largest requests came from Astra (13,620 satellites), Boeing (5,921 satellites), Telesat (1,969 satellites), and Hughes (1,440 satellites; Sheetz 2021). To make things even worse, in October 2021 Rwanda, a country that has no space industry to speak of and just one small satellite in orbit (UNOOSA, n.d.), announced that it had requested frequency rights for a LEO constellation of a staggering 327,000 satellites (Rwanda Space Agency 2021).[2]

Certainly, there are many areas on Earth where internet access is rare or not available at all, even in highly developed countries such as the United States. The provision of affordable and stable internet coverage for remote areas by multiple satellites by commercial companies could be considered as compliant with Article I of the 1967 Outer Space Treaty, which was adopted by the United Nations Committee on the Peaceful Use of Outer Space (UN COPUOS) and ratified by the vast majority of states. It provides, in part, that

> The exploration and use of outer space, including the moon and other celestial bodies, shall be carried out *for the benefit and in the interests of all countries*, irrespective of their degree of economic or scientific development, and shall be the province of all mankind.
>
> Outer space, including the moon and other celestial bodies, shall be *free for exploration and use* by all States without discrimination of any kind, on a basis of equality and in accordance with international law, and there shall be free access to all areas of celestial bodies [emphasis added].[3]

But is the availability of broadband internet worth the deterioration or loss of astronomical observations? Do we really need that many satellites? How can we manage the unbridled ambitions of "NewSpace" entrepreneurs? The Outer Space Treaty provides some guidance on how to handle private commercial space activity. Article VI places international responsibility for national space activities with the State, and that includes private commercial space activities. States are obliged to authorize and continuously supervise such activities, and they often do that through national space legislation. The treaty also addresses the issue of harmful interference with the activities of other States, in Article IX. It provides that states must have due regard for the activities of other states, and that in case of harmful interference, international consultation should take place to address the issue.

There is also an indication that the treaty recognizes the importance of science, in that same Article I:

There shall be *freedom of scientific investigation* in outer space, including the moon and other celestial bodies, and States shall facilitate and encourage international co-operation in such investigation [emphasis added].

Science is indeed hugely important to society, and we must realize that without astronomy there would be no space commerce.[4] Astronomy has brought immense knowledge to humankind. Moreover, observatories are often built and operated by public funding at national and international levels, and they must be protected. Astronomy has been used traditionally for navigating at sea and in the air. Astronomy also forms an important part of the cultural heritage of Indigenous people. Aboriginal astronomy for instance goes back tens of thousands of years. Surely all of these are at least as important as broadband internet? So, the million-dollar question is: can a way be found to accommodate both science and commerce in a fair and equitable manner, for the benefit of humankind? The answer is two-fold. On the one hand, states (and in particular the United States where most operators are based) should try harder to strike a balance among the interests of different stakeholders through the process of authorization and supervision, and states whose (astronomical) activities are harmed should use the options given to them by the Outer Space Treaty and request consultations. On the other hand, UN COPUOS, as the primary international body for space diplomacy, should play its role by providing a forum for the exchange of views among States as well as all other stakeholders, in order to raise awareness and reach solutions.

Technical Reactions from the Industry

Some companies have already engaged in discussions with the astronomical community and have taken technical measures to reduce the reflection of sunlight caused by their satellites. This shows that they are willing to seek solutions, although such voluntary action by some is not a sustainable solution in the longer term, and there is no way to make satellites completely invisible to astronomical observations.

SpaceX has launched several experiments, such as *Darksat*, where a Starlink satellite was covered in black paint in an effort to reduce reflectivity, with some positive results (Witze 2020; Foust 2020). *Visorsat*, where a sun visor was added to the satellites to block sunlight from the white parts of the main body and the antennas, was more successful and resulted in a noticeable, though still insufficient, reduction of light pollution (Zhang 2020; Mallama

2021). Similarly, discussions are ongoing with other commercial operators, such as OneWeb. There are also plans to build more space-based telescopes, which could solve part of the problem (Skran 2020; Ralph 2021). But space-based astronomy also has certain drawbacks, as space telescopes cannot easily be maintained or repaired,[5] and there are financial and technical limits on the mass and size that can be sent to space. Moreover, amateur astronomers will most likely not have access to such space-based observing platforms.

Astronomers' Initial Reactions and Why They Did Not Work

The initial focus of some astronomers has been on trying to protect the dark skies as a human right. A "Starlight Initiative" and "Starlight Declaration" were issued in 2007, claiming that "an unpolluted night sky should be considered an inalienable right of humankind," and promoting a "World Declaration on the Right to the Starlight as a common heritage of mankind" (The Starlight Initiative, n.d.). However, there is no human-rights instrument under current international law that recognizes or codifies a human right to unimpeded observation of the night skies.[6]

It has also been suggested that astronomers should bring a claim before the International Court of Justice (ICJ; Gallozzi et al. 2020).[7] But the ICJ only settles disputes between states which have recognized its jurisdiction; it cannot be seized by individuals or non-governmental organizations such as astronomical associations (O'Callaghan 2020).[8] Moreover, the United States has not recognized the jurisdiction of the Court, so a claim against the United States, from where the majority of constellations are being operated, is in any case not an option.

Yet another suggestion was to register the night skies as world heritage under the UNESCO World Heritage Convention.[9] Again, this will not work, because UNESCO can only receive proposals from member states to protect sites falling under their national jurisdiction and the night skies fall under no state's jurisdiction. Outer space is free for exploration and use by all states, as provided by Article I of the Outer Space Treaty.

This does not mean that UNESCO has not been active in the field of astronomy. It launched the "Astronomy and World Heritage Initiative" (AWHI) in 2004 to identify and preserve astronomical sites globally (UNESCO, n.d.). And in 2008, it created the "Portal to the Heritage of Astronomy" in cooperation with the International Astronomical Union (IAU), with which it concluded a memorandum of understanding (UNESCO 2008).

But UNESCO has been clear in rejecting the possibility of protecting the dark skies:

Taking into account the growing number of requests to UNESCO concerning the recognition of the value of the dark night sky and celestial objects, the World Heritage Centre made its first statement in 2007 underlining that the sky or the dark night sky or celestial objects or starlight as such cannot be nominated to the World Heritage List within the framework of the Convention concerning the Protection of the World Cultural and Natural Heritage.

The World Heritage Centre wishes to underline that the "Starlight" Initiative developed by a group of international experts is not part of the UNESCO Thematic Initiative "Astronomy and World Heritage".

[. . .] neither Starlight Reserves, nor Dark Sky Parks can be recognized by the World Heritage Committee as specific types or categories of World Heritage cultural and natural properties since no criteria exist for considering them under the World Heritage Convention. (UNESCO, n.d.)

More recently, an international petition titled "Safeguarding the Astronomical Sky" was launched and has been signed by more than two thousand astronomers (Gallozzi 2020). They request governments, institutions, and agencies around the world to, amongst others, provide legal protection to ground-based astronomical facilities, put on hold further launches of large constellations, and impose a moratorium on all technologies that can negatively impact astronomical space- and ground-based observations or scientific, technological, and economic investments in astrophysical projects. They further seek a right of veto for national and international astronomical agencies on all projects that can negatively interfere with astronomical facilities. Again, these wishes are not realistic, and have no legal validity.

Thus, however appealing all these ideas may seem, none of them is legally tenable. Are there other, better options?

More Realistic Action by the Astronomical Community

Since the early 2020s, various national and international astronomical associations and organizations have brought the issue of interference by satellites to light. Most commercial actors developing large satellite constellations are based in, and operate from, the USA, so it is not surprising that the American

Astronomical Society (AAS) was the first to ring the alarm bells, by organizing SatCon1 in June–July 2020, followed by SatCon2 in 2021. The report contains valuable recommendations, for instance that coordinated international regulation of the satellite constellation industry is needed, including oversight and enforcement. It also raises the idea of industry slowing down until meaningful solutions can be developed. Further, it calls on governments to conduct due diligence concerning the activities of commercial satellite operators, specifically regarding the impact of in-orbit operation of such activities.

In Europe, the European Astronomical Society (EAS) has organized sessions on the impact of large constellations on astronomy at its annual conference in 2020, and again in 2021 and 2022. The EAS established a Working Group on satellite constellations focusing on the concerns of astronomers and space scientists in Europe.

At the global level, the IAU, which has 13,000 individual members, issued a statement on satellite constellations in 2019, stating: "[W]e urge appropriate agencies to devise a regulatory framework to mitigate or eliminate the detrimental impacts on scientific exploration as soon as practical" (IAU 2019). It issued a further statement in 2020, stating: "[T]he IAU considers the consequences of satellite constellations worrisome; . . . will continue to initiate discussions with space agencies and private companies" (IAU 2020). During a presentation at COPUOS in 2020, a representative of the IAU noted that "currently there are no internationally agreed rules or guidelines on the brightness of orbiting manmade objects" and asked to include the subject on the agenda (Benvenuti 2020). He also spoke the following true words: "Space users should be continuously reminded that their satellites would not fly nor properly communicate without the essential contributions that astronomy and physics have made to celestial mechanics, orbital dynamics and relativity."

In October 2020, the IAU organized a conference in cooperation with the UN Office for Outer Space Affairs (UNOOSA), named "Dark and Quiet Skies" (DQS), followed by a second edition in October 2021. In April 2021, the IAU submitted a Conference Room Paper (CRP) to the Scientific and Technical Subcommittee (STSC) of COPUOS with the support of several COPUOS member states, summarizing its recommendations (Chile et al. 2021).[10] In 2022, concrete results started to emerge, as the STSC adopted a new agenda item, titled "General exchange of views on dark and quiet skies for science and society" (Scientific and Technical Subcommittee 2022), while the IAU established the "IAU Centre for the Protection of the

Dark and Quiet Sky from Satellite Constellation Interference" (IAU CPS 2022).[11]

What Should the US Government Do (And What If It Doesn't)?

As explained, Article VI of the Outer Space Treaty provides for international responsibility of states for the activities of their national entities, which must be implemented at the national level through a process of authorization and continuing supervision, usually by means of national space legislation and a licensing process. Currently, most constellation projects operate from the United States, and the Federal Communications Commission (FCC) is the agency in charge of licensing. It could be argued that the FCC should carry out more thorough assessments on the (environmental) impact of the full project before granting authorization for large constellations. A problem is that in terms of environmental impact, the FCC is exempted from applying the National Environmental Policy Act (NEPA). This exemption is currently being challenged in a court procedure regarding authorization by the FCC of the Starlink constellation without applying NEPA (O'Callaghan 2020; Wall 2021; Ellis 2021). Whatever the outcome, it can be debated whether applying NEPA would make a big difference, since the impact of constellations on science is not purely of an environmental nature.[12] The UN Guidelines on the Long-Term Sustainability of Space Activities, adopted by COPUOS in 2019 (UNOOSA 2019), might be more helpful. States are expected to implement these guidelines in their national legal regimes. The US government should fulfill its obligation under Article VI of the Outer Space Treaty in a meaningful manner, and that includes due consideration and protection of the interests of all stakeholders, including the astronomical community, and the need to preserve the long-term sustainability of the space environment.

If another state feels that the US does not adequately fulfill its duty of oversight and this harms its activities, it may be possible to take international legal action. Article IX provides for a mechanism of international consultations in case of harmful interference, allowing states which suffer from harmful interference with their activities to request international consultation. This has not often happened yet but could well be used increasingly in the future as states feel their interests and investments in science are damaged. In addition, the effect of "naming and shaming" in international fora such as UN COPUOS should not be underestimated.

What Should COPUOS Members and Permanent Observers Do?

As referenced earlier in this chapter, COPUOS can play a role by providing an international forum for raising international awareness of the need to ensure the protection of space science, without which space commerce would not exist. In addition to the new STSC agenda item specifically dedicated to the Dark and Quiet Skies discussion mentioned above, relevant agenda items in the COPUOS Legal Subcommittee are the "General exchange of views on the application of international law to small-satellite activities," as well as the "General exchange of views on the legal aspects of space traffic management." The STSC agenda item on the "Long-term sustainability of outer space activities" also remains highly relevant, and discussions will continue there with input by the IAU and support from COPUOS member states. As a permanent observer at COPUOS, the IAU can make interventions and give presentations to raise awareness and convince states of the need to find a balance among the interests of all stakeholders, rather than only focusing on commercial interests. In addition to the IAU, which represents individual astronomers, there are two international intergovernmental organizations in the field of astronomy which can potentially have more impact. The "European Organisation for Astronomical Research in the Southern Hemisphere" (ESO) is already a COPUOS permanent observer, whereas a new international organization named "SKA Observatory" (SKAO), created in March 2021, has also recently been admitted as a permanent observer (SKAO Observatory 2021). They can help to amplify the voice of the astronomical community in COPUOS.

Conclusion

Should commerce prevail over publicly funded science? Should we allow the desires of commerce to override the needs of science? Is it fair, equitable, or ethical to let commerce dictate how outer space is used? No, clearly not. "Free enterprise" must not be allowed to prevail over publicly funded science. The problem is that science does not have as strong a voice as commercial "pioneers" do, and the astronomical community needs help in clearly formulating its message and broadcasting it to the correct channels. It needs to underline that without space science there would be no space commerce, but it also needs to explain exactly what the issues are and how it would like to see them solved. Commercial enterprise is vital to take space exploration to a next level; after all, private industry made rockets reusable, and the prospect of using resources of celestial bodies would be much less realistic without commercial enterprise investing and pioneering. However, commerce needs to be regulated, so that the "benefit" and "equity" principles of the Outer Space Treaty are respected. The ongoing privatization and commercialization of space activities places a much bigger responsibility on states in implementing Article VI than ever before. In addition to action at the national level, awareness must also be raised at the international level, and UN COPUOS as the prime forum for international space diplomacy and law-making, must take center stage in that respect. The adoption of a new agenda item in COPUOS and the establishment of the IAU CPS Centre are good steps towards establishing a balance that is reasonable to the ambitions of commerce and the needs of science, to protect the dark skies as well as global connectivity for the benefit of all humankind.

Notes

1. They have been defined as "a series of shells or 'elements' each with a fixed height and inclination and a given number of orbital planes and number of satellites per plane" (McDowell, n.d.).
2. Apparently, this move is being masterminded by Greg Wyler, a serial entrepreneur who was also involved in Google, O3b, and OneWeb (Forrester 2021). This claim is reminiscent of an ITU filing for several coveted geostationary satellite slots by the tiny Pacific island Tonga in 1988 ("TongaSat," n.d.).
3. Treaty on Principles Governing the Activities of States in the Exploration and Use of Outer Space, including the Moon and Other Celestial Bodies, adopted on January 27, 1967, 610 UNTS 8843. This will be further discussed below. See the full text and status of ratifications of the Treaty at https://www.unoosa.org/oosa/en/ourwork/spacelaw/treaties.html. See generally on space law: Masson-Zwaan and Mahulena Hofmann (2019).
4. See on this subject: Schwartz (2020).
5. The famous Hubble Space Telescope was repaired in space several times.
6. See, for instance, the Universal Declaration of Human Rights (UN General Assembly 1948) or the International Covenant on Civil and Political Rights (UN General Assembly 1966).
7. The cited paper contains many legal inaccuracies.
8. The ICJ is the highest judicial organ of the UN.
9. UN Convention concerning the Protection of the World Cultural and Natural Heritage, 1972. Its mission is to encourage the identification, protection, and preservation of cultural and natural heritage around the world considered to be of outstanding value to humanity. Member States may identify sites on their territory that deserve protection.
10. The areas covered included: (1) The Impact of Satellite Constellations on the Science of Astronomy, (2) Protection of Dark Sky Oases, (3) Protection of Ground-Based Optical

Astronomy Sites and Related Science, (4) Protection of the Bio-Environment, and (5) Protection of Radio Astronomy Sites and Related Science. Several COPUOS delegations who were sympathetic to most of the content could not subscribe to the entire set of recommendations because some elements did not fall within the mandate of COPUOS.

11. https://cps.iau.org/ (IAU CPS 2022)
12. Editor's note: See Chapter 23 for William R. Kramer's discussion of NEPA in the context of space.

References

Benvenuti, Piero. 2020. "The Impact of Mega-Constellations of Communication Satellites on Astronomy." Presentation at UNCOPUOS Scientific and Technical Subcommittee. February 7, 2020.

Chile et al. 2021. "Recommendations to Keep Dark and Quiet Skies for Science and Society." UN Doc A/AC.105/C.1/2021/CRP.17. April 19, 2021. https://www.unoosa.org/oosa/oosa doc/data/documents/2021/aac.105c.12021crp/aac.105c.12021crp.17_0.html.

Ellis, Michael. 2021. "Keep Environmental Red Tape out of Outer Space." *The Heritage Foundation*, no. 288, August 6, 2021, 1–13. https://www.heritage.org/government-regulat ion/report/keep-environmental-red-tape-out-outer-space.

Forrester, Chris. 2021. "Wyler behind Rwanda's 300,000 satellite plan." *Advanced Television*, November 8, 2021. https://advanced-television.com/2021/11/08/wyler-behind-rwandas-300000-satellite-plan/.

Gallozzi, Stefano. 2020. "Appeal by Astronomers: Safeguarding the Astronomical Sky." Blog post. January 9, 2020. https://astronomersappeal.wordpress.com/2020/01/09/astronomers-appeal/.

Gallozzi, Stefano et al. 2020. "Concerns about Ground Based Astronomical Observations: A Step to Safeguard the Astronomical Sky." arXiv (preprint only), February 4, 2020, 1–16. https://arxiv.org/abs/2001.10952v2.

Foust, Jeff. 2020. "SpaceX Claims Some Success in Darkening Starlink Satellites." *SpaceNews*, March 18, 2020. https://spacenews.com/spacex-claims-some-success-in-darkening-starl ink-satellites/.

IAU. 2019. "Statement on Satellite Constellations." June 13, 2019. https://www.iau.org/news/announcements/detail/ann19035/

IAU. 2020. "Understanding the Impact of Satellite Constellations on Astronomy." February 12, 2020. https://www.iau.org/news/pressreleases/detail/iau2001/

IAU CPS. 2022. IAU Centre for the Protection of the Dark and Quiet Sky from Satellite Constellation Interference. https://cps.iau.org/

Mallama, Anthony. 2021. "Starlink Satellites are Fainter Now—But Still Visible." *Sky & Telescope*, January 22, 2021. https://skyandtelescope.org/astronomy-news/starlink-satelli tes-fainter-but-still-visible/.

Masson-Zwaan and Mahulena Hofmann. 2019. *Introduction to Space Law*. Kluwer.

McDowell, Jonathan. n.d. "Section 2: Constellation Models." Starlink Simulations. https://pla net4589.org/astro/starsim/con.html.

Mohanta, Nibedita. 2021. "How Many Satellites Are Orbiting the Earth in 2021?" *Geospatial World*, May 28, 2021. https://www.geospatialworld.net/blogs/how-many-satellites-are-orbit ing-the-earth-in-2021/.

O'Callaghan, Jonathan. 2020. "Legal action could be used to stop Starlink affecting telescope images." *New Scientist*, February 3, 2020. https://www.newscientist.com/article/2232324-legal-action-could-be-used-to-stop-starlink-affecting-telescope-images/.

O'Callaghan, Jonathan. 2020. "The FCC's Approval of SpaceX's Starlink Mega Constellation May Have Been Unlawful." *Scientific American*, January 16, 2020. https://www.scientificamerican.com/article/the-fccs-approval-of-spacexs-starlink-mega-constellation-may-have-been-unlawful/.

Ralph, Eric. 2021. "SpaceX CEO Elon Musk Talks Starship Space Telescopes, Artificial Gravity." *Teslerati*, July 7, 2021. https://www.teslarati.com/spacex-starship-telescopes-artificial-gravity/.

Rwanda Space Agency. 2021. Twitter post. October 20, 2021, 5:56am. https://twitter.com/RwandaSpace/status/1450762768601264137/photo/1.

Schwartz, James S. J. 2020. *The Value of Science in Space Exploration*. Oxford University Press.

Scientific and Technical Subcommittee. 2022. "COPUOS Report of the Scientific and Technical Subcommittee on Its Fifty-Ninth Session." Vienna, February 7-18, 2022. UN Doc A/AC.105/1258.https://www.unoosa.org/oosa/en/ourwork/copuos/2022/index.html.

Sheetz, Michael. 2021. "In Race to Provide Internet From Space, Companies Ask FCC for about 38,000 New Broadband Satellites." *CNBC*, November 5, 2021. https://www.cnbc.com/2021/11/05/space-companies-ask-fcc-to-approve-38000-broadband-satellites.html.

SKAO Observatory. 2021. Twitter post. September 3, 2021, 7:12am. https://twitter.com/skao/status/1433749617519255565.

Skran, Dale. 2020. "Space-Based Astronomy Is our Future." *Space.com*, March 14, 2020. https://www.space.com/space-based-astronomy-is-our-future-op-ed.html.

The Starlight Initiative. n.d. "The Starlight Initiative." https://starlight2007.net/index_option_com_content_view_article_id_234_itemid_78_lang_en.html.

"TongaSat." n.d. Wikipedia. https://en.wikipedia.org/wiki/TONGASAT.

UN General Assembly. 1948. "Universal Declaration of Human Rights." 217 A (III). December 10, 1948. https://www.un.org/en/about-us/universal-declaration-of-human-rights.

UN General Assembly. 1966. "International Covenant on Civil and Political Rights." 2200A (XXI). December 16, 1966. https://www.ohchr.org/en/professionalinterest/pages/ccpr.aspx.

UNESCO. n.d. "Astronomy and World Heritage Thematic Initiative." UNESCO World Heritage Convention. http://whc.unesco.org/en/astronomy/.

UNESCO. 2008. "UNESCO and the IAU sign key agreement on Astronomy and World Heritage Initiative." Press release. October 30, 2008. https://www.astronomy2009.org/news/pressreleases/detail/iya0803/.

UNOOSA. n.d. "Online Index of Objects Launched into Outer Space." United Nations Office for Outer Space Affairs. http://unoosa.org/oosa/osoindex/search-ng.jspx.

UNOOSA. 2019. "Long-Term Sustainability of Outer Space Activities." United Nations Office for Outer Space Affairs. https://www.unoosa.org/oosa/en/ourwork/topics/long-term-sustainability-of-outer-space-activities.html.

Wall, Mike. "Change to SpaceX's Starlink Internet Constellation Faces Legal Challenge." *Space.com*, June 3, 2021. https://www.space.com/spacex-starlink-megaconstellation-fcc-viasat-dish.

Witze, Alexandra. 2020. "SpaceX Tests Black Satellite to Reduce 'Megaconstellation' Threat to Astronomy." *Scientific American*, January 10, 2020. https://www.scientificamerican.com/article/spacex-tests-black-satellite-to-reduce-ldquo-megaconstellation-rdquo-threat-to-astronomy/.

Zhang, Emily. 2020. "SpaceX's Dark Satellites Are Still Too Bright for Astronomers." *Scientific American*, September 10, 2020. https://www.scientificamerican.com/article/spacexs-dark-satellites-are-still-too-bright-for-astronomers/.

23

Creating a Culture of Extraterrestrial Environmental Concern

William R. Kramer

Repeating Past Mistakes

A critical piece is missing from the puzzle of outer space exploration and exploitation, yet its absence has not been recognized by those leading government and commercial space actions. Without it, we will miss opportunities to increase the efficiency of space ventures and sustain resources and diminish options for future uses for decades or centuries. We will duplicate the same costly but avoidable mistakes made on Earth. We need to make a commitment to consider ways to minimize the foreseeable adverse effects of our actions on the extraterrestrial environments we will alter. Without awareness and precaution, our actions will result in a range of irreparable environmental damage that will affect future science, human habitation, commercial enterprise, and other futures. The issue was a foreseeable problem as early as 1986, when space ethicist Eugene Hargrove warned that space exploration "remains steadfastly focused on earthbound environmental issues" (1986, ix). Yet even now, well into the 21st century, most have not recognized this need. If we begin to significantly affect those extraterrestrial areas without environmental precautions, future government and private industrial/commercial projects "may simply produce a new environmental crisis that dwarfs our current one (on Earth)" (ix–x). Such outcomes are, however, avoidable. We have and use the analytical tools and procedures for achieving more favorable outcomes on Earth. We need to apply them to space.

Impact assessments required prior to undertaking major construction projects have proven to be effective in decreasing the costs of environmental damages (e.g., the costs of remediation, wasted resources, and other detriments) by creating efficiencies and supporting resource sustainability. In

William R. Kramer, *Creating a Culture of Extraterrestrial Environmental Concern* In: *Reclaiming Space*. Edited by: James S. J. Schwartz, Linda Billings, and Erika Nesvold, Oxford University Press. © Oxford University Press 2023. DOI: 10.1093/oso/9780197604793.003.0023 the US, the National Environmental Policy Act of 1970 (NEPA)[1] was enacted specifically to:

- fulfill the responsibilities of each generation as trustee of the environment for succeeding generations;
- attain the widest range of beneficial uses of the environment without degradation or unintended consequences; and
- preserve important historic, cultural, and natural sites.

Should we not seek the same goals in space? Absolutely! But there are essentially no requirements for impact analyses for actions outside low Earth orbits (LEO) in the US, Europe, or other spacefaring nations except for Belgium and France (Mustow 2018).[2] Article 7 of the Moon Agreement of 1979[3] states that nations shall take measures to not cause adverse changes to bodies in our Solar System or disrupt the balance of their environments, but among spacefaring nations, only France and India have signed the document. In the United States, for example, the federal agency responsible for administering NEPA has determined that although the act allows impact analysis for US actions in space, it will not be required for actions beyond LEO (Boling 2019). The policy demonstrates the political resolve to not impede space activities with regulations that may stifle near-term growth. But ignoring environmental impacts has proven to be short-sighted; it does not support longer-term, sustainable development and can lead to environmental disasters and significant social and other costs. There is no reason to expect different results in space. We will establish a precedent of disregard that will continue as we expand beyond Mars. It will encourage a pattern of destruction rather than enlightened creation (Kramer 2014).

Western Roots of Environmental Estrangement

An increasingly technological and industrial world that demands continued economic growth has altered Western cultures' relationships with the environment (Tarnas 1991, 362–363). Both renewable and nonrenewable resources have often been overutilized to secure quick profits with a minimum of consideration for their environmental impacts. A few examples include mining, deep wells that draw down millennia-old aquifers, and reliance on fossil fuels. All have contributed to maintaining Western economic, political, and cultural facets of society that demand growth. American historian and biographer Frederick Turner captured these relationships with the environment when he describes that "a feeling of American loneliness began to insist upon itself, a crucial, profound estrangement of the inhabitants from their habitat: a rootless, restless people with a culture of superhighways precluding rest and a furious penchant for tearing up last year's improvements in a ceaseless search for some gaudy ultimate." He continues, "This is an extraordinary phenomenon, and indications of it are to be found earlier than the political origins of the Republic" (1992, 5). Much earlier, he traces its roots to Old Testament periods where nature, in Abrahamic cultures, exercised "a cruel power over these wanderers, and they sought emancipation from it. . . . They sought to suppress the world of nature" (44). The gods ceased to be of the Earth, the palpable world we saw and touched, and were placed in the supernatural heavens. Perhaps this was where the Earth was first considered the "other," a force and personality independent of humans and a threat to be fought and defeated.

Aided by technology over the past 500 years, Western expansionism and exceptionalism have allowed increasing access to the globe's resources. Waves of European colonization encouraged acquiring lands cheaply, often without the free consent of Indigenous populations, opening areas to exploitation with no penalty for long-term environmental damages done (Diamond 1999; Mann 2011). Lands were put to their "highest and best use," where resources were considered worthless unless they created wealth. Valuation of landscapes failed to consider their other attributes, such as their role in resource renewal, water quality and quantity, climate moderation, and, importantly, their spiritual meaning. Conveniently, those privileged by policies of resource extraction purely for capital gain were those who defined "highest and best" and determined what was "unproductive." The physical and spiritual relationships of Indigenous peoples to the land and its resources were largely ignored. This model for accumulation of wealth, from the land through ever-increasing production, is basic to both capitalism and communism (with its communal rather than private ownership); the environmental consequences are the same. Regulations that might reduce profits or slow production, such as analyses of environmental impacts, are avoided unless overwhelming political or social pressures demand them.

During the second half of the twentieth century, the US public became more aware of the relationship between the environment and their quality of life. Oil spills, pesticide-related wildlife die-offs, environmental contaminants, the persistence of pathogenic industrial compounds, declining water quality, and other problems were frequently publicized. Simultaneously, evidence demonstrated that a continued growth model of development that ignored consideration of environmental effects was unsustainable.

Many industries that profited from a lack of regulation argued that the science did not support opponents' claims, that there was little proof that their actions had adverse impacts. Pesticide manufacturers reacted strongly against Rachel Carson's *Silent Spring* (1962), an analysis of pesticides' effects on non-target species, including humans. Her data and conclusions were challenged. An executive of the American Cyanamid Company, a major manufacturer of the agricultural chemicals cited by Carson as harming the environment, stated, "If man were to faithfully follow the teachings of Miss Carson we would return to the Dark Ages" (Weis 2014, 11). Carson was vindicated, however, as her analyses were proven valid, and all phases of the industry have since been regulated. With tobacco, research demonstrated a link between smoking and cancer as early as the 1950s. Again, industries challenged scientific findings and claimed their products to be (nearly) harmless. They employed tactics to resist regulation including public relations campaigns, buying scientific and other expertise to create controversy about established facts, funding political parties, hiring lobbyists to influence policy, using front groups and allied industries to oppose tobacco control measures, and corrupting public officials (Saloojee and Dagli 2000, 902). We are now witnessing a third, potentially devastating example of industries' reluctance to be regulated regarding the environment: global climate change. Although evidence of the relationship between greenhouse gases produced through human activities and climate is overwhelming, many in governments and affected industries still use the same tactics to stall regulation that were employed by the tobacco industry (Dunlap and McCright 2010).

Similar patterns are now emerging regarding impacts to extraterrestrial environments, but many government, private industry, and science-oriented space actors maintain their future actions will not harm extraterrestrial landscapes. Consider artists' depictions of human use of lunar or Martian landscapes. Nearly all (save dystopian science fiction imaginings) avoid images challenging a singular vision that human uses will have only benign impacts—no landfills or heaps of mining tailings; no evidence of subsurface ice polluted by what may drip from machinery or be dumped to cheaply dispose of chemical waste. We are to believe the space miner or builder is immune from error, poor judgment, or accident, and that by not considering environmental impacts, they will not materialize. Our experiences on Earth have proven that is a costly assumption.

Why should we expect better on the Moon or Mars than what we witness on Earth daily? Where is the incentive for better environmental behavior in extraterrestrial environments? Perhaps it is easier to focus on dreams of space exploration unencumbered by the reality of our potential to adversely affect those landscapes. That vision certainly supports the continued growth model, the Western pattern of landscape domination, and a denial (or ignorance) of the potential for environmental damage. How such relationships are evolving in space is evidenced in the colonizing language we frequently hear when space is defined as an adversary that must be fought, conquered, tamed, and "civilized" (Kramer 2014; Billings 1997). It creates the dichotomy of we and them, the terrestrial and the extraterrestrial. Such an approach to space creates not only a rationale for not regarding outer space landscapes as worthy of conserving, but an obligation to dominate and pillage them.

Rachel Carson quotes essayist E. B. White in the introduction to *Silent Spring*: "Our approach to nature is to beat it into submission. We would stand a better chance of survival if we accommodated ourselves to this planet and viewed it appreciatively instead of skeptically and dictatorially" (1962). While White is decidedly referencing the Earth, it is applicable to space. We have established an adversarial relationship with the Moon and, especially, Mars. By characterizing other worlds as a threat, we are justified, as he observes, to beat it into submission. The National Geographic Society's television miniseries *Mars* (2016) blended a fictional story of the founding of a Mars settlement with interviews and commentary from actual scientists and others prominent in governance and commercial enterprises. One of the fictional characters, Hana Seung, refers to Mars as a "vicious planet" and that "Mars would kill us in any of a thousand ways." Immediately afterwards, Casey E. Dreier, the hardly fictional Director of Space Policy of the Planetary Society, states, "Mars itself is your enemy. You have a shared common enemy of Mars trying to kill you every day." Giving Mars such a malicious and menacing personality not only allows decimation of its landscape but imposes a duty to subdue it, to despoil it. This language abets colonizers' claims to property and diminishes care for long-term environmental consequences. A different approach is needed as we voyage into space, and non-Western traditions offer much to consider.

Ruling Authorities and Indigenous Perspectives

The Outer Space Treaty of 1967[4] is clear in Article I in asserting that "the exploration and use of outer space . . . shall be carried out for the benefit and in the interests of all countries . . . and shall be the province of all mankind." Similarly, Article 4 of the Moon Agreement reiterated the standard in stating, "the exploration and use of the Moon shall be the province of all mankind." Their intent is clear: all should benefit from outer space exploration and exploitation. It follows that to achieve the documents' objectives, all must participate in describing desired outcomes. This would include representation of the world's philosophies, cultures, and traditions, not just the Western, frequently colonial, perspectives that have been dominant from initial research on rocketry to the emergence of "new space."[5]

In a critique of the assumption of "the Colonial Mindset" of the West, Danielle Wood, Massachusetts Institute of Technology professor and specialist in societal development, writes, "whoever has the technology, economic means, and the will to do so, has the right to claim property, territory, and resources, regardless of . . . claims of other people and the claims of environment. . . . This Colonial Mindset is already built into the fabric of thought as space agencies, engineers, scientists, entrepreneurs and explorers contemplate future human activity on the Moon, Asteroids, Mars, and beyond" (2020).[6] While there are certainly many advantages of Western capitalism in elevating the human condition, when applied to space, extraterrestrial environments are being commodified to be exploited for continued growth at the expense of sustainability, landscapes, and should endemic life be present, ecological balance (Shammas and Holen 2019). Unfortunately, environmental sustainability has been provided little status in current visions of extraterrestrial development; sustainability of profits is paramount.

Indigenous cultures' spiritual and philosophical relationships with landscapes are certainly not uniform, and the generalities expressed here make no claim to be universals. Their cosmologies are as varied as the cultures themselves (Monani and Adamson 2016). But while non-Western voices are excluded from conversations on the futures of space, they must be considered in building policies affecting extraterrestrial environments, if we wish to balance the prevalent pretensions to objectivity of the Western view of science and the cosmos with Indigenous perspectives, which are often dismissed as quaint and of no relevance (Asselin 2015; Baird 2012). As expressed by sociologist Victor L. Shammas and independent scholar Tomas B. Holen, "there is an expedient conflation of capitalist interests with a universalizing notion of the interests of humanity" (2019, 3). Exclusion, or worse, pandering by reference to Indigenous thought in dialogues citing outer space "for all mankind" without including them in decision-making, is hypocritical.

Some Indigenous cultures have traditions of viewing landscapes and humans as a cooperative whole, a relationship where the line distinguishing human and landscape is blurred. They tend to be more holistic as opposed to the West's predominantly mechanistic and materialistic associations with the land and its characteristics. As such, Indigenous thought may contribute to greater respect for and a wiser use of space environments. They may be crucial in structuring a sustainable relationship with extraterrestrial landscapes in that they often offer "alternative ways of seeing ourselves in relationship to the natural world" (Young 1987, 270). Rather than having an adversarial relationship with the environment, many non-Western traditions seek to deepen a cooperative spirit with landscapes by recognizing that humans are an intrinsic part of that environment; quite literally, "we are the land, and the land is us" (Korff 2016). Such a bond is not only with plants and animals, but can extend to the inanimate, such as mountains, the sky, stones, water, and even celestial bodies (Hollabaugh 2017; Capper 2020). In Australia, Ambelin Kwaymullina, an Aboriginal woman, explains,

> For Aboriginal peoples, country is much more than a place. Rock, tree, river, hill, animal, human—all were formed of the same substance by the Ancestors who continue to live in land, water, sky. Country is filled with relations speaking language and following Law, no matter whether the shape of that relation is human, rock, crow, wattle. Country is loved, needed, and cared for, and country loves, needs, and cares for her peoples in turn. Country is family, culture, identity. Country is self. (Korff 2016)

In North America, many Indigenous peoples developed strong identities with celestial objects. The Skidi Pawnee's world, for example, centers on the stars and the interaction of the Earth and the cosmos. "The role of astronomy is extensive, for it appears as an organizing principle for all other aspects of Skidi life—political hierarchy, village layout, agricultural practices, socioreligious activities, and the embellishment of material culture. (All combine to form) a complex star theology" (Chamberlain 1982, 207). None of this infers that Indigenous humans do not alter landscapes to sustain their health, welfare, and longevity as a society (Mann 2005; Crosby 2004). Some routinely burn prairies to encourage the species they hunt or divert water for agriculture (Abrams 2020). But those actions shape landscapes; they don't destroy them.

In Hawaii, Indigenous traditions of creation hold the mountain Maunakea to be an ancestor, sharing genealogical ties with Kānaka Maoli, native Hawaiians. It is a sacred part of the landscape. As such, they are protesting the installation of a thirty-meter telescope on its summit. A spokesperson for the group explained, "We have always revered Maunakea as our sacred mauna. In fact, it is part of our cosmology, the very beginning of Earth from which man descends, so for us it's a very spiritual matter" (Bartels 2020). In New Zealand, Māori activism resulted in granting the Whanganui River legal personhood status (Hsiao 2012). In North America, many features of the landscape have spiritual status that deeply affect native cultures and traditions. And, as Daniel

Capper, Professor of Philosophy and Religion at the University of Southern Mississippi, cites in Chapter 16 of this book, Buddhist traditions include inanimate (from a Western perspective) objects such as mountains as an intrinsic part not only of the landscape but as instructive for us all.

These few examples speak to an Indigenous sense of place not generally found in Western approaches to landscapes. If our moon and Mars are considered as nonplaces without spiritual meaning and as enemies to be conquered without consideration of what is described by Brad Tabas (Professor of Space Humanities at the École Nationale Supérieure de Techniques Avancées Bretagne) as "human entanglement," they remain environmentally at risk, not worthy of even an afterthought (2021; Mitchell et al. 2020).

A Pathway to a Solution

Two options would help in averting environmental failures in space. The first, national or international legal regulation through "hard law" (similar to NEPA), would clarify space actors' responsibilities (Kramer 2014; Mustow 2018). However, enacting laws or treaties seems remote given political inertia and current competition to advance space enterprises. A second option is for space industries themselves to create, administer, and enforce industrial standards, practices, codes of conduct, and protocols for identifying and reducing adverse environmental impacts. Such self-regulation has been practiced for centuries by artisanal and other groups, such as trade guilds and manufacturing associations. This could be implemented quickly, would be responsive to industries' needs, and could be modified to adapt to the new challenges of working in space without the need for a lengthy legislative process (Kramer 2017). While I believe it is in space industries' best interest to be proactive, this option, too, seems unlikely given industries' past record of resistance to environmental regulation. Action might be taken only after there is wide recognition of the issue and a demand for an effective response, and that may occur only after an extraterrestrial environmental disaster or after significant profits are made from space resources. Should large profits be made at the expense of alien environments, there may be a call from "all humankind" both for sharing the wealth and caring for those environments. However, as with the environmental issues precipitated by European colonization, it may take decades for extraterrestrial damages to trigger regulations that slow, if not stop, that harm. And as we have witnessed on Earth, many of those harms will be irreversible.

Although some Indigenous voices have been invited to participate in space policy discussions, it is debatable whether the results of those dialogues have influenced capitalist approaches to space exploration and exploitation. There are programs that extend space "science" to Indigenous peoples as part of overall outreach programs, such as STEM (science, technology, engineering, and mathematics) education, but has the reciprocal education of Westerners in Indigenous ways of thinking, especially regarding relationships with landscapes, happened? If so, were responses taken seriously in the formulation of policy? It is foreseeable that Indigenous groups may need to demand consideration of the fate of extraterrestrial environments, whether through hard law, international treaties, industrial standards, or some other protocol. Representatives of those groups should strongly consider organizing a united effort to make their perspectives heard, recognizing that the extraterrestrial environment is our shared environment, that as we are a part of Earth's landscapes, we are also part of the landscapes of the Moon and Mars and the universe. Non-Western perspectives may be our best hope for achieving those goals.

Conclusion

To summarize, I would like to highlight four conclusions presented in this chapter. First, extraterrestrial actions will result in adverse environmental impacts that may reduce options for future generations' use of those landscapes, decrease sustainability of resources, increase costs of development, and cause human health problems and other negative outcomes. Second, we have the tools to assess environmental impacts and mitigate or avoid harms. Third, space-faring governments and commercial ventures have avoided discussions of extraterrestrial environmental impact assessment, perhaps believing that knowledge of potential environmental harms may stifle commercial development. This is short-sighted and counterproductive to sustainable use. And finally, Western philosophies regarding landscapes tend to be adversarial— "man against nature." Many Indigenous traditions view humans as an integral part of the landscape, a coexistence that is not only mutually favorable but essential to their physical and spiritual existence. These perspectives need to be heard to promote the wise use of space in keeping with the sentiments of the Outer Space Treaty, Moon Agreement, and similar documents.

Notes

1. NEPA describes national policy to create conditions under which man and nature can exist in productive harmony, for present and future generations. It directs the preparation of environmental assessments.
2. Belgium's Law on the Activities of Launching, Flight Operation, or Guidance of Space Objects of 2013 requires consideration of the impact on both the Earth and any celestial body affected. The French Space Operations Act includes requirements for those proposing actions to list measures to avoid, reduce, or mitigate the adverse effects on Earth and outer space.
3. Agreement Governing the Activities of States on the Moon and Other Celestial Bodies. While specifically naming the Moon, the Agreement applies to all celestial bodies in our Solar System other than the Earth. Eighteen parties have ratified the Agreement, four have signed (including France and India). Source: https://www.unoosa.org/res/oosadoc/data/documents/2021/aac_105c_22021crp/aac_105c_22021crp_10_0_html/AC105_C2_2021_CRP10E.pdf.
4. Treaty on Principles Governing the Activities of States in the Exploration and Use of Outer Space, including the Moon and Other Celestial Bodies. Source: https://www.unoosa.org/oosa/en/ourwork/spacelaw/treaties/introouterspacetreaty.html.
5. New space refers to the recent growth of private commercial space ventures such as SpaceX, Blue Origin, and others that have replaced predominantly scientific and political space activities with commercial and other economic purposes.
6. Editor's note: See Chapter 9 of this volume for Danielle Wood et al.'s discussion of "Opportunities to pursue liberatory, anticolonial, and antiracist designs for human societies beyond Earth."

References

Abrams, Marc D. 2020. "Don't Downplay the Role of Indigenous People in Molding the Ecological Landscape." *Scientific American Online*. August 5, 2020. Accessed September 3, 2021. https://www.scientificamerican.com/article/dont-downplay-the-role-of-Indigenous-people-in-molding-the-ecological-landscape/.

Asselin, Hugo. 2015. "Indigenous Forest Knowledge." In *Routledge Handbook of Forest Ecology*, edited by Kelvin S.-H. Peh, Richard T. Corlett and Yves Bergeron, 586–596. Routledge.

Baird, Melissa F. 2012. "'The Breath of the Mountain Is My Heart': Indigenous Cultural Landscapes and the Politics of Heritage." *International Journal of Heritage Studies* 19, no. 4: 327–340.

Bartels, Mehgan. 2020. "The Thirty Meter Telescope: How a Volcano in Hawaii Became a Battleground for Astronomy." *Space.com*. March 23, 2020. Accessed September 14, 2021. https://www.space.com/thirty-meter-telescope-hawaii-volcano-maunakea-opposition.html.

Billings, Linda. 1997. "Frontier Days in Space: Are They Over?" *Space Policy* 13(3): 187–190.

Boling, Edward. 2019. Associate Director for NEPA US Council for Environmental Quality. Personal communication. April 10, 2019.

Capper, Daniel. 2020. "American Buddhist Protection of Stones in Terms of Climate Change on Mars and Earth." *Contemporary Buddhism* 21, no. 1–2: 149–169.

Carson, Rachael. 1962. *Silent Spring*. New York, NY. Ballentine Books.

Chamberlain, Von D. 1982. *When Stars Came Down to Earth: Cosmology of the Skidi Pawnee Indians of North America*. Berkeley, CA: Ballena Press Anthropological Papers.

Crosby, Alfred W. 2004. *Ecological Imperialism: The Biological Expansion of Europe, 900–1900*. Cambridge: Cambridge University Press.

Diamond, Jared. 1999. *Collapse: How Societies Choose to Fail or Succeed*. New York: Penguin Group, USA.

Dunlap, Riley E. and McCright, Aaron M. 2010. "Climate Change Denial: Sources, Actors and Strategies." In *Routledge Handbook of Climate Change and Society*, edited by Constance Lever-Tracy, 270–290. New York: Routledge.

Hargrove, Eugene C. 1986. "Beyond Spaceship Earth." In *Beyond Spaceship Earth: environmental ethics and the solar system*, edited by Eugene Hargrove, ix. San Francisco: Sierra Club Books.

Hollabaugh, Mark. 2017. *The Spirit and the Sky: Lakota Visions of the Cosmos*. Lincoln, NE: University of Nebraska Press.

Hsiao, Elaine C. 2012. "Whanganui River Agreement." *Environmental Policy and Law* 42, no. 6: 371–375.

Korff, Jens. 2016. "Meaning of Land to Aboriginal People." *Creative Spirits*. Accessed September 3, 2021. https://www.creativespirits.info/aboriginalculture/land/meaning-of-land-to-aboriginal-people.

Kramer, William R. 2014. "Extraterrestrial Environmental Impact Assessments: A Foreseeable Prerequisite for Wise Decisions Regarding Outer Space Exploration, Research and Development." *Space Policy* 30: 215–222.

Kramer, William R. 2017. "In Dreams Begin Responsibilities: Environmental Impact Assessment and Outer Space Development." *Environmental Practice* 19: 128–138.

Mann, Charles C. 2005. *1491: New Revelations of the Americas Before Columbus*. New York: Random House.

Mann, Charles C. 2011. *1493: Uncovering the New World Columbus Created*. New York: Alfred A. Knopf Inc.

Mitchell, Audra, et al. 2020. "Dukarr Lakarama: Listening to Guwak, Talking Back to space Colonization." *Political Geography* 81(102218): 1–10.

Monani, Salma and Adamson, Joni. 2016. *Ecocriticism and Indigenous Studies: Conversations from Earth to Cosmos*. New York: Routledge Press.

Mustow, Stephen Eric. 2018. "Environmental Impact Assessment (EIA) Screening and Scoping of Extraterrestrial Exploration and Development Projects." *Impact Assessment and Project Appraisal* 36: 467–478.

National Geographic TV. 2016. *MARS*. Television miniseries. Episode 3, "Pressure Drop." Director, Everado Gout.

Saloojee, Yussuf and Dagli, Elif. 2000. "Tobacco Industry Tactics for Resisting Public Policy on Health." *Bulletin of the World Health Organization* 78, no. 7: 902–910.

Shammas, Victor L and Holen, Tomas B. 2019. "One Giant Leap for Capitalistkind: Private Enterprise in Outer Space." *Palgrave Communications* 5, no. 1: 1–9.

Tabas, Brad. 2022. "On Earthlings and Aliens: Space Mining and the Challenge of Post-Planetary Eco-Criticism." *Resilience* 9, no. 3: 26–61.

Tarnas, Richard. 1991. *The Passion of the Western Mind: Understanding the Ideas That Have Shaped Our World View*. New York: Ballantine Books.

Turner, Frederick. 1992. *Beyond Geography: The Western Spirit Against the Wilderness*. New Brunswick, NJ: Rutgers University Press.

Weis, Judith S. 2014. *Physiological Development and Behavioral Effects of Marine Pollution.* New York, NY: Springer.

Wood, Danielle. 2020. "On Indigenous People's Day, Let's Commit to an Anticolonial Mindset on Earth and in Space." *MIT Media Lab.* Accessed July 12, 2021. https://www.media.mit.edu/posts/Anticolonial_Mindset_Earth_and_Space/.

Young, M. Jane. 1987. "Pity the Indians of Outer Space: Native American Views of the Space Program." *Western Folklore* 46, no. 4: 269–279.

PART 5

VISIONS OF THE FURTHER FUTURE

24

Desire, Duty, and Discrimination

Is There an Ethical Way to Select Humans for Noah's Ark?

Evie Kendal

Introduction

There is a common ethical dilemma used in philosophy classrooms to explore different views on the value of human lives. A group of people are stranded on a lifeboat, but will all drown unless a certain number are thrown overboard. The activity is then to determine the most ethical way to decide who stays. Do we prioritize the doctor who might save the lives of others if the survivors end up shipwrecked on an island awaiting rescue? Do we draw lots and let fate decide? Do we refuse to choose some lives over others and all sink together? In short, are some lives more worth saving than others, and if so, why? One motivation for expanding human civilization beyond our own planet is to preserve the species should the worst happen, e.g., catastrophic climate change, asteroid or cometary impact, nuclear war, etc. But given limited resources, creating such a "lifeboat" for the species will require hard decisions of a similar nature to our classical philosophical thought experiment. If we can only save a small portion of the human population, who do we choose?

This chapter explores some alternative methods for selecting candidates for our species lifeboat, including random lottery, targeted skilled migration, and representative sampling, paying particular attention to the ethical implications for people of diverse genders and sexual orientations. It concludes that the goal of saving humanity cannot be separated from the goal of protecting cherished human values, including reproductive autonomy and cultural diversity.

Evie Kendal, *Desire, Duty, and Discrimination* In: *Reclaiming Space*. Edited by: James S. J. Schwartz, Linda Billings, and Erika Nesvold, Oxford University Press. © Oxford University Press 2023. DOI: 10.1093/oso/9780197604793.003.0024

Assumptions for a Lifeboat Ethics of Global Catastrophe

Before any serious discussion of ethically selecting candidates to save in the case of a global catastrophe is possible, it is first necessary to accept the assumption that humans are worth saving at all. This is by no means an uncontroversial claim, especially given the level of destruction this one species has caused on the countless others that make up Earth's biodiversity. After considering various methods of selecting who might make the cut, it is possible that the conclusion will be nobody, or that in the absence of an agreed-upon ethical standard for making such decisions, we must instead choose to all die together, equality intact. However, assuming there is value in saving the species, there are several features from which this value is typically thought to derive. Chief among them are humanity's rationality, cognitive ability, language capabilities, and sociocultural development. The question then becomes, when faced with an extinction-level disaster, can we preserve these unique characteristics with a small sample of survivors sent off-world, or would choosing some people to survive while the majority perish undermine the humanity we are trying to save? Would such an endeavor dissolve the concepts of human dignity and fairness for those left behind, or would they cherish the fact that at least some of their number might survive to carry on their legacy? To defend our lifeboat, we must therefore also assume species preservation is theoretically possible through this means, and that at least some among the species consider this desirable.

Lifeboat ethics refers to a dire situation in which some lives must be sacrificed lest all lives be lost. In bioethics, this kind of dilemma most frequently pops up in resource allocation decisions and disaster triage. Choices that would usually be considered morally repugnant, such as withholding life-saving treatment from some patients or engaging in "mercy killings" of victims of disaster who can't be evacuated,[1] are thus rehabilitated in a state of exigency as the very definition of a "necessary evil." The COVID-19 global pandemic provides ample examples where healthcare providers have had to choose which patients get access to limited mechanical ventilators or intensive care support. As in the case of the sinking ship in our thought experiment, these choices are justified due to the harsh reality that we cannot save everyone (James 2020). However, physician Jeffrey Hall Dobken (2018), ethicist Tom Koch (2012), and philosopher Naomi Zack (2009)[2] all note that this defense often overlooks the proximate causes of "lifeboat making" scenarios. Exploring the original lifeboat dilemma illuminates their position.

In 1841, the US ship *William Brown* hit an iceberg and sank, taking thirty-one passengers with her. Captain George Harris, eight crew, and one passenger escaped in a small jollyboat, while First Mate Francis Rhodes, eight other crew, and thirty-two passengers were loaded into a poorly maintained longboat only designed to carry half so many. Both were auxiliary vessels intended for small scale activities, such as ferrying passengers to and from the shore, and not designed for extended use as rescue boats. As the longboat took on water, Rhodes ordered the crew to toss half the passengers overboard, so the boat would not sink (Dobken 2018). After those remaining were rescued, a case was brought against one of the crew, generally considered a politically expedient scapegoat, with *United States v. Holmes* ultimately resolving that the ship's captain, crew, and owners were not to blame for the loss of life (Koch 2012). However, a recurrent concern in reporting on the incident was that Rhodes did not attempt to "cast lots" to decide the fate of all aboard the longboat, a standard typically applied in maritime disasters. This had been the method used in 1820 by the much-valorized Captain George Pollard Jr. of the American whaling ship *Essex*, which sank following an unexpected whale attack, leaving twenty crew stranded on leaky whaleboats with insufficient rations.

In 1974, eugenicist and "eco-conservative" Garrett Hardin used the lifeboat analogy to argue against providing humanitarian aid to developing nations or opening immigration, proposing that food should neither be brought to the poor nor the poor to the food (Angus and Butler 2011). His argument was that Earth had finite resources and to attempt to share them equally would only serve to impoverish everyone. Philosopher Michael K. Potter (2011) criticizes this view, noting there is enough food produced on the planet to feed all of its inhabitants "if it were equitably distributed" (2011, 661). This echoes the objections of Dobken (2018), Koch (2012), and Zack (2009), who all note the scarcities cited in lifeboat defenses are often artificially created. Regarding healthcare rationing, Dobken claims bioethicists concern themselves with designing resource allocation methods, while ignoring the "social, political, regulatory, and financial planning that created scarcity" (2018, 121). Meanwhile, Koch (2012) notes the *William Brown* example is used in ethical discussions "as if its exigencies were natural and thus inevitable. Lifeboat ethics assumes the only question is who gets thrown overboard. It never asks whether anyone need be drowned at all" (2012, 98–99). Had the *William Brown* not been speeding through treacherous waters chasing financial reward, and had its longboat been properly maintained and sufficiently large to cater to all passengers, the supposed "necessity" of tossing sixteen people overboard would never have occurred. Here the difference between the *William Brown* and the *Essex* is important for our own analogy: if the sinking of the *William Brown* represents catastrophic climate change, a predictable, preventable, "lifeboat making" situation, the whale attack on the *Essex* perhaps stands in for less foreseeable or avoidable possibilities, like an asteroid collision. This is relevant when determining a lifeboat ethics of global catastrophe, as it dictates the degree of disaster planning and mitigation required in advance to justify any lifeboat-style decisions made in the future. In other words, to develop an ethical method of selecting candidates for our lifeboat, we must first strive to avoid the need for one.

Having outlined our main assumptions here, namely, that preserving humanity by selecting a small number to transport off-world is morally permissible, theoretically possible, at least minimally desirable, and that all reasonable measures have already been taken to avoid the need for such a drastic intervention, the next step is determining how places will be allocated. Whether our proposed lifeboat will be interstellar or interplanetary, roam space in desperate search of rescue, travel to an already established human settlement or research post, or just orbit the Earth for future return, fall outside the scope of this chapter, but for the purposes of the upcoming debate, we should assume a scenario in which a place on the lifeboat equates a chance at survival, while exclusion carries the certainty of death.

Common Ethical Arguments for Rationing Resources

Lifeboat scenarios cast the conflicts between consequentialist and nonconsequentialist ethical frameworks into sharp relief (Ryberg 1997). The former define moral rightness according to the consequences of actions, while the latter judge on other factors, e.g., intention. While appeals to the classical utilitarian goal of "greatest good for the greatest number" are highly influential in typical resource allocation dilemmas, in the case of planetary-wide destruction it is difficult to decide who and what should be counted. Will animals, insects and plants be afforded space in our lifeboat? For each human that is saved, how many other lifeforms could have been preserved? Do the qualities discussed earlier, e.g., cognition, communication, etc., justify prioritizing human life, especially if the survival of *any* other species may be dependent on our care?[3] Or is this an example of what philosopher Charyl E. Abbate calls an "elitist, masculine bias" toward impartial reason as a source of moral worth (2015, 2)? Similarly, preference-based utilitarianism aims to maximize preference satisfaction in a way that often favors human interests, at least for those aware and able to articulate their preferences. Zack claims that when faced with emergencies, rapid consequentialist moral reasoning often prevails, but whether this is caused by the exigencies of the situation or "merely bring out our underlying consequentialist moral principles" is unclear (2009, 34). She claims that history demonstrates that when survival of a group demands the sacrifice of a minority among them, this sacrifice will usually be made (40). She suggests that the predictable nature of many disasters means the grounds for choosing who will be sacrificed should be fairly determined in advance, under a principle of "Fairly Save All Who Can Be Saved with the Best Preparation" (126). For the *William Brown* and climate change catastrophe analogies, this again supports the idea that lifeboat ethics can only justify saving the few if reasonable prevention strategies were in place to mitigate the foreseeable disaster. As Koch notes, "[b]y the time lifeboat ethics is engaged all good solutions are gone" (2010, 99).

Going beyond consequentialist frameworks, other popular resource allocation models include philosopher Daniel Callahan's prioritization of the young as those who "haven't had their chance yet," sometimes called the "fair innings" framework, and influential bioethicist and political theorist Norman Daniels' argument for expending public resources to promote each citizen's enjoyment of the "normal opportunity range" for someone in their situation (Koch 2012, 100–101). While neither of these is likely to assist in selecting between similar *young* people, they will both functionally exclude the elderly from accessing extremely limited resources. James (2020) notes that this aligns with Hardin's harsh lifeboat ethics, where the aged were discounted, alongside the poor, as unworthy of receiving support. Physician Chadd K. Kraus et al. (2007) further note that in triage decisions, "expected outcome in survival and function" is the dominant influence, which has implications for ageism and disability discrimination.

Another popular method of choosing between lives in a disaster is casting lots, as seen in the maritime examples discussed earlier. Regarding the *William Brown*, Zack claims that if "the members of the longboat had drawn lots, sympathy for the survivors would be stronger" (2009, 39). This claim seems to be supported by the fact the captain of the *Essex* was deemed to have done the "honorable" thing by including himself in a lottery, despite the fact he was not ultimately called upon to sacrifice himself for his crew. Reports of the *William Brown* suggest Rhodes had initially discussed the possibility of a survival lottery with Captain Harris, but when faced with a sinking boat "brooked no discussion of drawing lots" (Koch 2012). The issues here are further complicated by the fact that the passengers were mostly poor Irish emigrants, whose deaths were considered unfortunate but of low significance.

As Koch states: "One nineteenth-century newspaper report gave the tenor of the age when it chronicled the loss of 'twenty souls and 240 emigrants.' While the former were lamented the latter were, well, just emigrants" (2012, 88). This highlights that lifeboat ethics decisions have traditionally been inherently classist and racist, and in the case of Hardin's formulation, grossly lacking in recognition of how some people, in his case citizens of wealthy countries, came to be on the "better" boats in the first place (Okyere-Manu 2016).

Hardin (1974) did have a solution for the "conscience-stricken" who felt guilty for surviving at the expense of others, though: they could give up their space on the lifeboat in exchange for someone else. This might align with a virtue ethics perspective which considers the right action as that which a virtuous person would do. As seen in Captain Pollard's actions, this need not lead to a situation where only the *un*virtuous survive, as *willingness* to sacrifice oneself for the good of the collective might satisfy our ethical requirements here. In fact, the alternative, where only those unwilling to promote the good of the species at personal expense retain their spaces in our lifeboat, is likely to undermine the whole project of preserving those things about humanity we most value. Hardin also claimed that a Marxist or Christian ethic of refusing to choose some lives over others and instead trying to save everyone would lead to "[c]omplete justice, complete catastrophe" (393). However, for some virtue ethicists and people ascribing to deontological (duty-based) ethical systems, allowing everyone to die may be less ethically problematic than choosing some to live. For example, moral philosopher Immanuel Kant's famous "categorical imperative" is unlikely to permit the ends (of preserving humanity) to ever justify the means (of developing a system that values some human lives higher than others). The intrinsic value of all human lives afforded in the Kantian framework opposes a consequentialist defense that some must be sacrificed for the good of others, irrespective of scale.

Another method that can incorporate elements of casting lots and the "fair innings" framework is applying political philosopher John Rawls' "veil of ignorance" to deliberations predetermining our lifeboat decisions. The Rawlsian model asks people to design a fair resource allocation system for a hypothetical society in which they are unaware of their own position, e.g., where they cannot know whether they will personally draw the short straw. In such a society, do they still feel overall that a lottery is the best way to make lifeboat-style decisions? When each member is unaware of what their own age will be, will the group still preference the young to be saved?

The final ethical defense considered here is appealing to the doctrine of double effect. By this model, selecting candidates for the lifeboat is following an ethically permissible goal of preserving humanity, which unfortunately carries the foreseen, but unintended side effect of letting all other potential candidates die. In a disaster setting, Zack notes that euthanizing dependents who cannot be rescued, including frail humans and pets, can be seen as preventing suffering, rather than killing, according to the same logic (2009, 43). However, this defense is only valid when non-lethal methods would have been chosen in a context in which they could achieve the same objectives; in other words, preserving humanity in a way that sacrifices the majority is only permissible if chosen exclusively in the absence of an option that does not call for this sacrifice. Again, our ethical duty to first protect our existing population and planet are affirmed.

Additional Ethical Considerations for Noah's Ark and Promoting Diversity

While all of the ethical considerations considered in this chapter so far are relevant in an extinction-level disaster, what we are discussing in this chapter goes beyond a lifeboat intended to save as many *individuals* as possible, to a Noah's Ark intended to save the *species*, at least in some form. This is where discussions regarding fertility come into the debate. If the existence of the lifeboat is justified on the grounds of preserving humanity, does this imply the occupants have a duty to procreate to rebuild the human population on the other side? If so, this would imply physiological and social fertility might be deciding factors in who is selected. The remaining sections of this chapter argue that establishing such a duty is not only unethical but counterproductive. By systematically excluding anyone outside of their reproductive years or who may be otherwise physiologically or socially infertile, including people of diverse genders or sexual orientations, such a proposal would fail to save humanity in any meaningful sense, inasmuch as it would strip us of some of the very diversity that makes our species unique. Even more concerning is the potential this "procreative duty" would create a sub-class of physiologically fertile, cis-gender women essentially living in reproductive slavery.

Reproductive autonomy is a highly cherished human value, as it relates to intimate partnerships, family creation, and bodily integrity. Likewise, cultural diversity in the human species is one of the elements that differentiates us from many other species. As such, we must ask ourselves, is diversity the kind of thing we can sacrifice in the face of extinction, or would sacrificing it *be* our extinction? Likewise, if the freedom to exercise our human rights, including reproductive liberty, is central to a flourishing human experience, how can we be said to save humanity while undermining these?

On a more practical note, if we are promoting a lifeboat predominantly aimed at preserving human lives, in preference over many other animal species, and using our advanced cognitive and communication abilities to justify our special place in this system, this suggests we should be aiming to maximize those abilities, for example, through conserving as many different cognitive styles and languages as possible. This requires a maximally heterogeneous sample of humans for best effect, with diverse representation across age groups, cultures, genders, etc. The benefits of this would also be conferred to any future generation of settlers, as if we excluded, for example, queer individuals from the lifeboat on the grounds they may have reduced reproductive potential, this would lead to a scenario in which queer children are born into a community where they lack representation. Given the pronatalism evident in such a policy, this further suggests these children would experience hostility based on their own real or perceived reproductive potential. A similar problem would arise should people with disabilities be excluded from lifeboat selection, with future discrimination against those with congenital or acquired disabilities likely to compound this original discriminatory policy.

Possible Selection Criteria

This final section considers potential methods of allocating limited spaces in our lifeboat, assuming an objective of preserving the species without sacrificing cherished human values.

User Pay

The first method, which involves abandoning both the goal of preserving human diversity and any concept of fairness, would allow the super wealthy to purchase spaces on the lifeboat. This option is included here only to be categorically rejected on several grounds. Billionaires, by and large, are a fairly homogenous group, and age considerations aside, relying on this sub-population to preserve humanity would likely lead to failure, both philosophically and practically. Given the intrinsically immoral state of being super wealthy in a world in which people die of starvation every few seconds, allowing a situation in which only these candidates survive the global apocalypse would make for a very bleak future indeed, and violate principles of virtue ethics and the "fair innings" framework. Furthermore, Rawlsian "veil of ignorance" experiments consistently reject systems that allow the fortunate to hoard benefits at the

expense of the rest, favoring more equal resource distribution. While the consequentialist might argue that any human survival is better than none, using
Earth's climate as precedent, we can already predict what the super wealthy
will do in any new human settlement, thus rendering their rescue antithet-
ical to the purpose of preserving the species.[4] According to ecological feminism, unfettered capitalism and its patriarchal value system are inextricably
linked to our current state of climate disaster, due to extractivism and the desire for "domination of nature" (Davion 2002, 14). Allowing a concentration
of the world's greatest resource hoarders and environmental destroyers into
our lifeboat merely diminishes the possibility our species preservation project
will succeed, both in terms of literal survival and the endurance of cherished
human values.

While there may not be a case for allowing the wealthy to buy their way
onto the lifeboat, there may be compelling grounds to do the reverse, and prioritize those from lower socio-economic backgrounds and developing nations. Ethicist Beatrice Okyere-Manu (2016) notes that the ecological crisis
facing many African nations, for example, are in part the result of economic
and environmental injustice perpetrated by the wealthy, including through
imperialism, the slave trade, and mineral extraction. Prioritizing the interests
of these groups in the case of a global disaster might go some way toward providing reparation for these injustices, while also preserving diversity and a
potentially less extractivist culture for any new human settlement. Applying a
"fair innings" framework to the development of nations, rather than individuals, might also support this option.

Global Lottery or Random Stratified Sample

As demonstrated in the maritime disasters discussed earlier, casting lots is a
popular method for making lifeboat-style decisions, as it promotes equality
and fairness for all candidates. But another aspect that is intuitively appealing
is that we don't have to actively choose between lives. However, there is a risk
that, given the relative size of the global population to our lifeboat, a random
lottery will fail to select some roles that are crucial to the survival of the group,
e.g., a truly random sample may lack a single doctor. It could also produce a
sample with unsustainably few physiologically fertile candidates.

An alternative would be a random stratified sample, where the global population would be separated into strata according to relevant characteristics to
guarantee, for example, a certain number of doctors or women of childbearing
age. However, choosing potential *characteristics* would not be that dissimilar

to choosing potential *people* to preserve, and thus some of the ethical appeal of casting lots would be reduced. A "veil of ignorance" experiment to determine relevant strata could help manage some of these practical concerns.

Reproductive "Inherent Requirements"

To avoid accusations of discrimination on the grounds of fertility status, while also ensuring sufficient reproductively capable and willing candidates make it onto our lifeboat, we could establish a quota of specified "roles" for which fertility would be considered an "inherent requirement." This would be analogous to the "job" of a commercial surrogate, for which physiological fertility is a necessary selection criterion. Those chosen for these roles could be purposively selected or drawn from a pool of potential candidates, as in our random stratified sample.

While this might appear a good solution on the surface, this option poses numerous threats to reproductive autonomy. What if, after being evacuated, a survivor changes their mind about wanting children? What if after falling pregnant they seek a termination? Will they be compelled to procreate according to the conditions under which they were recruited, in the name of the greater good? Such a dystopian future of reproductive slavery may seem farfetched, but this quote from an interview with astrobiologist Charles S. Cockell effectively demonstrates why this concern should be taken seriously:

> Space is an inherently tyranny-prone environment . . . You are living in an environment where the oxygen you breathe is being produced by a machine." On Earth, he notes, governments can rob their people of food and water, "but they can't take away your air, so you can run off into a forest and plan revolution. (Shermer 2019)

It is not just space that is tyranny-prone in this situation, but also the social processes of reproduction itself. In modern medicine, there is a presumption that intrusion into the body without consent is a violation, meaning forced treatment or tissue donation are not ethically or legally permissible. Pregnancy is often the exception though, with many women forced to continue unwanted pregnancies, or undergo medical interventions, transfusions, and/or surgical procedures, including caesarean sections, to preserve the life of the fetus. Reproductive technology law specialist Judith F. Daar claims this suggests that "once a woman becomes pregnant, she sheds her right to make an informed choice about her medical care" (1992, 836). She relates the US case of *Fosmire v. Nicoleau* where a woman was given a transfusion against her will due to postpartum hemorrhage. The court's decision was that this violated the patient's autonomy but included a statement that had the transfusion been needed to protect the fetus, "the State's interest, as *parens patriae*" would consider the wellbeing of the child "to be paramount." For Daar, this statement confirmed the court would "treat a woman one way if she were pregnant and another if she were not" (1992, 837). In other words, women are already routinely denied their rights in light of their reproductive capacity, and transposing this situation into the "tyranny-prone" space environment during a time where the species is at risk of extinction is only likely to increase this threat.

Antidiscrimination and Family Plans as "Protected Characteristics"

One way to avoid the risk outlined above is to follow existing antidiscrimination laws that prohibit disadvantaging candidates based on protected characteristics, such as family plans. While there are many examples where these protections fail to achieve their objective on Earth, excluding fertility status and sexual orientation from consideration when selecting lifeboat candidates, through random or targeted allocation, might provide some protection against discrimination. The preservation of the species would then depend on whether this sample contained enough members willing and able to gestate the next generation.

Targeted Skills Migration Points-Based System

Another alternative to random allocation or preferential selection based on fertility status might involve a points-based system like in skilled migration. This could either exclude protected characteristics, including disability, fertility status, gender identity, etc., or attach certain point weightings to these. Candidates would be competitively selected according to the needs of the mission, with the possession of certain skills "earning" them a place on the lifeboat. This method could also enhance the diversity of the final sample, e.g., given our desire to preserve as many languages and cultures as possible, a polyglot would be a highly desirable candidate. However, this system also carries significant risk of discrimination. As with all supposedly meritocracy-based admission procedures, a points-based system might be biased against minority groups. As satirical news reporter Andrew P. Street (2021) notes, "i you insist that success is about merit and then notice that men are disproportionately achieving it then it's possible that you might have *accideliberately* mistaken 'merit' for 'wangsmanship.' " Again, a "veil of ignorance" exercise might help minimize overt sources of bias here, but unconscious bias is likely to remain, to the detriment of the diversity of our lifeboat cohort.

Resituate Gestation

Another option for promoting both procreation and diversity is to provide a gestational option that doesn't depend on the bodies of the lifeboat survivors, such as an artificial womb. While pregnancy would still be available to those who wished it, neither the desire nor ability to gestate would be a factor in candidate selection. This would place people of all genders, ages, fertility status, and sexual orientation on equal footing regarding reproductive potential, as all could serve in the communal raising of a new settlement's children. This system could also promote much more diversity in the future community, by transporting many embryos as well as the small complement of grown humans.

When considering whether this would negate the need to select *any* humans to transport off-world, it is important to remember the overarching goal of our Noah's Ark. It is not just to promote survival of human beings, but human cultures and values. Therefore, human socialization will be essential for the next generation to carry on this legacy, and thus the parenting role could not be substituted by an artificial intelligence. Also, the highly prized goal of preserving humanity's unique communication skills is best achieved through children acquiring language through dialogue, which, as anyone engaging in an app-based language program can attest, is not replaceable with technology.

All things considered, given the current absence of an alternative to physical gestation, the risks of discrimination based on reproductive capacity are so great that, at least for now, a global lottery or stratified random sample seem the most viable strategies for promoting reproductive autonomy in our extreme Noah's Ark scenario.

Conclusion

Developing an ethical method for selecting humans for Noah's Ark requires that we first try to avoid the disasters that might necessitate such a project, then work to develop fair selection criteria that promote diversity, reproductive autonomy, and survival potential. This might involve consideration of the skills and reproductive potential needed to sustain a fledgling human settlement. Ensuring that not just human life but human values survive requires that we avoid unfair discrimination in our selection of candidates and carefully consider how we might prioritize historically disadvantaged groups for inclusion to promote diversity and restorative justice.

Acknowledgments

Thanks to Dr. Ben Beccari and the collection editors for discussion and feedback.

Notes

1. Lifeboat arguments have been referenced regarding lethal injections given to patients abandoned without hope of rescue during Hurricane Katrina (Zack 2009; Kraus et al. 2007). They have also been used when allocating scarce medication supplies, including edavarone for ALS patients, where some clinics used lotteries, others had strict inclusion and exclusion criteria, and others allowed "differential access based on personal finances" (Breiner et al. 2020, E320).
2. Unless otherwise stated most of the theorists cited in this chapter are bioethicists, including some who are also health practitioners.
3. At the time of writing my own feline overlord, Proletariat, is making it clear she believes a better use of this human's skills would be opening tins of tuna for her.
4. Note some cultural groups are place-based and may opt out of displacement from heritage lands, even in the context of a global disaster.

References

Abbate, Cheryl E. 2015. "Comparing Lives and Epistemic Limitations: A Critique of Regan's Lifeboat from an Unprivileged Position." *Ethics & the Environment* 20, no. 1: 1–21.

Angus, Ian and Simon Butler. 2011. *Too Many People? Population, Immigration, and the Environmental Crisis.* Chicago, IL: Haymarket Books.

Breiner, Ari, Lorne Zinman and Pierre R. Bourque. 2020. "Edavarone for Amytrophic Lateral Sclerosis: Barriers to Access and Lifeboat Ethics." *CMAJ* 192, no. 12: E319–E320.

Daar, Judith F. 1992. "Selective Reduction of Multiple Pregnancy: Lifeboat Ethics in the Womb." *Davis L. Review* 25, no. 4: 773–844.

Davion, Victoria. 2002. "Ecofeminism, Lifeboat Ethics and Illegal Immigration." *Global Dialogue* 4, no. 1: 114–124.

Dobken, Jeffery Hall. 2018. "Physician-Assisted Suicide (PAS)/Physician-Assisted Death (PAD): The Rise of Lifeboat Ethics." *Journal of American Physicians and Surgeons* 23, no. 4: 121–124.

Hardin, Garrett. 1974. "Lifeboat Ethics: The Case Against Helping the Poor." *Psychology Today* 8: 38–43.

James, Paul. 2020. "Engaged Ethics in the Time of COVID: Caring for All or Excluding Some from the Lifeboat?" *Bioethical Inquiry* 17, no. 4: 489–493.

Koch, Tom. 2012. *Thieves of Virtue: When Bioethics Stole Medicine*. Cambridge, MA: MIT Press.

Kraus, Chadd K., Frederick Levy, and Gabor D. Kelen. 2007. "Lifeboat Ethics: Considerations in the Discharge of Inpatients for the Creation of Hospital Surge Capacity." *Disaster Medicine and Public Health Preparedness* 1, no. 1: 51–56.

Okyere-Manu, Beatrice. 2016. "Overpopulation and the Lifeboat Metaphor: A Critique from an African Worldview." *International Studies in the Philosophy of Science* 30, no. 3: 279–289.

Potter, Michael K. 2011. "Lifeboat Ethics." In *Encyclopedia of Global Justice*, edited by Deen K. Chatterjee, 660–662. New York: Springer. https://link.springer.com/referenceworkentry/10.1007/978-1-4020-9160-5_318.

Ryberg, Jesper. 1997. "Population and Third World Assistance: A Comment on Hardin's Lifeboat Ethics." *Journal of Applied Philosophy* 14, no. 3: 207–219.

Shermer, Michael. 2019. "Our Mission on Mars: Obey the Lessons of Mutiny on the Bounty." *Quillette*, December 22, 2019. Web. https://quillette.com/2019/12/22/our-mission-on-mars-obey-the-lessons-of-mutiny-on-the-bounty/.

Street, Andrew P. 2021. "The Never-ending Contradictions of Scott Morrison." *The Big Smoke*, November 4, 2021. Web. https://www.thebigsmoke.com.au/2021/11/04/the-never-ending-contradictions-of-scott-morrison/?fbclid=IwAR2Ks20QR0xSUJeOorHEE4QMqmbZkfaMoYA4cPRZmqOWM0dRQFVNSqdFGLk.

Zack, Naomi. 2009. *Ethics for Disaster*. Lanham, MA: Rowman & Littlefield Publishers.

25

Deconstructing and Reprivileging the Education System for Space

Janet de Vigne

There is considerable discussion among educators today about decolonizing the curriculum (Arshad 2021; Muldoon 2019). This means that the growing recognition that our current systems privilege certain ways of knowing over others is challenging deeply held assumptions and perceptions of what it means to teach and learn in communities worldwide, not only in our aim to educate everyone on the planet, but also in consideration of future education. Discourses of the past have used military, economic and political might to enforce one worldview—theirs—and to privilege the ways of knowing they use over those of the conquered. History is written by the winners, or so it is said.

We are familiar today with the discourse of neoliberalism—a product of the West—and educators are trying to enable students to recognize its "master plot" (Ritivoi 2009) and how this informs political decision making. Given that many criticize neoliberalism for its monetary focus (del Cerro Santamaria 2019) and absence of focus on issues of justice and equality (Klees 2008), are these critiques represented in our imagining of the education of the future? How do Western ways of thinking dominate such systems? What models of education might we develop to take into the future—to take into space—that are not based on exclusively Western ways of knowing? Any discussion of decolonizing the curriculum must be pluridimensional in nature—it must look to the wrongs of the past; the ideologies, inequalities, and prejudices of the present; and a systemic recentering of future practice. Our own stories tell us that Western models are unsustainable. Let's reexamine who we are as one human race through these lenses—acknowledging who and what we have been, looking to where we are going—and working toward what we could be.

Janet de Vigne, *Deconstructing and Reprivileging the Education System for Space* In: *Reclaiming Space*. Edited by: James S. J. Schwartz, Linda Billings, and Erika Nesvold, Oxford University Press. © Oxford University Press 2023. DOI: 10.1093/oso/9780197604793.003.0025

Models of Education in Science Fiction

Human imagination is a powerful thing, not least because it can, in some cases, result in innovation and even invention of the thing imagined. People have already imagined educational curricula in science fiction. Perhaps we could look at Robert Heinlein's 1952 novel *Space Family Stone* (published in the US as *The Rolling Stones*) —a work very much of its time—for an example of this. Here, teenage male twins Castor and Pollux refuse to go to school, specifically to the United States Military Academy at West Point in all but name— the model that inspired *Star Trek*'s Starfleet Academy. Then there is Madeleine L'Engle's 1962 *A Wrinkle in Time*, ground-breaking in its female protagonist, and still number two on the top 100 best children's books (according to the US School Library Journal [Bird 2012]). Here, school is viewed perhaps as it still is in many contexts—unhelpful, not able to equip children with what they need to overcome a particular challenge. It is a necessary evil required by society, but irrelevant to children's lives outside. Both of these portrayals—Heinlein's and L'Engle's—are ideas we are familiar with; they still exist today. Arguably, the discussion in both these works could be described as social commentary.

For a reimagined educational model, we might then choose to consider the 1985 novel *Ender's Game* by Orson Scott Card, where a child is pushed through a series of ever more violent computer-based activities preparing him for war. This is an allegory of current educational practice in some ways: fear is the motivator (some suggest that economic competitiveness drives educational decision making at the political level [Jessop, Fairclough, and Wodak 2008]) where violence is symbolic (in that it indirectly subordinates individuals and causes them to accept the ideas that oppress them [Connolly and Healy 2004]) but no less actual. However, in *Ender's Game*, competitiveness is present because of war and the "win or die" imperative, plus violence is present and enacted in the reality of Ender's day-to-day life—fighting fellow classmates, fighting in virtual reality. Another type of violence (space combat) is removed from direct experience when it takes place through a computer screen—the juxtaposition of the actual and the virtual. Violence is therefore pluridimensional, in that it moves through and in many dimensions, and ever present.

It is not until Ender realizes how he has been exploited both by the cruelty of war and also through his brutal education that he is able to seek something approaching redemption, or at least to instigate redemptive action. Ender shares a capacity for the development of deadly strategic and fighting skills— dimensionally speaking virtual and actual—with other dystopian characters. Two of these might be: Thomas Cale, the 1995 protagonist of Paul Hoffman's dystopian fantasy novel, *The Left Hand of God*, educated in an equally brutal system, and Katniss Everdeen, victim, vanquisher, and agent of Suzanne Collins' *The Hunger Games*, where education serves to reproduce the status quo for the disadvantaged class, of which she is a member.

Ender's education is harsh: he is removed from his family because of a talent recognized by the authorities, and the exploitation of that talent is key to his success in winning a war—through playing a computer game that is not, as he thinks, a simulation.

Ender's Game: Twenty-First Century Perspectives and Militaristic Models

The Ender novels are still being used by military organizations as demonstrations of the effectiveness of strategy. They could be used to show the threats, human and existential, to the perpetrators, as well as the victims that they so clearly portray: "Marines reading Ender's Game as part of their professional military education (PME) would be remiss if they did not reflect on the moral and strategic questions posed by the end of the novel and Card's follow-on works" (Hovey 2019). These questions include the human cost of Ender's strategic decisions as well as the moral questions around using a teenager to wage war. There are questions around the way he is deceived by the authorities in imagining this as a computer game, when in fact the conflict is "real." His own culpability is another question—how responsible is he for his actions? The risk, it seems, even now, is that the focus will remain, for some, only on the winning. What type of education is this? Imperialist thinking is still with us where readers of *Ender* are concerned: "We're adapting to the future and growing up like Ender. We're getting a head start by maturing early on so we'll be ready to face anything and everything. . . . In the end, it's our dollhouses and our Lego castles, our freedom and our imagination, that create the true building blocks of an empire" (Kaufman 2014). Written in the Georgetown University student newsletter *La Hoya*, this comment is rather worrying, if it is an example of human exceptionalism. Do humans really consider themselves to be the only beings in the universe worth consideration? Are we to dominate and subjugate, rather than get alongside? The consideration of human superiority is value-laden in ways unconscionable to only some of us. Should this be the case? Should we attempt to decentralize ourselves, and if so, would this process erase the horror, inequality, and injustice associated with our historic experience of empire? Or is Kaufman referring to an empire of the mind—the enlargement of the internal space that enables us to construct ourselves rath than impose our ideals onto something else? Would we simply transfer all of this baggage as one human race onto an off-world alien ecology (assuming that, like the Klingons, "they" are equally warlike and self-seeking?).

The Starfleet Academy vision is both better and worse than this. It's not species-ist, perhaps, but it divides the mind and the heart in a deterministic manner reminiscent of Robert Louis Stevenson's Dr. Jekyll and Mr. Hyde (except that here it is feelings vs. cognition rather than goodness vs. badness). *Star Trek*'s Spock, half human and half Vulcan, is engaged in a perpetual struggle with emotions, his own or the lack of them as attributed to him by his almost exclusively human colleagues. This binary division is unfair. The so-called scientific/rational vs. emotional method is very much still present in the discourse of science: "[T]here's always an effort to minimise emotion and intelligence in other species," according to Tony LaCasse, director of public relations at the New England Aquarium (Montgomery 2015, 11). There is perhaps a justifiable fear among scientists around anthropomorphization—attributing human feelings to nonhuman beings. However, in a time when we are questioning the meaning of sentience (the London School of Economics has just recommended that the UK government ban farmed octopus on the basis that they are sentient (Devereux 2021), i.e., that these animals possess feelings and intelligence), it may indeed be time to move more of our thinking from a human exceptionalist perspective to one which is more egalitarian—to move from ego to eco, viewing humanity as a part of, not masters and oppressors of, a system. Even, as ecofeminist scholar Donna Haraway might say, by going into the chthulucene—delving into the earth, the soil, the mud, to get down with the bugs without which we would not exist (Haraway 2016).

Global citizenship is one thing; universal citizenship may be another—examining the values of citizenship in our own contexts may well be enabled by the humble stance toward such a developing understanding of creatures other than human. We need to enable a language of emotion in order to articulate this, and understand how emotions filter experience.

Understanding and Using Story

Stories are so important for humans—they are where we explore who we are and how we find meaning: "We tell ourselves stories in order to live" (Didion 1979). We need new stories now. Different stories will require a deeper awareness of the language we use, and a critical examination of the influences of our society, assumptions, and ideologies on it.

Haraway speaks to this in her discussion of the stories necessary for the post-human age—where we are moving from an ego- to an eco-perspective, placing humans within an ecosystem, rather than above or at the center of it. She states that the Anthropocene (the age of the human—a word, incidentally, used only by the privileged, not those affected adversely by the decisions made by governing structures alien to them, according to her) has too top-heavy a social structure. But revolution—for example, along the Hunger Games model (Katniss Everdeen becomes the figurehead for a violent rebellion that, she recognizes, will simply reproduce itself) will not work: "Revolt needs other forms of action and other stories for solace, inspiration and effectiveness" (Haraway 2016). If we need to avoid violence—perhaps just because we need to preserve life in space rather than self-destruct—perhaps a more liberating educational system will succeed where excessive control has failed.

The Affordances of Technology

The Personalized Learning Environment (PLE) described in the postcyberpunk (Huereca 2010) novel, *The Diamond Age: Or, A Young Lady's Illustrated Primer* (by Neal Stephenson, 1995) presents a "third" way. Doing away with teachers altogether, algorithms in this novel design and plot an educational trajectory for a young girl. Designed for a member of the elite in a highly structured and class-divided society, an illegal copy of the book intended for the protagonist's daughter falls into the hands of a disadvantaged child, with revolutionary consequences. Learning is configured as a series of problems to be solved—it is experiential, rather than cerebral. This resembles the "web-quest" concept of education in some ways, where students are sent out on an enquiry, an information hunt.

Could this idea be described as decentered in terms of an educational system? Is the PLE as described here any less Western in its ideas? There is a central problem: the technology. Even this is value-laden, as a character in the novel, whose group will not use it, explains: "Yong is the outer manifestation of something. Ti is the underlying essence. Technology is a yong associated with a particular ti that is Western, and completely alien to us . . . [W]e could not open our lives to Western technology without taking the Western ideas" (Stephenson 1995, 432).

The issue of technology has had a demonstrable effect on cultural practices today. For instance, modern Chinese no longer reads in up and down columns, as the Western configured keyboard can only cope with left to right or right to left horizontally placed text. Such requirements produce an effect on cultur identity. The teaching of English in Terran classrooms on- and off-line today results in mashups in street advertising and other texts using English words in non-English syntax. Where it has always been the case that languages adopt and adapt words and ideas from other languages, we live in a time when, because of technology, the process is faster and more invasive than before.

Technology and Reprivileging Ways of Knowing

Could any of the ideas in this chapter—the use of algorithms to direct education, recognition and direction of emotion, child-centered planning, be usefully adopted? Should we, could we abandon the teacher in favor of algorithm-directed experiences? There are some problems and some advantages here. If teaching could be less directive and more enabling, as in the inquiry model (where students find the answers to questions by exploring the topic, rather than being told what to do by a teacher—they are guided rather than directed), children might be more engaged. The materials they access, and the tools of inquiry—hard and software as well as the internet—would need to be re-privileged, because these are the things that carry (and promote) dominant ideologies: "[T]he selective and biased knowledge to which students are exposed often hinders their development of the kind of global perspective that can critique ethnocentric and stereotypical ways of reading the world" (Subedi 2013).

Binaya Subedi, human ecologist and educator, discusses the deficit approach to the stored knowledge to which we have access, in that this has historically represented cultures other than white European as lacking "better" cultural values. A deficit model assumes one-way communication, where certain knowledge and ways of knowing are considered 'better', because they originate with one particular group—usually the dominant one. She also explains the danger of exceptionalist (here, the assumption that one group's ideas are better than another's) perspectives that "reduce cultural traits to 'boxes'", and frame them as "neatly distinct" in the world (Subedi 2013). In the process of privileging one narrow view, there is no sense of ethical recognition or engagement "with what the Other may know" (Subedi 2013) or how they may know. Knowledge is therefore lost.

Imagine how it might be to land on a new world and learn to sing it. How much could be learned from watching the flora and fauna with no exploitative motive, considering instead how the new landscape might shape us and how we could work with rather than against it? We would need to develop language that works with, not against, our surroundings. Language is the essential tool of human tools: the foundation of all channels through which we communicate. The signs and signifiers of clothing, make-up, the body, the senses; the discourses of chemistry, history, the languages of home, school, the legal system demonstrate the pluridimensionality of language. In any cases, particularly English, language is no longer considered to be a bounded system belonging to a particular group. Linguistic diversity is the norm. "Plurilingual repertoires" means that speakers possess diverse linguistic repertoires "of a fluid and partly fragmentary nature" (Gogolin and Duarte 2017). Our educational systems have not yet taken this into account, and still prioritize some languages over others in their curricula. Does the teaching in classrooms today reflect the diversity of its students? The answer is "not yet."

We could start this development by examining nondominant education systems on Earth and explore and value their ways of knowing. This would require us to analyze the language we use and become aware of linguistic choices—lexical (vocabulary) and syntactical (agent versus acted upon). How far a possible decolonization could be achieved through such analysis and by technological means is an interesting question.

In his research exploring the perspectives of South African students (on strike for the decolonization of their curriculum), educator Lawrence Meda found that students valued technology as a means of expediting the decolonizing process: "Decolonising the curriculum is a subject that has come at a time when technology has advanced and it can be used to make the process smooth and efficient" (2020). Meda's students did not want to eradicate Western educational processes, but to incorporate Indigenous Knowledge (IK) and give it equal status: "They are not advocating for the eradication of Western knowledge in the curriculum, but for decentring to take place. Decolonisation is about decentring a Western curriculum with IK" (Meda 2020).

Such an approach will include logistical issues with choices about what people should learn, where, and how. A nation might decide on a mixed curriculum, hopefully to the benefit of its people, but with an eye on the world, its place in the world and its contextualized needs. Mbembe and wa Thiong'o address the issue of defining the process: "decolonising is about having a liberated perspective that allows people to see themselves clearly in relationship to themselves and to other selves in the universe" (wa Thiong'o 1986).

Critical Awareness and Pervasive Ideologies

Kenyan writer and academic Ngugi wa Thiong'o's perspective is a critical one, in the Freirean sense (Freire 1970). Paulo Freire, the celebrated Brazilian

educator and philosopher, wrote the *Pedagogy of the Oppressed*, a work introducing the concept of criticality to education: seeking to liberate through
opening students' eyes to the ideologies and powers that position and oppress
them. Ira Shor, professor of composition and rhetoric and friend of Freire,
suggests that taking such a stance in the classroom might mean the teacher
reengineers the learning to encourage students to become critical and active
rather than passive recipients of knowledge, particularly where the current
curriculum reproduces stratification in society (i.e., keeping people in their
place) and "politically retards alternative thought" (Shor 1980).

Achille J Mbembe, Professor of history and political science at the University
of Witwatersrand, South Africa, also takes a critical perspective, while recognizing the "waking up" process as continuous and social: the process requires
togetherness, a "becoming" —it is dynamic, not static: "[D]ecolonisation is
not viewed as a once off exercise, but an ongoing process of people seeing
themselves clearly; emerging out of a state of either blindness or dizziness"
(Mbembe 2016).

Neither of these perspectives rids us of militaristic inclinations or the in-
sidious effects of technology. The process is of disentangling and making
choices about what to keep and what to reject. Going into space will require
us to disentangle a lot, particularly with a (hopefully) diverse group of people.
We won't be able to control how these people develop or what might happen
when they reproduce; we can only make suggestions and hope that they do
not repeat the mistakes of our Terran past. Suggesting that they carry with
them the practices of decolonization, holding different ways of knowing as of
equal value, aiming for understanding first before oppression, will be the way
forward.

Education: Visions of the Future

Looking at the process of learning across various systems, some things appear
to be held in common. We might accept that learning is a social activity and
that therefore knowledge is distributed: "Learning, in other words, occurs in
communities, where the practice of learning is the participation in the community. A learning activity is, in essence, a *conversation* undertaken between
the learner and other members of the community" (Downes 2010). Stephen
Downes, pioneer in learning technology, writes in celebration of E-learning
2.0, by which he means the move towards open conversations between learner
and teacher, and beyond—the social revolution enabled by the Web, personal, and syndicated learning. However, Otto Scharmer, Senior Lecturer at

the Sloan School of Management, Massachusetts Institute of Technology and organizational development specialist, takes this vision forward to Education 4.0. Where, for Scharmer, Education 2.0 leaves us in a fairly reductionist output- and efficiency-centric space (the conveyor-belt system), 4.0 moves us into a cocreative and ecosystem-centric model. Here, deep sources of learning are activated, sources of wellbeing are strengthened, food and its production are regarded as a medium for healing the planet and the people, our focus on finance moves from extractive capital to generative, reforming the system, and governance comes from awareness-based collective action (Scharmer 2019). There is no space for hierarchical jockeying for position here, or unilateral militaristic ambition. It's a huge move from where we are now, but it might solve some of the problems in the seismic shifts occurring in education. Education 4.0 also takes us into the type of creative space where we might solve the problem of overreliance on the Internet (the foundation for Downes' E-learning 2.0 theory). In space, the World Wide Web ceases to be as dynamic as it is on Earth. Deep sources of learning must be relevant and generated from local contexts, not imposed for ideological, oppressive and or extractive purposes. How might we move forward?

James Paul Gee, the father of conversation analysis (a method of analyzing the way we talk, looking at the cultural and social constructions behind interaction) and the "godfather" of games-based learning, would suggest that we use games. Online gaming might be regarded as a useful and entertaining pursuit of education, but it is also laden with Western values. Learning may also be vicarious rather than experiential here—and there are dangers as well as benefits in adopting different identities and learning through games, not least that of addiction (Siddall 2019). Gaming software and hardware is also expensive to produce and therefore expensive to access, thus limiting its effectiveness as a learning medium. Decolonization—if we wish to practice this through developing computer games—might also mean "not-for-profit" in terms of enabling access to equipment and games worldwide. The PLE of *The Diamond Age* predicts a synthesis of a game played pluridimensionally, but in real time and places—not vicariously. These experiences are actual in their reality and come at a cost to the protagonist, although they are designed to help her "overcome."

Where Do We Want to Go?

We need to consider knowledge, then—whether acquired through everyday life experience, or in a space designed for its acquisition, or gain vicariously: What do we want our children to get from knowledge? What do we want them to know? Where do we want them to go for it? And how much do we want them to remember? Arguably, education has always been a mechanism of social control. In Aldous Huxley's *Brave New World*, regarded by some as "eerily prescient" (Miltimore 2017), learning occurs through hypnosis as the children (whose societal status has already been chemically predetermined prebirth) sleep. There is no liberation here—people are trained to be content with their lot—until their ideology is disrupted by the "Savage." The question for us remains—do we want to perpetuate our systems, or might it be possible for humanity to evolve beyond, to a place where learning ignites an open mind, heart, and will? I suggest that history shows, time and time again, that the type of social reproduction depicted in *Brave New World* and elsewhere is doomed to fail us as a species.

The PLE might be viewed as enabling in that it views its students with no prejudice. It adapts to any neurodiversity issues by changing the learning paradigms. It utilizes positive psychological concepts to encourage and enable its students to overcome obstacles, partly self-created but also in manipulating the dominant ideology for the students' benefit. Algorithms know no binaries. This type of PLE might work towards addressing distributive, recognitional, and associational issues of injustice (Gewirtz 2006) by working inside the system to re-privilege the student. Do we need an educational system like this? The danger is that we could only program it with known quantities on earth—and it would need to remain dynamic rather than static in an off-world context, drawing on unpredictable scenarios. The machine is less flexible than a person living through an experience.

Could we manage without teachers? Sugata Mitra, pioneer in educational technology, thinks we might—his work, along the lines again of web-inquiry and so dependent on the technology and its values, is enabling any students to gain an education in places where teachers won't go, and reprivileging the education process by making it child-led: "They engineered a system that was so robust that it's still with us today, continuously producing identical people for a machine that no longer exists. The empire is gone, so what are we doing with that design that produces these identical people, and what are we going to do next if we ever are going to do anything else with it?" (Sugata Mitra, in Moino, 2015).

It would be good to encourage, therefore, a move away from Western-centric educational ideas in considering the space curriculum. Can we balance the use of technology and its values with its benefits? Can we raise up different ways of knowing and still subject these to critical rigor? Will the missions on which humanity embarks carry forward the militarist or the ethnographer—the soldier or the social scientist? "It matters what stories make worlds, what worlds make stories" (Haraway 2016, 12). Let's engage in speculation for a new system where not just the human Terrans, but the species and the environment exist in a mutual becoming-with. That, surely, is at the heart of a curriculum re-centered around valuing being—and ultimately, survival through collaboration and cooperation.

References

Arshad, Rowena. 2021. "Decolonising the Curriculum: How Do I Get Started?" *Times Higher Education*, September 14, 2021. https://www.timeshighereducation.com/campus/decolonis ing-curriculum-how-do-i-get-started.

Bird, Elizabeth. 2012. "Top 100 Children's Novels #2: A Wrinkle in Time," *School Library Journal*, June 28, 2012. https://blogs.slj.com/afuse8production/2012/06/28/top-100-childr ens-novels-2-a-wrinkle-in-time-by-madeleine-lengle/.

Connolly, Paul, and Julie Healy. 2004. "Symbolic Violence, Locality and Social Class: The Educational and Career Aspirations of 10–11-Year-Old Boys in Belfast." *Pedagogy, Culture & Society* 12 (1): 15–33.

del Cerro Santamaria, Gerardo. 2019. "A Critique of Neoliberalism in Higher Education." *Oxford Research Encyclopedia of Education*, September 30, 2019. https://oxfordre.com/ education/view/10.1093/acrefore/9780190264093.001.0001/acrefore-9780190264 093-e-992.

Devereux, Charlie. 2021. "Cruelty Claim as World's First Octopus Farm Poised to Open in Spain." *The Times*. December 21, 2021. https://www.thetimes.co.uk/article/81cf8c10-61c2-11ec-b279-fa13aec304af.

Didion, Joan. 1979. *The White Album*. London: 4th Estate.

Downes, Stephen. 2010. "Learning Networks and Connective Knowledge." In *Collective Intelligence and E-Learning 2.0: Implications of Web-Based Communities and Networking*, edited by Harrison Hao Yang and Steve Chi-Yin Yuen, 1–27. New York: Hershey IGI Global.

Freire, Paulo. 1970. *Pedagogy of the Oppressed*. New York: Continuum.

Gewirtz, Sharon. 2006. "Towards a Contextualized Analysis of Social Justice in Education." *Educational Philosophy and Theory* 38, no. 1: 69–81.

Gogolin, Igrid, and Joanna Duarte. 2017. "Superdiversity, Multilingualism, and Awareness." In *Language Awareness and Multilingualism: Encyclopedia of Language and Education*, edited by J. Cenoz, D. Gorter, and S. May, 375–390. Berlin: Springer.

Haraway, Donna. 2016. *Staying with the Trouble: Making Kin in the Chthulucene*. Durham, NC: Duke University Press.

Hovey, Eric. 2019. September 1, 2019, Comment on "Call to Action: Ender's Game," blog post, published June 10, 2019. https://mca-marines.org/blog/2019/06/10/call-to-action-enders-game/.

Huereca, Rafael Miranda. 2010. "The Age of 'The Diamond Age': Cognitive Simulations, Hive Wetwares and Socialized Cyberspaces as the Gist of Postcyberpunk/'La era de' La Era del Diamante: Simulaciones Cognitivas, Wetware En Enjambre y Ciberespacios Socializados Como La Esencia Del Pos." *Atlantis* 32, no. 1: 141–154.

Jessop, Bob, Norman Fairclough, and Ruth Wodak. 2008. *Education and the Knowledge-Based Economy in Europe*. Rotterdam: Sense Publishers.

Kaufman, Hannah. 2014. "Education and the Pressure to Grow Up." *The Hoya*, June 30, 2014. https://thehoya.com/education-and-the-pressure-to-grow-up/.

Klees, Steven J. 2008. "Neoliberalism and Education Revisited." *Globalisation, Societies and Education* 6, no. 4: 409–414.

Mbembe, Achille Joseph. 2016. "Decolonizing the University: New Directions." *Arts and Humanities in Higher Education* 15, no. 1: 29–45.

Meda, Lawrence. 2020. "Decolonising the Curriculum: Students' Perspectives." *Africa Education Review* 17, no. 2: 88–103.

Miltimore, Jon. 2017. "'Brave New World' Predicted Today's Schooling Trends." *Foundation for Economic Education*, July 14, 2017. https://fee.org/articles/brave-new-world-predicted-tod ays-schooling-trends/.

Moino, Isabel. 2015. "The Future of Learning." Blog post. Published May 15, 2015. Accessed October 1st, 2021. https://medium.com/@isabelmoino/let-f1a3053148ee.

Montgomery, Sy. 2015. *The Soul of an Octopus*. London: Simon and Schuster.

Muldoon, James. 2019. "Academics: It's Time to Get Behind Decolonising the Curriculum." *The Guardian*, March 20, 2019. https://www.theguardian.com/education/2019/mar/20/academ ics-its-time-to-get-behind-decolonising-the-curriculum.

Ritivoi, Andreea Deciu. 2009. "Explaining People: Narrative and the Study of Identity." *Storyworlds: A Journal of Narrative Studies* 1: 25–41.

Scharmer, Otto. 2019. "Vertical Literacy Reimagining the 21st-Century University." *Presencing Institute*, April 16, 2019. https://medium.com/presencing-institute-blog/vertical-literacy-12-principles-for-reinventing-the-21st-century-university-39c2948192ee.

Shor, Ira. 1980. *Critical Teaching and Everyday Life*. Chicago: University of Chicago Press.

Siddall, Liv. 2019. "My Land of Make Believe: Life after The Sims." *The Guardian*, October 6, 2019. https://www.theguardian.com/games/2019/oct/06/my-land-of-make-believe-life-after-the-sims-video-games-computers .

Stephenson, Neal. 1995. *The Diamond Age: Or, A Young Lady's Illustrated Primer*. London: Penguin.

Subedi, Binaya. 2013. "Decolonizing the Curriculum for Global Perspectives." *Educational Theory* 63, no. 6: 621–638.

wa Thiong'o, Ngugi. 1986. *Decolonising the Mind: The Politics of Language in African Literature*. Portsmouth: Heinemann.

26

Astrobioethics Considerations Regarding Space Exploration

Octavio Chon-Torres

Astrobiology is the study of the possibility of life in the universe, but also of our expansion into it (Hays 2015). However, not everything is limited to the natural sciences as far as this field of study is concerned. We also have the participation of the social sciences and humanities, where analyses of the cultural implications of finding life elsewhere in the universe or settling in other planetary environments play very important roles in our becoming aware of our place in the universe. Questions such as whether or not we are the guardians of life in the universe, or whether other nonterrestrial life forms have equal rights considerations, are questions that cannot be directly answered by the natural sciences (Chon-Torres 2018a). In fact, it is essential to have a complex view, a diverse view, which characterizes astrobiology. Astrobiology, being transdisciplinary in nature (Santos et al. 2016; Chon-Torres 2018b), requires an interconnected way of looking at the world, and in this particular case, of examining the possibility or expansion of life in the cosmos. While we are advancing more rapidly at the technological level, however, at the moral or ethical level, we are not keeping pace. Seen in this light, we have the pending task of thinking about our moral role in the expansion or treatment of life outside Earth.

Astrobioethics is a philosophical approach within astrobiology that examines the repercussions or moral implications of the research and actions we take regarding life in the universe. Some of the key questions about space exploration with respect to astrobioethics are: (1) Are we the guardians of life in the universe? (2) Can we be morally empathetic to possible extraterrestrial life forms? (3) Is the discovery of life elsewhere in the universe a key factor in the improvement of the human condition? (4) Can becoming multi- or interplanetary beings mean that we finally come together as humanity? (5) Should we avoid at all costs any possible attempt to communicate with other possible

Octavio Chon-Torres, *Astrobioethics Considerations Regarding Space Exploration* In: *Reclaiming Space*. Edited by: James S. J. Schwartz, Linda Billings, and Erika Nesvold, Oxford University Press. © Oxford University Press 2023. DOI: 10.1093/oso/9780197604793.003.0026 civilizations in the universe? Each of these questions will be discussed in this chapter.

Are We the Guardians of Life in the Universe?

The fact that so far, we have not found life elsewhere in the universe makes us wonder about our role with respect to existing life on Earth, and whether this role is to prolong this life as much as possible. If humans were the only intelligent living forms in the universe, then that would highlight our role in taking care of it. This reflection arises from the fact that it could be objected that nobody has given us the right or permission to expand in the universe. By what right do we pretend to establish human colonies on Mars or the Moon? Do we think we are the experts in the universe? Do we think we are so important as to think that we are the consciousness of the universe in action?

In the situation that humans are the only conscious, technologically advanced life form so far, we would have an ethical obligation to maintain and preserve life. If it is the case that we are the only self-aware living form in the universe capable of protecting life via deliberate action, that should make us aware that if we do not do something to preserve or maintain life, it will become extinct and may never appear again. In this scenario, if we never do anything to maintain life in the universe, we would be condemning it to disappear completely. It would be difficult to think of not doing anything to perpetuate and expand it in the universe, unless some kind of technological limitations do not allow us to do so. If we speak of life as such, it is essential that it continues to exist in the universe. Whatever cultural or religious considerations humanity may have about life, it is essential that life spreads throughout the universe. What happens if we do not? Inevitably, this expansion will have to happen if life is not to meet its end. That is, assuming, for example, that humanity does not meet its extermination in the very long term, we will have to move to other planetary environments, which means we have no choice. Let us keep in mind that this is only possible if humanity does not become extinct, whatever the cause.

Someone might say that it may not have to be humankind that maintains life in the universe, but a posthuman form based on artificial intelligence. Again, whatever intelligent life form it is, if it survives long enough, it will have to confront that at some point, the planet of origin of Earth's life forms will not be able to sustain them so that they can maintain their existence where they always have. Therefore, if a life form has the necessary intelligence and tools to maintain life, it will have to do so.

However, there are two strong arguments that may limit us, in terms of time, from spreading out into the universe. The first argument is that if we find life on the planet we are targeting, such as Mars, then necessary precautions would need to be taken in order to preserve local Martian life—such as developing new care policies for the benefit of extraterrestrial life and all that this entails, which would entail investments of time. The second argument concerns the technology we need to be able to move to other planetary environments. For example, even with the right technology to travel to and terraform a planet like Mars, it would take hundreds of years and many generations, and humanity may not survive to see it happen along the way. In both cases, certain amounts of time are required to overcome these obstacles.

Of course, achieving all of this takes more than just time, but in a hypothetical case where we face a life-or-death situation for the entire human species, or even for all life on Earth, time is crucial. However, if humanity or some intelligent post-human form continues to survive, at some point we will find the solution and move forward. In any scenario, as long as some intelligent life form persists and has the means to do so, it will have to move off Earth in order to perpetuate itself. Considering this—considering ourselves as some form of guardians of life in the universe—becomes an understatement on a pure survival issue. Do we have a moral burden to be able to preserve life in the universe? Yes, for otherwise we would all be dooming ourselves. Is it too early to think about it? It is never too early, since we are talking about time scales that do not reach beyond a few generations. Does that mean we should abandon the care of life on Earth? Neither, because at the end of the day, we would be watching over the life of all. All these reflections lead us to the following consideration.

Can We Be Morally Empathetic to Possible Extraterrestrial Life Forms?

This question is: if we were to find some form of microbial life while exploring space, would we be obliged to respect it? There is nothing in Western cultures that obliges us to respect forms of life found outside the Earth, except our ethical frameworks of action, which is why it is important to extend ethical analysis to the case of extraterrestrial life. Beyond the utilitarian value that we can project toward this potential life form, even in a scenario where its potential utilitarian value is very high, we nevertheless need to consider a morality that manages to overcome biogeocentrism.

According to astrobiologist Julian Chela-Florez (2001; 2009; Aretxaga 2004), biogeocentrism is a perspective of life in the universe (as well as its value) based on the only reference we have so far, which is terrestrial life. Since we have no evidence of life outside Earth, it is impossible to overcome biogeocentrism at this time. However, the same is not true if we examine the value of extraterrestrial life on a conceptual level, through philosophical speculation. For instance, there is the idea of teloempathy, which, according to astrobiologist Charles Cockell (2005a, 2005b), is a position humans can take in relation to forms of nonterrestrial life where humans recognize the right of non-terrestrial life not to be annihilated. If we find extraterrestrial microbial life forms, and they cannot defend themselves from a possible threat from us, a teloempathetic response would consider the interests that these life forms have in not dying. The fact that they would be singular life forms with an origin outside of Earth and no possibility of defending themselves is what makes teloempathy so important.

Up to this point, however, someone could object that it is not necessary to invent so many words to refer to situations already known in different environments. But that would be a mistake, since it would limit the capacity and developmental potential for the elaboration of arguments and, following that logic, we would not even have the conceptual baggage that we now use to talk about everything we know. We conceive the world thanks to the concepts that emerge as new constructs that allow us to shed light on the changing scenarios we face. Still, another other question may arise: If we agree we should empathize with exterrestrial microbial life forms, should we extend the same consideration to microbes on Earth?

Consider a Martian bacterium versus a terrestrial bacterium. Both are alive, but the former presumably has an origin in a second genesis of life, and examining it would answer whether we are not alone in the universe, as well as giving us clues for a potential astrobiocentric theory of life in the universe. At the level of scientific interest, the Martian bacterium represents a unique opportunity to understand the emergence and evolution of life in other planetary environments. In this way, it has a different value than a common Earth bacterium. However, our critical interlocutor may insist that, when talking about the rights of Martian microbial life, we should also talk about the rights of terrestrial microbial life. Terrestrial microbial life forms also have interests in avoiding annihilation, but the same would be applicable for any other form of life. If this meant we had an obligation to protect terrestrial microbial life, we would be obligated to become vegans, and we would even have to leave alone the harmful bacteria that may affect our human health. If we take the logical consequences to the letter by defending terrestrial bacterial interests, we end up in an unrealistic situation. For this reason, the notion of teloempathy belongs and makes sense primarily in the context of astrobioethics, since its narrative and substance are framed by the importance of finding, studying, and understanding life in the universe. In the event of confirmation of evidence of extraterrestrial life, the treatment given to it will not be at all the same as that given to terrestrial bacteria. This may lead us to the next question, for if such a finding is confirmed, we would need to discuss even the possibility that our own way of considering life on Earth will change, as a direct effect of our consideration of it in the universe. This leads us to the following question.

Is the Discovery of Life in the Universe a Key Factor in the Improvement of the Human Condition?

It is interesting to think about this possible consequence, to think that humanity might actually turn around in its thinking and actions primarily because of confirming evidence of life on other worlds. Of course, such a change would be different following a discovery of microbial life as opposed to a discovery of extraterrestrial intelligent life forms. If we consider specifically a scenario where we have confirmed the presence of non-terrestrial microbial life, would that also mean that we would achieve world peace and the unification of nations? Unfortunately, that does not seem to be the destiny of humankind should such a finding ever be confirmed. If we take as a reference the situation generated by another globally impactful event, the SARS-CoV-2 pandemic, we observe that economic and political interests did not disappear in a considerable way. Even the World Health Organization (WHO) pronounced the possibility of a catastrophic moral failure with regard to the vaccination process in developing countries. The head of the WHO claimed that healthy people in rich countries could be vaccinated much earlier than sick people living in poor countries (WHO 2021). The fact that the life expectancy of a sick person in a poor country is very low is not new, but in the context of a pandemic, this gives us much to think about. Considering that we have not been able to act collectively in ways that really give us signs of any possibility of overcoming political differences, even in the face of an invisible common enemy, suggests that confirming the presence of extraterrestrial life in the universe will not make us better people. In the beginning, there may be reactions and strategies that lead improvements in the human ethical condition, but overall, our mistakes and imperfections as human beings may win out over us.

What will happen to religion? What has always happened to religion when new discoveries are made could happen: a process of adaptation. In this case, it would be a process involving a narrative of the meaning of the universe where the notion of the *imago Dei* (image of god) can embrace other forms of life (Pryor 2020). And yes, new religious movements will emerge, and some sects will also emerge from already established religions, but the key factor that characterizes it, *religare* (religious ties or binds) will give a new meaning for a good part of humanity.

However, it is not necessary to wait for confirmation of life elsewhere in the universe to see if there is a substantial change in us. What if we are the extraterrestrials for other worlds as we begin to colonize them? Therefore, the following question deserves to be raised.

Could Becoming Multi- or Interplanetary Beings Mean That We Finally Unite as Humanity?

The very idea of being a multi- or interplanetary species has inspired the imagination of many, so much so that there is no shortage of film and literary production on the subject. However, practically speaking, does it make sense to talk about any species becoming a multiplanetary species? In a literal sense, this would only be appropriate if the species to which we refer has considerable genetic changes compared to one which remained on its planet of origin. So, if we stick to the idea of species at the biological level, a new species emerges if its genetic variations are sufficient to differentiate it from its predecessor (Gohd 2009; Chon-Torres and Murga-Moreno 2021), and thus a new "multiplanetary" species would only arise following the onset of sufficient new genetic variations acquired on new planets. But that does not seem to be the popular idea when one imagines humanity outside the Earth. That is, when one imagines humanity as a multi- or interplanetary species as billionaire and space colonization advocate Elon Musk might do, he has in mind that the current version of humanity will be the one that will actually reach the multiplanetary/interplanetary stage.

On a rhetorical level, the notion of becoming a species from the stars may sound appealing; if we use literary and flexible language, it sounds nice. On a concrete, biological level, referring to humanity as a multi- and interplanetary species implies essential changes in what it means to be human beings. So it may not even be possible to speak of a multiplanetary species and human species at the same time, in the traditional sense of the word, that is, the current humanity inhabiting multiple locations outside Earth. The importance of differentiating the terminology between being a multiplanetary species and a multiplanetary *sub*species, then, is that the former would directly allude to genetic improvements or modifications that take place in order to inhabit other worlds. In the second case, we do not need to think of a humanity with genetic variations derived from survival in another planetary environment as a new species in its own right. The consequences of alluding to one or the other will lead us to different discussions.

This being so, the most concrete proposal would be to refer to humanity as a multi- *and* interplanetary subspecies (Chon-Torres and Murga-Moreno 2021). The issue that remains is to identify the conceptual differences between being multiplanetary and being interplanetary. A multiplanetary humanity would eventually be able to properly *inhabit* different environments outside the Earth through a process of terraforming. The idea of being multiplanetary sounds too much like being limited only to inhabiting planets, but in practice it could refer to human inhabitation of any extraterrestrial location, for example, artificial habitats on a moon, even on our own Moon.

On the other hand, if we refer to humanity as an interplanetary subspecies, the matter is more limited to the question of being able to *travel* to other environments outside of Earth. Becoming an interplanetary subspecies refers directly to humanity's technological ability to travel to other planetary bodies. That being the case, the way we are exploring space currently is more likely to make humanity an interplanetary subspecies than a multiplanetary species. However, the ethical issue of being an interplanetary subspecies remains to be discussed.

In the hypothetical case of being able to travel to other worlds, of being able to come and go to the Moon, or being able to transport ourselves to Mars, would that mean that humanity has achieved or realized a moral achievement? What about the idea of being planetary human beings—would a multiplanetary humanity end up becoming a humanity fragmented into each environment it manages to inhabit, as we already do on this planet? If we are going to judge the moral capacity of a civilization by its technological development, we would not do well. Developing technology does not guarantee ethical progress. Therefore, a dystopian scenario where humanity is interplanetary is possible, for instance, a scenario where forms of repression of freedom are allowed on Mars under the idea that the atmospheric conditions merit it. To think of humanity traveling to other worlds only from the perspective of a researcher, or thinking that it will only be trained or prepared people who will do so, would be inadequate. Human beings are political beings by nature, they are social beings, and this implies a complexity that cannot be foreseen or predicted. If we were to travel off Earth, this complex sociality would follow us like a shadow. The few times that people come together to overcome their differences is when we have a common enemy, and yet we eventually return our division. But what if that threat were permanent, such as being contacted by an advanced civilization? This brings us to the following astrobioethics question.

Should We Avoid at All Costs Any Attempt to Communicate with Other Civilizations in the Universe?

This question may reflect the fears we may have should we make contact with an intelligent life form that turns out to be hostile. It is not for nothing that the late physicist Stephen Hawking also expressed this concern when he thought that an alien species could themselves be colonizers, in which case the destiny of humanity would be a grim one (Cofield 2015). Therefore, it might be thought that the best thing to do would be to avoid all forms of potential contact, or to arm ourselves in order to be prepared for possible hostile activity. However, neither of these two alternatives seem to be adequate in the short and long term.

First, if humanity decided to avoid any form of contact, it would be wasting an opportunity to answer the age-old question of whether or not we are alone in the universe. Moreover, the fact that on Earth this type of experience of encounter between different worlds and cultures has led to hostilities and cultural assimilation does not mean that these negative consequences will be replicated in cosmic contexts. We cannot readily extrapolate from a biogeocentric perspective to a scenario with our potential galactic companions. Besides, eventually, if there is intelligent life with sufficient technology out there, and if humanity or post-humanity continues to exist, we will have to find them, or they will have to find us. Of course, certain conditions would have to be in place for contact to happen, because if it is the case that humans are the only intelligent living beings with technology, then no matter how much we expand into the cosmos, we will not find signs of advanced technological life. We might encounter environments with microbial life forms, but they would pose us no threat. In the hypothetical case that we continue to exist as a species—in all its variations—it is inevitable that we will encounter something out there, and regardless of whether it is hostile or not, that contact will have to happen. According to interdisciplinary researcher Seth Baum (2019), if humanity follows an *astronomical trajectory*, that is, if it survives everything that poses a threat to our species, our destiny will be to inhabit other worlds by means of, for example, terraforming. Now, if we extract this idea and contextualize it in a situation where, over thousands of years in the future, humanity may travel beyond the Solar System, either it is likely to encounter some extraterrestrial life forms, or it may never do so.

On the other hand, assuming that we decide to arm ourselves—militarize space—in order to prevent any form of hostile activity against Earth by a threatening civilization, we would end up making an unsuccessful attempt. If there is a sufficiently advanced civilization, technologically speaking, then our weapons, whatever we have, will not be able to cope with such a tremendous military capability. We may hold out a little at first, but eventually we will have to give in to their greater force. Of course, this is a scenario where we are the visited ones, and where we are at the mercy of our visitors. Does this mean that, in general, we should do nothing to defend ourselves against extraterrestrial threats? We have no absolute guarantee that our intergalactic visitors will be benevolent or not, but militarizing space would make no sense, unless it is used as a pretext for geopolitical purposes. In this geopolitical case, it may be that space becomes militarized under the pretext of guaranteeing the survival of humanity, generating influence and pressure on other countries, leading to a kind of arms war or cold war taken to outer space—but where the potential enemies are ourselves.

Conclusion

Taking up and giving proposals to the questions posed, I could say that: (1) Are we the guardians of life in the universe? Yes, as long as it is a matter of preserving the life we know and protecting any extraterrestrial life that cannot defend itself. (2) Can we be morally empathetic to possible extraterrestrial life forms? Yes, developing a teloempathetic perspective can bring us closer to an astrobiocentric view. (3) Is the discovery of life in the universe a key factor for the improvement of the human condition? Not necessarily, because the improvement in humanity's moral quality does not depend on what we discover or not outside, but inside ourselves. (4) Can becoming multi- or interplanetary beings mean that we finally unite as humanity? Neither, because technological development does not necessarily represent any form of moral advancement. (5) Should we avoid at all costs any attempt to communicate with other civilizations in the universe? In the scenario where we develop enough technologically to find other life forms, they will detect us or we will detect them; in the scenario where they visit us, we will have no way to defend ourselves against such technological advancement, if they pose any threat at all.

References

Aretxaga, Roberto. 2004. "Astrobiology and Biocentrism." In *Life in the Universe*, edited by Joseph Seckbach, Julian Chela-Flores, Tobias Owen, and François Raulin, 345–348. Dordrecht: Kluwer Academic.

Baum, Seth D., Stuart Armstrong, Timoteus Ekenstedt, Olle Häggström, Robin Hanson, Karin Kuhlemann, Matthijs M. Maas, James D. Miller, Markus Salmela, Anders Sandberg, Kaj Sotala, Phil Torres, Alexey Turchin, and Roman V. Yampolskiy. 2019. "Long-Term Trajectories of Human Civilization." *Foresight* 21, no. 1: 53–83. https://doi.org/10.1108/FS-04-2018-0037.

Cockell, Charles. 2005a. "Planetary Protection: A Microbial Ethics Approach." *Space Policy* 21: 281–292.

Cockell, Charles. 2005b. "Duties to Extraterrestrial Microscopic Organisms." *Journal of the British Interplanetary Society* 58: 367–373.

Chela-Flores, Julian. 2001. *The New Science of Astrobiology: From Genesis of the Living Cell to Evolution of Intelligent Behavior in the Universe*. Dordrecht: Kluwer Academic.

Chela-Flores, Julian. 2009. *A Second Genesis: Stepping Stones Towards the Intelligibility of Nature*. Singapore: World Scientific.

Chon-Torres, Octavio A. 2018a. "Astrobioethics." *International Journal of Astrobiology* 17: 51–56. https://doi.org/10.1017/S1473550417000064.

Chon-Torres, Octavio A. 2018b. "Disciplinary Nature of Astrobiology and Astrobioethic's Epistemic Foundations." *International Journal of Astrobiology* 20, no. 3: 186–193. https://doi.org/10.1017/S147355041800023X.

Chon-Torres, Octavio A. and César Murga-Moreno. 2021. "Conceptual Discussion around the Notion of the Human Being as an Inter and Multiplanetary Species". *International Journal of Astrobiology* 20, no. 5: 327–331. https://doi.org/10.1017/S1473550421000197.

Cofield, Calla. 2015. "Stephen Hawking: Intelligent Aliens Could Destroy Humanity, But Let's Search Anyway." *Space.com*, July 21, 2015. https://www.space.com/29999-stephen-hawking-intelligent-alien-life-danger.html.

Gohd, Chelsea. 2019. "Can We Genetically Engineer Humans to Survive Missions to Mars?" Space.com, November 7, 2019. https://www.space.com/genetically-engineer-astronauts-missions-mars-protect-radiation.html.

Hays, Lindsay, ed. 2015. *The Astrobiology Strategy 2015*. NASA.

Pryor, Adam. 2020. *Living with Tiny Aliens: The Image of God for the Anthropocene*. New York: Fordham University Press.

Santos, Charles Morphy D., Leticia P. Alabi, Amânio C. S. Friaça, and Douglas Galante. 2016. "On the Parallels Between Cosmology and Astrobiology: A Transdisciplinary Approach to the Search for Extraterrestrial Life." *International Journal of Astrobiology* 15: 251–260.

WHO. 2021. "WHO Director-General's Opening Remarks at 148th Session of the Executive Board." January 18, 2021. https://www.who.int/director-general/speeches/detail/who-director-general-s-opening-remarks-at-148th-session-of-the-executive-board.

27

Greening the Universe

The Case for Ecocentric Space Expansion

Andrea Owe

Introduction

We live in a peculiar time, and I have a peculiar job. I am neither a space scientist, nor a STEM scientist. I am an artist turned environmental moral philosopher working in the cheerfully named field of *global catastrophic risk*.[1] Essentially, my job has two major parts. The first is to figure out how we may avoid going extinct, or at least avoid being harmed to such an extent that life and civilization as we know it might never recover. It is examining and avoiding these trajectories that define the field of global catastrophic risk. The other part is studying the moral potential of the long-term human future that could be realized if we overcome these risks, and how we may realize this potential. Current humans do not necessarily have the answers to these moral questions, but we should strive toward ensuring they can be answered and realized in the future. Therefore, we must think quite far ahead—hundreds, thousands, sometimes millions of years into the future. Such a future is likely to entail space expansion.

As an environmental philosopher, I differ from most of my colleagues in that I define catastrophic risks as they pertain to the total Earth-life-system, not just the human species. By the Earth-life-system, I mean the living and evolving ecosphere that characterizes planet Earth. I further differ from some of my colleagues in that my intuitions about a morally optimized long-term space future neither point toward some space technotopia nor the maximization of intelligence and pleasure only (Bostrom 2008). Rather, they reflect a broader view on the entire world we are a product of—a rich and ever-changing ecosphere with an extraordinary story to tell.

In this chapter, I take the liberty to be ambitious and idealistic, as I bring together my perspectives on environmental ethics, global catastrophic risk,

Andrea Owe, *Greening the Universe* In: *Reclaiming Space*. Edited by: James S. J. Schwartz, Linda Billings, and Erika Nesvold, Oxford University Press. © Oxford University Press 2023. DOI: 10.1093/oso/9780197604793.003.0027 and a long-term space future to make a deeply environmental case for space expansion.

An Argument for Ecocentric Space Expansion

An environmental motivation for space expansion holds a question of what *nonhuman* entities and values humans should aim to protect and promote throughout spacetime. As an *ecocentric* environmental philosopher, I see Earth's ecosphere and its continued story as the primary moral object to protect and promote.[2] Ecocentrism is grounded in the sciences of the Earth system and sees moral value holistically in the systems that make up the natural world, of which humans, all other life, and abiotic natural elements are part, through myriad interdependent and symbiotic relationships.[3] Ecocentrism further entails a historical perspective that sees the evolutionary Earth-story as a whole, stretching from its past, through to the present, and further into its potential future. Independent of any utility for humans, all of this has intrinsic value—value in its own right.[4,5] Ecocentrism is also value-pluralistic. While the highest level of intrinsic value is the ecosphere with its continued story, subecospheric things, such as the well-being of the different organisms within the ecosphere, also have intrinsic value.[6]

Inherent in the ecocentric view is a rejection of ontological and ethical anthropocentrism—ontology being about the nature of the world, whereas the ethical concerns how the world should be. This means a rejection of the beliefs that humans are distinct from, more important, and/or better than the rest of nature, solely on grounds of being human. While the ethical rejection is expanded on in the next section,[7] modern science provides the ontological rejection. There is no human and nonhuman world; there is one ecosphere. Humans are part of ecosystems and the animal kingdom alike. We are all subject to the same physical laws. Anthropocentrism is a modern illusion, making humans behave in ways that destroy our own home, other living beings, and ultimately ourselves.

Why Ecocentrism?

While I endorse ecocentrism, it is important to note that when we talk about moral value in the far future, it is unlikely that present humans can identify all future values accurately. Ecocentrism is what I propose as a viable strategic starting point for optimizing the potential for moral value in the long-term future, by ensuring maximal optimization at both the ecospheric and sub-ecospheric levels. By optimization, I mean making something be the best it can be (discussed further toward the end of the chapter). There are several reasons why I advocate considering intrinsic value at these levels.

First, I am sympathetic to moral realism[8] and find ecocentrism to accurately reflect natural history and science. Second, I find all instances of life, evolution, and nature's creative force to be extraordinary and of moral value. Importantly, I find the whole world to be very much alive, and recognize that all life, both individual organisms and holistic entities such as species and ecosystems, have an innate drive to live and flourish, they have interests, and they have well-being. I find all this morally relevant. I also strongly suspect that sentience/consciousness/intelligence[9]—commonly considered of moral significance—permeates much of, if not the entire, natural world.[10] Third, I recognize the inherently relational and symbiotic structure of the Earth-life-system, which makes it difficult to single out certain entities, qualities, or properties as the sole container of moral value. Intuitively, the holistic view on Earth's entities and components has always made sense to me, whereas the hierarchical one never has. Ecocentrism reflects my actual experience of the world.[11] Fourth, I find moral value in the diversity of these instances of evolution, life, and creativity, as central to the ecosphere's realization of potential. The fifth is pragmatic. I believe it is in the best interest of humans and nonhumans alike that humans value the natural world in this holistic manner. I think the failure to do so is central to our current socio-environmental predicament, as well as to more subtle harms, such as to our characters and loss of meaning. The sixth is the simplest reason: as far as we know, what has happened on Earth—an astonishingly diverse and living world—is so cosmically extraordinary that there is good reason to value its totality rather than singling out only a fraction of it. Going further, moral uncertainty (MacAskill et al. 2020) and the prospect of moral progress (Sauer et al. 2021) give us reasons to promote a "radically" inclusive ethic, especially when considering long-term trajectories. For example, just as moral consideration of nonhuman animals has gained ground over past decades, moral consideration across the plant kingdom could also be regarded as common sense in a century or two (I hope sooner).

In short, this holistic perspective avoids the pitfall of missing the forest for the trees and is precautious, open-minded, and humble toward the unknown future. Focusing moral effort on making the ecosphere be the best it can be entails embracing that the world and its myriad past, present, and future inhabitants do not exist for human purpose, that we are all part an intricate, everchanging web, the value and potentiality of which goes far beyond that of our own species, and that all are better off living in a world (or worlds) humans respect and cherish, including into our distant cosmic future.

What Does This Mean for Space Expansion?

Unfortunately, standard accounts of space exploration and expansion carry strong bonds to anthropocentrism. For example, the idea that biological humans can detach from the rest of terrestrial nature and survive alone is ontological anthropocentrism. The idea that humans are destined to conquer the universe, and entitled to do so, is ethical anthropocentrism. Consequently, many environmentalists oppose the ongoing pursuit of a space future.[12]

However, in a cosmic perspective on Earth, space expansion is not about one particular species in its current form. It is about the continuation of the extraordinary story that is the *story of life on and from Earth*. This story holds a broad range of moral values, including beauty, knowledge, diversity, subjective experiences, creativity, love, and life itself. They are all worthy of promotion. A truly nonanthropocentric perspective on the pursuit of a space future will not only consider what humans exclusively think and feel about nature. It will also consider the perspectives, interests, and values of nonhumans. Simultaneously, it must recognize that the Anthropocene human is indeed distinct, in that our collective agency now has long-term, evolutionary, planetary, and possibly multiplanetary implications. We have, therefore, an equally encompassing responsibility. The time when the proper environmental ethic was to "leave nature alone" is over. In its place must come active aid and assistance—on Earth and beyond.

Indeed, an ecocentric vision of space expansion leaves behind anthropocentric ideas of a cosmic manifest destiny and interstellar subjugation. It contrasts with the ideas of safeguarding the human species alone, of escaping the "Earth cradle," of leaving Earth for nonhumans, and in the process fleeing from our destruction of Earth. Ecocentric space expansion entails the core motivation being the protection and promotion of the totality of the story of life on and from Earth, where the optimization of moral value is built upon the broadest possible foundation of bio- and ecosystem diversity, even if "life" and "ecosystems" will come to be fundamentally different from how they are today.[13] The idea is not to preserve the exact current state of the ecosphere— there is no *status quo* in nature—but to create the largest possible opportunity space for ecospheric value: if many elements within the ecosphere have intrinsic value, then the future potentiality of each element together accumulate in the greatest opportunity space.

As the only species currently capable of space expansion, as well as contemplating the ethics of planetary and cosmic trajectories, humans have a unique instrumental role in the continuation of this story.[14]

What Does This Mean for Humans in Space?

Importantly, an ecocentric space expansion equally has humans' and any posthuman descendants' best interests in mind, as the ecocentric project is to reconcile human culture with the reality of the world we come from, toward flourishing for all, instead of for the few (Curry 2011). The physical, psychological, and moral distancing that modern economic and social paradigms have enforced between humans and the rest of nature has already caused much self-harm, including the onset of anthropogenic catastrophic risks (more on catastrophic risk in the following section). Evidence shows that humans not only instrumentally depend on a healthy ecosphere but find in it much meaning and moral value (Bruskotter et al. 2015; Johansson-Stenman 2018; Berry et al. 2018). To pursue an even more existentially lonely way of being human beyond this planet, therefore, appears not only exceptionally risky but deeply undesirable.

Most importantly, we must ask ourselves why we would *want* to separate ourselves from what is by all known accounts the most extraordinary phenomenon in the universe. Even if it should prove possible to eventually cut the human-ecosphere umbilical cord (Holt 2021), how can an astronomical future with only one form of being (human? posthuman? AI?) be better than a future filled with trillions of them? How can such a future be better for humans? How can the potential of what humans can become and create be greater in this narrow cone of light, than in one that illuminates the full creative and moral potential of the Earth story?

Space expansion that takes the terrestrial ecosphere as its starting point will enable future worlds that can sustain and optimize human flourishing, that are rich in value and meaning *for* humans, and where the potential human impact across space can be as positive as possible.

An Ecocentric Argument for Space Expansion

Many reasons for pursuing space expansion have been proposed, some more convincing than others. Here, I present three interrelated reasons that apply to all Earth-originating entities and show what they imply for ecocentric space expansion.

Global Catastrophic Risk

One reason for space expansion comes from a practical observation of current global catastrophic risks on Earth. Some catastrophic risks are natural, such as supervolcanic eruptions or Earth getting hit by a large asteroid, while others are human-caused, such as nuclear war or unaligned artificial intelligence.[15] While the unfolding of any global catastrophe would not necessarily entail the complete annihilation of life, it would likely involve mass extinction, the destruction of ecosystems, and a massive blow to human civilization. The exact window of time for spreading beyond Earth in terms of catastrophic risk is unknown. Space expansion is therefore, generally speaking, a viable risk mitigation strategy. By spreading beyond Earth, we would become more resilient to risk. This applies to all Earth-originating entities (Tonn 2007; Baum 2010). For the ecocentric case, this means that all life and the total ecosphere have interests in avoiding catastrophic risk and pursuing a long-term future.

Importantly, this strategy relies on there being no or minimal risk correlation between the different locations. For example, a Mars habitat dependent on Earth provisions is vulnerable if catastrophe hits Earth, whereas a fully self-sustaining habitat is more resilient. Currently, anthropogenic risks are the most pressing. Therefore, in the immediate future, this first of all means that great effort should be taken toward mitigating anthropogenic risks on Earth, so we have enough time to achieve the conditions necessary to pursue space expansion in a risk-resilient manner. For example, if the present human generation fails to act on climate change and the destruction of the living world, then many things worthy of protection and promotion could be irreversibly lost by the time space expansion is undertaken. We could bring the underlying causes of the environmental crisis with us to space, or civilization could be left incapable of space activities all together. This is vital for all terrestrial life as space expansion requires a technologically advanced civilization.

Still, overcoming immediate risks is not enough. It is highly unlikely that we can reach a sustainable state of zero risk while remaining on Earth only. If nothing else, Earth will eventually become uninhabitable due to the expansion and warming of the sun (Wolf & Toon 2015). In light of catastrophic risk, any long-term future necessarily entails space expansion (Baum et al. 2019).

Equality across Spacetime

A second reason is motivated by an ethical principle of equality across space and time. This means that someone or something of moral value is of the same value independent of the time in which they happen to exist (Cowen & Parfit 1992; Tonn 2018), and the location they happen to be in (Smith 1998). This matters for the future in general, such as morally accounting for future generations (Tonn 2018). However, it also matters for space expansion. By expanding in space, and by effect mitigating the risks inherent in remaining on Earth only, we open an exceptionally larger temporospatial frame that allows for astronomical opportunities to advance moral value. These opportunities are vastly larger than those available for the present and the near-term future, and in trajectories where life remains restricted to Earth (Baum et al. 2019). In effect, the act of ensuring we can expand into space has a compounding effect on moral value realization.

Therefore, on the ecocentric account, pursuing space expansion might prove the equivalent of keeping the door open to a multitude of possible futures for *all moral values inherent in the ecosphere*, whereas not expanding into space would effectively close the door on all of them. Despite our past and present misconduct toward the rest of Earth's life, it would be even more immoral to deliberately shut the door, including on behalf of future beings that could be more moral than present humans. In the same way that it is deeply unethical to blame all humans equally for the ongoing environmental crisis, it is deeply unethical to condemn all future life for what a few generations of one species have done. If there is a possibility of continuing the incredible story of terrestrial life into the far future by expanding into space, then it is deeply antienvironmental, and deeply anthropocentric, to deny the pursuit of this possibility.

Optimization of Moral Value

Thirdly, in order to realize the potential for moral value at great scales, we must go beyond just ensuring the expansion of Earth life into space. We must work to create conditions necessary for not only the continuation, but the flourishing, of life and civilization. We want the future to be better than the present and, ideally, we want the future to be as good as it possibly can be. Ensuring that life exists in some form in some cosmic locations might be good, but ensuring that life and civilization can flourish, continuing to advance and improve moral goods at great scales and into the distant future, is much better (other things being equal).

This difference can be understood as that between sustainability and optimization (Owe & Baum 2021). Optimization is more demanding and provides reason to pursue space expansion sooner rather than later. For example, sustaining life could potentially be secured through relatively simple measures, such as planetary seeding, where terrestrial microorganisms are deliberately sent to other celestial bodies. Optimizing the flourishing of a diversity of life, on the other hand, requires the establishment of complex and resilient large-scale environments.

For ecocentric space expansion, this means the most promising methods are either to make other planetary bodies suitable for Earth-life (while leaving billions of others as they are), such as through terraforming,[16] or constructing Earth-like environments from scratch, mimicking a symbiotic whole.[17,18] The latter may be less adaptationally demanding for living beings, whether naturally or through biotechnology, than terraforming existing planets with challenging planetary conditions. Constructed habitats could have further benefits such as less moral loss from uprooting existing planets, better safety and risk resilience from mobility, and potentially less cost and effort to make ideal habitats. In either case, as the source of creative and moral potential in our world is the entire ecosphere, trying to make terrestrial life start over by adapting to abiotic extraterrestrial environments is a much poorer starting point than advancing the full preconditional sphere for ecospheric flourishing. After all, the ecosphere is a "billions of years stress-tested form of adaptive complexity" (Holt 2019). Even in the trajectory of natural evolution merging with or being replaced by technological design, the total available environment will determine its potentiality. Quite possibly, an astronomical future holds myriad hybrid beings and systems very different from the life we know today, ideally also incorporating extraterrestrial nature for increased diversity.[19]

Significant efforts have been made to expand human environmental considerations to space environments *as they are* (Rolston 1986; Milligan 2015; Schwartz 2018), including a proposed cosmocentric ethic (Lupisella 2020). I fully endorse these efforts and many of their sentiments. However, while ecocentric space expansion can incorporate extraterrestrial nature, it necessarily entails bringing terrestrial nature into space nature in the best interest of humans and nonhumans alike, and, in cases of trade-offs, prioritizing terrestrial over extraterrestrial nature. An ecocentric ethic that considers the perspectives, interests, and values of nonhumans and the greater terrestrial ecosphere has its primary moral obligation to the Earth-life story, not to everything that exists in the universe, nor to the universe itself.

A Universe of Weird and Beautiful Earths?

An ecocentric account of space expansionism leaves us a tentative objective that considers a universe of myriad flourishing Earth-inspired worlds as morally good. Each of these worlds would be as extraordinary as the original, each distinct in their own right, but all part of the overarching story of life that came from Earth. If the ecosphere itself and its story is of utmost moral value, then a diversity and optimization of and within Earthly ecospheres throughout spacetime is of even greater value. This ecocentric account also presents a common policy and ethics compass for humans as morally responsible stewards of the total Earth story.

Naturally, there are many challenges toward such an idealistic objective, including those that arise from disunity among people, and between people and the rest of Earth's life. A significant task is to develop the adequate environmental and technological expertise. Establishing and investing in educational programs in Earth design is a good start. Other challenges include creating comprehensive collections of seeds, microorganisms, DNA, etc., and minimizing suffering in Earth reproductions.

The long-term future requires ambitious goals. As we are potentially at the very beginning of a vast opportunity space, the observations laid out in this chapter give strong incentives to prioritize a long-term space future, and to approach it by our most optimistic visions. Still, I hold that in our search for meaning and value across the universe, we need not look so very far, certainly not inward only. Importantly, we need to leave the door open to the ecospheric potential, even if that potential is unrecognizable to us today.

Space expansion grounded in the reconciliation of humans (and our technologies) with the rest of nature, rather than the further ostracism and condemnation of humans from nature, will not only allow our mutual continued existence, but lay a solid foundation for our flourishing and realization of moral potential. While humanity's greatest immediate challenge is to survive the next century or two, our greatest achievement will be eventually greening the universe and bringing it to life.

Acknowledgments

Thanks to Lauren A. Holt, Seth D. Baum, and Tyler T. Barrott for comments on an earlier version of this chapter.

Notes

1. Also called *existential risk*. See, e.g., Bostrom and Ćirković (2008) and Ord (2020).
2. Others who have argued similar cases in a space context include Margulis and West (1997), Tonn (2002), and Randolph and McKay (2014).
3. On ecocentrism, see, e.g., Curry (2011) and Washington et al. (2017).
4. That is not to say all things within the ecosphere are morally good.
5. The concept of intrinsic value has many interpretations and is subject to much debate, but I will not entertain these matters here.
6. Importantly, well-being/suffering are defined differently for different entities. We should be careful not to consider well-being/suffering by conceptions grounded exclusively in human experience.
7. There are many ways to reject ethical anthropocentrism. For example, I reject the sanctity of human life (only) also on grounds of speciesism, but this is not unique to ecocentrism.
8. Moral realism is the view that moral goods and truths can be derived from objective features of the world.
9. However, and to what extent, you want to distinguish the three.
10. See, e.g., Reber (2019), Calvo et al. (2020), Parise et al. (2020), and Goff (2017).
11. I spend time in wild nature on a daily basis. In my opinion, I would have no business calling myself an environmental philosopher if I did not.
12. See the *Futures* 2019 special issue "Human colonization of other worlds" for a variety of arguments https://www.sciencedirect.com/journal/futures/vol/110/suppl/C.
13. Such as through biology-technology hybrids and postbeings. Editor's comment: See also Chapter 28 of this volume, "Will Posthumans Dream of Humans? *A Message To Our Dear Post-Planetary Descendants*," by Francesca Ferrando.
14. In my opinion, it is this capacity that morally denies us the option to be self-centric, in contrast to other lifeforms.
15. Value-aligned AI is the problem of ensuring that advanced AI does what we want it to do, or otherwise is benevolent.
16. To modify the conditions of a celestial body to make it habitable for Earth-life.
17. This can be contrasted with directed panspermia/planetary seeding, advocated by panbiotic ethics, as that entails intrinsic value of life alone.
18. Inherent in this symbiosis is a principle of equity. Extreme inequities are discouraged because they are unsustainable for the total system. The needs and interests of all parties must be met to a roughly equal extent. This also applies within social animals, which includes humans and at least some other animals, who psychologically experience such inequities as unjust, which can encourage social unrest, mistrust, etc. (de Waal 2006; Beckoff & Pierce 2009; Rowlands 2012). Therefore, the cargo for space expansion cannot be a random collection of organisms, nor a disproportional favoring of certain system elements over others, whether human or nonhuman.
19. Possibly, such a trajectory could even be an important part of the human-nature reconciliation (Morton 2016). It could, of course, also entail utter horror.

References

Baum, S. D. 2010. "Is Humanity Doomed? Insights from Astrobiology." *Sustainability* 2, no. 2: 591–603.

Baum, S. D. et al. 2019. "Long-Term Trajectories of Human Civilization." *Foresight* 21, no. 1: 53–83.

Beckoff, M. and J. Pierce. 2009. *Wild Justice: The Moral Lives of Animals*. Chicago: University of Chicago Press.

Berry, P. M. et al. 2018. "Why Conserve Biodiversity? A Multi-National Exploration of Stakeholders' Views on the Arguments for Biodiversity Conservation." *Biodiversity and Conservation* 27, no. 7: 1741–1762.

Bostrom, N. 2008. "Letter from Utopia." *Studies in Ethics, Law, and Technology* 2, no. 1: 1–7.

Bostrom, N. and M. Ćirković. 2008. *Global Catastrophic Risks*. Oxford: Oxford University Press.

Bruskotter, J. T., M. P. Nelson, and J. A. Vucetich. 2015. "Does Nature Possess Intrinsic Value? An Empirical Assessment of Americans' Beliefs. *The Ohio State University*. DOI: 10.13140/ RG.2.1.1867.3129.

Calvo, P., M. Gagliano, G. M. Souza, and A. Trewavas. 2020. "Plants Are Intelligent, and Here Is How." *Annals of Botany* 125: 11–28.

Cowen, T. and D. Parfit. 1992. "Against the Social Discount Rate." In *Justice Between Age Groups and Generations*, edited by P. Laslett and J. S. Fishkin, 144–161. New Haven, CT: Yale University Press.

Curry, P. 2011. *Ecological Ethics. An Introduction* (2nd ed.). Cambridge: Polity Press.

De Waal, F. 2006. *Primates and Philosophers: How Morality Evolved*. Princeton, NJ: Princeton University Press.

Goff, P. 2017. *Consciousness and Fundamental Reality*. Oxford: Oxford University Press.

Holt, L.A. 2019. "Why the 'Post-Natural' Age Could Be Strange and Beautiful." *BBC Future*, May 3, 2019. https://www.bbc.com/future/article/20190502-why-the-post-natural-age-could-be-strange-and-beautiful.

Holt, L.A. 2021. "Why Shouldn't We Cut the Human-Biosphere Umbilical Cord?" *Futures* 133: 102821.

Johansson-Stenman, O. 2018. "Animal Welfare and Social Decisions: Is It Time to Take Bentham Seriously?" *Ecological Economics* 145: 90–103.

Lupisella, M. 2020. "Meaning and Ethics." In *Cosmological Theories of Value*, edited by M. Lupisella, 171–194. Springer.

MacAskill, W., K. Bykvist, and T. Ord. 2020. *Moral Uncertainty*. Oxford: Oxford University Press.

Margulis, L. and O. West. 1997. "Gaia and the Colonization of Mars." In *Slanted Truths*, edited by L. Margulis and D. Sagan, 221–234. Springer.

Milligan, T. 2015. *Nobody Owns the Moon: The Ethics of Space Exploitation*. Jefferson: McFarland and Company.

Morton, T. 2016. *Dark Ecology: For a Logic of Future Coexistence*. New York: Columbia University Press.

Ord, T. 2020. *The Precipice: Existential Risk and the Future of Humanity*. New York: Bloomsbury Publishing.

Owe, A. and S. D. Baum. 2021. "The Ethics of Sustainability for Artificial Intelligence." *Proceedings of AI for People: Towards Sustainable AI, CAIP'21*.

Parise, A.G., M. Gagliano, and G. M. Souza. 2020. "Extended Cognition in Plants: Is It Possible?" *Plant Signalling and Behavior* 15, no. 2: 1710661.

Randolph, R. O. and C. P. McKay. 2014. "Protecting and Expanding the Richness and Diversity of Life, an Ethic for Astrobiology Research and Space Exploration." *International Journal of Astrobiology* 13, no. 1: 28–34.

Reber, A. 2019. *The First Minds: Caterpillars, 'Karyotes', and Consciousness.* Oxford: Oxford University Press.

Rolston III, H. 1986. "The Preservation of Natural Value in the Solar System." In *Beyond Spaceship Earth: Environmental Ethics and the Solar System*, edited by E. Hargrove, 140–182. San Francisco: Sierra Club Books.

Rowlands, M. 2012. *Can Animals be Moral?* Oxford: Oxford University Press.

Sauer, H., C. Blunden, C. Eriksen, and P. Rehren. 2021. "Moral Progress: Recent Developments." *Philosophy Compass* 16, no. 10: e12769.

Schwartz, J. S. J. 2018. "Where no Planetary Protection Policy Has Gone Before." *International Journal of Astrobiology* 18, no. 4: 353–361.

Smith, D. M. 1998. "How Far Should We Care? On the Spatial Scope of Beneficence." *Progress in Human Geography* 22, no. 1: 15–38.

Tonn, B. E. 2002. "Distant Futures and the Environment." *Futures* 34, no. 2: 117–132.

Tonn, B. E. 2007. "Futures Sustainability." *Futures* 39: 1097–1116.

Tonn, B. E. 2018. "Philosophical, Institutional, and Decision Making Frameworks for Meeting Obligations to Future Generations." *Futures* 95: 44–57.

Washington, H., B. Taylor, H. Kopnina, P. Cryer, and J. J. Piccolo. 2017. "Why Ecocentrism Is the Key Pathway to Sustainability." *The Ecological Citizen* 1: 35–41.

Wolf, E. T. and O. B. Toon. 2015. "The Evolution of Habitable Climates under the Brightening Sun." *Journal of Geophysical Research: Atmospheres* 120, no. 12: 5775–5794.

28

Will Posthumans Dream of Humans?

A Message to Our Dear Postplanetary Descendants

Francesca Ferrando

Introduction

This chapter is written in the form of a message to our postplanetary descendants. Its stylistic approach is a mesh of poetry, fiction, and nonfiction. There are different paths to choose from: a letter, a poem, and short essayistic intermissions. They all convey the same message: we, humans, posthumans, and nonhumans, have always been in space, and we are all related. The notions of future posthumans, and posthumanities,[1] refer to the progenies of those humans who may migrate to space in the near future, and whose biologies and technologies would adapt, generation after generation, to conditions in space, eventually evolving into different species and beings. Yet, at the existential level, they will be not so radically different from us, humans of the twenty-first century, as to be an absolute enigma: the awareness of self-realization exceeds any individual, species, and spatiotemporal limitations. The theoretical frame which sustains this chapter is posthumanism[2]—in its philosophical, and more specifically, existential[3] take. Posthumanism approaches humans (in all of our diversities) as part of the entire cosmos: nets of ecological and technological emergencies, outside of any primacy or supremacy.

The title of this chapter evokes the dystopian science fiction novel *Do Androids Dream of Electric Sheep?* (1968) by North American writer Philip K. Dick (1928–1982). The book takes place in a dark near future; according to the plot, by 1992[4] a nuclear world war has killed millions of people, causing the extinction of entire species and forcing survivors to migrate to Mars and to other colony planets. Nonhuman animals, almost extinct, have become a rare commodity; androids are almost indistinguishable from humans, and humans have not changed much: they are still lost in human-centric discriminations and patriarchal stereotypes—the sexualization of the gynoid characters in the book is a clear example.

Francesca Ferrando, *Will Posthumans Dream of Humans?* In: *Reclaiming Space*. Edited by: James S. J. Schwartz, Linda Billings, and Erika Nesvold, Oxford University Press. © Oxford University Press 2023. DOI: 10.1093/oso/9780197604793.003.0028

Unlike in Dick's novel, in this chapter's posthuman scenario, originating at the dawn of the twenty-first century, there is no apocalypse to rely upon, nor any ultimate catastrophe. Earth's life has not been endangered by a nuclear global war; the sixth mass extinction has been brutally banal[5] and normalized, but not generally accepted. While some humans have uncritically approached this environmental crisis as the new normal, many others have realized that our actions, habits, and mindsets have all to do with it; that the well-being of the Earth is our own well-being; that (human and nonhuman) discriminations are detrimental to all; and that ecology and technology must be acknowledged in existential dignity. In the following two paragraphs, and throughout the text, we will present some possible histories of the futures, in order to contextualize both the letter and the poem. Multiple outcomes will be offered, as humans are many and they are, and will be, manifesting different stories. These hypothetical histories of the futures rely on the current human condition; they should not be taken as predictions nor prophecies. More radically, they are calls for action to be aware of our legacies. Let's then present some possible scenarios.

As of now (early twenty-first century), some humans are still unwilling to reverse the perverse dynamics of the Anthropocene; in the salvation tale of high-tech capitalism, technology is presented as the final solution to everything, even to our own destruction. Based on such a trajectory, this text presumes that, in the near future, instead of changing habits, some human societies might continue polluting and destroying natural habitats, to build more, buy more, have (and be) "more." Following this, some humans may eventually move to space with the same human-centric mindsets; in that case, the exhaustion of other planets, satellites, and asteroids would necessarily follow. More extensively, if we base our existential principles on the myth of "more," we will never achieve self-realization, neither as individuals nor as a species. In fact, the plus (or "more") can only exist through the minus (or "less"). One term necessitates the other in an inextricable attachment of meanings: in order to be the plus, someone—or something—must become the minus. From an existential perspective, one cannot reach satisfaction by following values based on quantity: ultimately, "more" is never enough (you can always add a zero). Only "enough" is enough, and that is a subjective concept: we are many.

Human life is diverse and dynamic. This is why it is important to underline that history cannot be summarized and simplified in the actions of a specific group of people sharing similar spatiotemporal, as well as embodied, physiological, and sociocultural conditions. While some humans may be lost in the illusions of high-tech salvation and anthropocentric masteries of the planets, others have already realized that technology is not separated from human intentions; that we are the planets (and the environments) in which we manifest; and that we cannot understand ourselves through an absolute separation from human, and non-human, others. For instance, human beings rely on breathing air, eating food, drinking water. We are generated by other humans, and we need human and non-human presence to communicate and survive. We are not alone. This applies to outer space as well. In this chapter, we will underline that humans cannot "go" to space, since they have always been *in* space. This is also the reason why space exploration and migration cannot be reduced to an economic endeavor and simplified as a search for new resources, ignited by the ongoing depletion of planet Earth. Journeying in space is based on an existential stance: it's a search for meaning.

The Message

Welcome

Let's introduce a brief history of the future to explain the genesis, developments, and outcomes of the message. By the end of the twenty-first century CE, humans on Earth became more and more used to space debris falling on the ground. And thus, some space artists and hackers (that is, artists working with space products and design) decided to reverse the trend, inscribing messages in space objects that would eventually fall on other planets and habitable asteroids. After pluralistic debates on what should be written, the agreement was reached on the universal message of existential relationality. No matter who you are, where you are from, and what embodiment you are manifesting through and in, "I" can understand who "You" are, because "We" are One and Many.

And still, these artists and hackers were aware of the fact that the message, written in the form of a letter, was expressed within the traditional frame of rational arguments and historical facts, and that logic and reason were not going to be characteristics featured by all posthumanities. They were aware that different posthuman species would descend from the human, and that some could evolve by engaging, more directly, with their inner selves. To convey the same message in a different voice, they chose to end it with a poem, in order to tap into other aspects of posthuman consciousness, such as insight and intuition.

The message was eventually discovered by future posthumanities, turning into a key astroarchaeological document which allowed our postplaneta descendants to realize that their human ancestors had also been "civilized" and "intelligent." Until then, the most common theory was that Earthly humans were brutish and barbarian animals without self-consciousness, capable of killing and purposely harming each other and all other beings. This is the document which became the standard to sustain the theory that twenty-first-century humans were already conscious and civilized people:

A Letter to Our Dear Postplanetary Descendants

Dear Posthumans,

Who may be reading this letter from a place other-than-Earth, that you call home. You were born on Mars, on an asteroid, on a moon, on a habitable exoplanet. You may even have been born on a spaceship—which happens sometimes, in those long journeys to the galaxies, where the ancient human experience of being part of a tribe becomes, again, a reality: a small group of people traveling together, physically and existentially, space nomads of the futures reconnecting to their ancestries (as a matter of fact, humans—from whom You descend—lived nomadically for most of their history). You may dream of black holes, dwarf planets and . . . biological humans; you then wake up and think: "*Was that real?*" You may wonder: "*Where do we come from?*" or: "*How was life on Earth?*" and "*Why did our ancestors leave?*" I am writing to You from planet Earth, in the twenty-first century CE. We are family, on some level. I am a human, genetically and genealogically related to You. I am aware that history is eventually forgotten and that, time passing, conflicting hypotheses flourish around important subjects.

The question of our origins (where we, as a species, come from), has been, and will always be, haunting us, because there is no beginning. Once we understand who we are, we realize that there is only constant transformation: "in the beginning, there was no beginning." You, living in the age of the spiritual machines,[6] may have already integrated this ultimate revelation, realizing that consciousness is everywhere, and that the biological and the technological realms are constantly merging into each other, in the vastness of Being. This awareness may have brought your *ethos* beyond any bio-, techno-, and species-centrism(s). After all, You, born of cyborgs, are the result of this lack of ontological borders. Your species evolved to specific conditions that were not previously found, and that will not occur again, in those exact material and spatiotemporal terms, in the future: even planets change, and are changed, by the life they sustain. A constant process of existential poiesis, cocreation, and devolution. And still, with posthuman life expanding outside of planet Earth, the memories of Earthly experiences may have faded away, but they are a valuable source of understanding that may unravel the roots of some of your reiterated species' habits and experiences. I will thus elucidate some queries that you may encounter in your path of self-enquiry, as posthuman beings and species.

Beyond Earth: Space Is the Place

Humans, and Posthumans, are made of stardust.

This is something that we, as a species, have always known. To date, there is no scientific consensus on how life on Earth originated, but the most credited hypothesis refers to this process as an abiogenesis, that is: biological life would have arisen from inorganic matter through natural processes.[7] Other hypotheses, such as exogenesis, claim that life did not originate on Earth, but somewhere else in space. Ultimately, these theories perpetuate the absolute separation of Earth/Space, as if Earth was not in space.[8] Other investigations reveal a different truth: *Homo sapiens* have always been aware of the stars. According to new archeological findings from the Upper Paleolithic time, "detailed solstice observances were common among complex hunter-gatherers, often associated with the keeping of calendars and the scheduling of major ceremonies" (Hayden and Villeneuve 2011, 331). Ritual centers of most ancient human civilizations had the sky at the center of their inquiry: their knowledge of astronomy was profound. This is why you need to realize that you did not leave "home," but rather, you are always home. The cosmos is our womb; it is in constant expansion and will welcome you, no matter what, because: *You are It.*

More specifically, migrating to space was not "against" human nature, as some Earth-centered humans claimed. The history of humans is a history of migrations. You descend from *Homo sapiens*, who throughout history have migrated from Africa to all continents on Earth. Migrating to space was not only based on the need for resources, as some of your knowledge-keepers may underline. Some of your ancestors left the Earth to create different ways of living, not strangled by millennia of human-centric habits. In fact, space colonization (that is, going to space for human interests and profits) was performed by some, but was not supported by many humans on Earth. For instance, according to the "Outer Space Treaty" (1967), a key document ratified by 104 countries in the twentieth century, celestial bodies shall be used "for peaceful purposes" and shall not be contaminated.[9]

By the time you read this letter, the universe might have expanded so much that you may be too far to visit the Earth, but you must know that it is, indeed, a beautiful planet. Agents of great rocks, carved by the winds and powerful waters, fruitful green lands and majestic deserts, shaping each other's discontinuous embodiments, like sacred serpents uncoiling spacetime. Once some people migrated out of the Earth, no longer overpopulated, nature took over the human-made ecological disasters that I am sure you have heard about: "because humans destroyed the Earth, humans had to leave." This is one of the hypotheses that you have probably learnt on your human origins. They call it the "Anthropocenic Hypothesis," which locates human migration to space in the attempt to escape a ruined planet, caused by human-centric habits. This is not entirely true.

If humans only left planet Earth because of the ecological collapse they created, then once in space, they would have repeated the same catastrophic behaviors, thus causing the collapse of the new places they called home, finally causing their own extinction. Your existence is proof that that is not the case. As you know, many planets and celestial bodies did eventually collapse due to human greed and carelessness. Yet, not all humans followed the same mindless trajectories. While some humans looked at space as the new "El Dorado"—a place to conquer and exploit by gaining resources for the benefit of a small pocket of humans—other humans, deeply aware of not just "living" on this planet, but being part of this planet, took care of the Earth as part of their ancestry. Indigenous nations and other communities, for instance, had a profound effect in reversing the anthropogenic-caused damages. You have to realize that humans cannot be simplified under a unified stereotype: there have been many different ways of existing and manifesting as *humans*. The fact that *You, posthumans,* are alive today, is the result of this multigenerational care for the cosmos, not of its ultimate destruction.

Beyond Human: We Are, the Dreams of the Dreamers

Dear Posthumans, not born of Earth,

I wonder what you see, when you dream of humans: do you envision them as females, males, others? What are the colors of their skins? Are they young or old? Are they dressed or naked? What languages do they speak? Probably your language. And still, I wonder about all of this, because humans were many, and their differences were crucial to what they were exposed to. They were female, male, and beyond; of many different ethnicities, ages, habits, cultures, and eras. I wonder what your ideas of humans are; what depictions you may have, if any. Now that your embodiments are gender-fluid, race-fluid, age-fluid, and even body-fluid (as you can manifest through different and multiple bodies), you may wonder: "How could humans discriminate based on physiological diversity?" It happened in different forms and at different levels. You may be able to relate to it, or not: discrimination is a social disease rooted in partial human perspectives, and sustained on economic interests, power, and social stratifications.

In your world, discrimination, if still existing, may be drastically different. For you, even the notion of "species" may be hard to tackle, as posthumans exhibit unlimited, and radically different, types of appearances and constitutions; but for millions of years, biological beings evolved mainly by means of natural selection and adaptation, which was very slow and efficient. Humans generated out of Earth and adapted to it: Earth was a constitutive element of our physiological systems and the planet we were most suited for. Genetic engineering and the technological shift to evolution caused an acceleration in realizable manifestations, from the individual to the species level; they were not always successful but opened many possibilities. By the time you read this letter, discrimination may be still around, in one form or another—for instance, by considering people of specific types of embodiments, or from particular planets, less than others. Be very careful with such views: they alter integrity and are real obstacles to full existential awareness. In fact: You Are, and always have been, Everything.

Beyond Spacetime: We Are

I was here when migration to Mars happened. I was here, at the beginning of human-traced history, when Paleolithic people would place massive rocks in circles to connect to space. I was here when You were born. I am here. I can see You, gazing at the thousand moons, absorbed; ecstatically dancing on the rocks with your fluid bodies that I cannot even detect, for they move too fast for my human brain. But I can relate to You, because You are like me, and like that owl that flew low and slow during a night of full moon in the Pleistocene, and like the rovers, shining robots sent to Mars before any human beings during the Infocene: You, like all of us, partake in the existential quest. We look at the sky, we howl at the mystery of existence; some of us can fly, physically and metaphorically. Some of us dream. I do not know what you dream of; if you lay down or if you prefer darkness to sleep, now that your asteroid may be constantly in shadow. But I know that you are manifesting new archetypes of existence. And I know that You Are, in the Self-Realized Ocean of Being—in This, We Are: One and Many.

A Poem to You, Born Out of Earth

Dear Ones, Born Out of Earth:
I am talking to You, Posthumans, Transhumans, Not-humans at all.
I am watching the Sky, and I know You are There, right Now. Because Time
is a cycle, and the Futures are already Here.

I am talking to You.
I don't know how You look, what You do, or who You want to be.
But I Know You,
because: We, Are.

You may not have been where I live; I may not have been where You live.
But We have always been in Space, as We All Are.
We may not have glared at the same Sun.
And still, Light is Light. Darkness is Darkness. They both are part of
    Existence.

I have not met You, but I know You. And You know me.
Why?
Because We ask the same question: "Who Am I?"
And We Both, know the answer.

We Are Inter-beings born of Stardust, to that We Return.
We Are Eternal Recurrences:
Spacetime liquefies in our Consciousness.

I am Aware of Who You Are.
Because I may be You, one day.
And You are already me.

Genetics is not a compromise nor a destiny;
It's a language.
And from the Past of your Ancestors,
I inscribe my message in our DNA code.
And I am not afraid.
Or maybe, just a little. To fly too high and melt my wings, like Icarus.
But We are not a Labyrinth: We cannot escape from Ourselves.

I am listening to You,
Sound of Space,
Dance of Celestial Bodies.
Too far to collapse into a Black Hole.
Too close to act like You have never been Human.

If You were an Angel, I would not be surprised.
If You were a Demon, it would also be fine.

You Are Who You Are.
You are Always the Same;
Always, Changing.

Our Actions: Re-Actions. Our Progenies, our Seeds.
We are Who We Are.
Manifesting the Galaxies, Among other Dimensions.
The Multiverse: our Home.

Welcome,
Posthuman Family.

Nothing really changed,
Just a little Spacetime between us.
Just, Everything.

You have always been Aware of Our Resonance.
In Romance with Existence;
Playing the Cosmic Game, of Birthing and Dying;
Creating new Archetypes, each and every time:
Full Existential Awareness.

It does not matter Who You Are Now.
Because You are Me, and I am You.
And We have always been Together.
Just a million years apart;
Nothing. Just a blink of an eye.
You may not get this metaphor,
Now that telepathy has replaced vision.
But think it this way:
"Do You Ever Dream of Biological Beings?"
If so: I am one of them.

I am talking to You,
and You are no longer Dreaming:
We Are, Fully, Awakened!

Conclusions

The end of the human, the beginning of the posthuman: these are not dichotomies, nor linear evolutions, but cosmic patterns of spiral manifestations. Planet Earth is not our ultimate place of origin, but part of who we are: humans are of planet Earth, therefore, have always been in space. Posthumanities will be many; some of them will be searching for their biological origins, and thus, will eventually tap into our historical presence, habits, and knowledges. At the core of the existential journey, in which We all Are, beyond species, time, and space, there is the yearning for Self-Realization. The journey to understand who "I am" begins by understanding that "I am Who We Are;" and also, that: "I am," "Where I am," "What I am," and "With Whom I am" (among others) are intra-connected terms. Ultimately, cosmic posthumanities will be not so drastically different from ourselves, humans of the 21st twenty-first century, that we cannot even relate to them: such absolute dualisms are not sustained in the merging texture of spacetime. We are always, necessarily, related to all that manifest in this dimension. This is a message to all of us, who come from the pasts, the presents and the futures of the (post-)humans. Be aware: we are who We are, who We were, and who We will be. "We" are One and Many.

Acknowledgments

My Gratitude to Thomas Roby, for his Cosmic Presence in my Life and for his important comments to this article.

Notes

1. For a deeper understanding of this concept, see Roden (2015).
2. See, among others: Braidotti (2013) and Ferrando (2019).
3. This approach is explained more clearly in my entry "Existential Posthumanism" (Ferrando, in press).
4. Originally 1992, the year was changed to 2021 in later editions.
5. Here, I am referring to Jewish philosopher Hannah Arendt's use of the term "banality" in relation to the trial of Adolf Eichmann, a Nazi high official involved in the planning and execution of the Holocaust. In "Eichmann in Jerusalem: A Report on the Banality of Evil" [1963] (2006), Arendt writes: "For when I speak of the banality of evil, I do so only on the strictly factual level . . . It was sheer thoughtlessness—something by no means identical with stupidity—that predisposed him to become one of the greatest criminals of that period. And

if this is 'banal' and even funny, if with the best will in the world one cannot extract any diabolical or demonic profundity from Eichmann, this is still far from calling it common-place . . . That such remoteness from reality and such thoughtlessness can wreak more havoc than all the evil instincts taken together which, perhaps, are inherent in man" (277–278).

6. For a further understanding of this notion, see Kurzweil (1999).
7. Soviet biochemist Alexander Oparin, for instance, famously developed the hypothesis of the "primordial soup" [1936] (1953).
8. For further investigation on the origins of life from an astrobiological perspective, see, among others, Hazen (2005); Seckbach (2012).
9. This is how the Treaty reads: "States shall avoid harmful contamination of space and celestial bodies" (1967).

References

Arendt, H. [1963] 2006. *Eichmann in Jerusalem: A Report on the Banality of Evil.* New York: Penguin.

Braidotti, R. 2013. *The Posthuman.* Cambridge: Polity.

Dick, P. K. 1968. *Do Androids Dream of Electric Sheep?* New York: Penguin Random House.

Ferrando, F. 2019. *Philosophical Posthumanism.* London: Bloomsbury.

Ferrando, F. 2022. "Existential Posthumanism: A Manifesto." In *More* Posthuman Glossary, edited by R. Braidotti E. Jones and G. Klumbyte, 47–49. London: Bloomsbury.

Hayden, B. and S. Villeneuve. 2011. "Astronomy in the Upper Paleolithic?" *Cambridge Archaeological Journal* 21: 331–355.

Hazen, R. M. 2005. *Genesis: The Scientific Quest for Life's Origin.* Washington: Joseph Henry Press.

Kurzweil, R. 1999. *The Age of Spiritual Machines: When Computers Exceed Human Intelligence.* New York: Penguin.

Oparin, A. I. [1936] 1953. *The Origin of Life.* New York: Dover Publications.

Roden, D. 2015. *Posthuman Life: Philosophy at the Edge of the Human.* London: Routledge.

Seckbach, J., ed. 2012. *Genesis—In The Beginning: Precursors of Life, Chemical Models and Early Biological Evolution.* Cellular Origin, Life in Extreme Habitats and Astrobiology Vol. 22. Berlin: Springer.

Treaty on Principles Governing the Activities of States in the Exploration and Use of Outer Space, including the Moon and Other Celestial Bodies. 1967. United Nations Office for Outer Space Affairs. Accessed December 30, 2021. http://www.unoosa.org/oosa/en/ourwork/spacelaw/treaties/introouterspacetreaty.html

Index

For the benefit of digital users, indexed terms that span two pages (e.g., 52–53) may, on occasion, appear on only one of those pages.
Tables and figures are indicated by *t* and *f* following the page number